The Stability and Evolution of Genes and Genomes

The Stability and Evolution of Genes and Genomes

Editors

Luigi Viggiano
Renè Massimiliano Marsano

MDPI

Basel • Beijing • Wuhan • Barcelona • Belgrade • Novi Sad • Cluj • Manchester

Editors
Luigi Viggiano
Department of Biosciences,
Biotechnology, and Environment
University of Bari "Aldo Moro"
Bari
Italy

René Massimiliano Marsano
Department of Biosciences,
Biotechnology, and Environment
University of Bari "Aldo Moro"
Bari
Italy

Editorial Office
MDPI
St. Alban-Anlage 66
4052 Basel, Switzerland

This is a reprint of articles from the Special Issue published online in the open access journal *Genes* (ISSN 2073-4425) (available at: www.mdpi.com/journal/genes/special_issues/Stability_Genes_Genomes).

For citation purposes, cite each article independently as indicated on the article page online and as indicated below:

Lastname, A.A.; Lastname, B.B. Article Title. *Journal Name* **Year**, *Volume Number*, Page Range.

ISBN 978-3-0365-9803-1 (Hbk)
ISBN 978-3-0365-9802-4 (PDF)
doi.org/10.3390/books978-3-0365-9802-4

Cover image courtesy of Alessia Viggiano

Contents

About the Editors

Luigi Viggiano

Luigi Viggiano is an Assistant Professor at the Department of Biosciences, Biotechnology, and Environment of the University of Bari "Aldo Moro". Viggiano's research interests include neurobiology, transposable elements, and human genetics.

Renè Massimiliano Marsano

Renè Massimiliano Marsano is an Associate Professor of Genetics at the Department of Biosciences, Biotechnology, and Environment of the University of Bari. Marsano's research interests include transposable elements, transposon-based technologies, and heterochromatin.

Preface

The intricate dance between stability and evolution within the realm of genes and genomes is a captivating saga that unfolds across the vast tapestry of life. In this Special Issue, aptly titled "The Stability and Evolution of Genes and Genomes" we embark on a journey into the dynamic interplay between evolutionary forces and the delicate maintenance of genetic information, a narrative that shapes the existence of diverse species on our planet.

This compilation assembles a harmonious ensemble of perspectives found in eight original research papers and four reviews, each exploring unique aspects of genetic stability and evolution. Covering the foundational influences of mutations, transposition, and natural selection, as well as delving into the subtle complexities of epigenetic effects, karyotype variability, and the evolution of intricate behavioral traits, the contents of this Special Issue shed light on the diverse and intricate landscape of genetic processes.

Luigi Viggiano and Renè Massimiliano Marsano
Editors

genes MDPI

Editorial

The Stability and Evolution of Genes and Genomes

Luigi Viggiano * and René Massimiliano Marsano *

Dipartimento di Bioscienze, Biotecnologie e Ambiente, Università degli Studi di Bari "Aldo Moro", 70121 Bari, Italy
* Correspondence: luigi.viggiano@uniba.it (L.V.); renemassimiliano.marsano@uniba.it (R.M.M.)

The existence of current species can be attributed to a dynamic interplay between evolutionary forces and the maintenance of genetic information. Genes and genomes are constantly evolving entities, shaped by a multitude of forces that maintain their stability while allowing for necessary changes. These forces include (but are not limited to) mutations, transposition, and natural selection.

Together, these forces ensure a delicate balance between preserving genetic information and generating crucial variability for species survival.

While mutations are the primary drivers of evolution, there are cellular mechanisms that counterbalance excessive variation and contribute to the stability of genes and genomes and preserve the faithful pass down of the genetic material from generation to generation.

In this Special Issue of *Genes*, titled "The Stability and Evolution of Genes and Genomes", we have collected eight original research papers and four reviews. These contributions aim to explore various fields, such as the role of transposable elements, the epigenetic effect of heterochromatin on gene expression, karyotype variability, evolution of complex behavioral traits, the stabilization of genetic information in organelles, and viral genome evolution.

Determining what primordial forms of life looked like is a debated topic for which several theories have been proposed [1]. Among the unanswered questions is whether primordial bacteria had a single or a double envelope membrane. In their article, Léonard et al. [2] addressed the significant question of the identity of the last bacterial common ancestor, advancing the suggestive hypothesis that bacteria might have evolved from a common ancestor with a monoderm cell wall architecture. They suggested that the appearance of the outer membrane was not a unique event in evolution and that selective forces have led to the repeated adoption of such an architecture.

Viruses and bacteriophages are also among the earliest form of life [3]. Due to their fast replication, they are considered as exceptional models to study the evolution and the stabilization of genetic information. In particular, understanding how viruses generate new genetic information during evolution and how this information is stabilized and modified in a limited genomic size is currently a debated subject and a relevant issue in public health. In a highly interesting review [4], Pavesi discussed the origin, evolution, and adaptive conflict of overlapping genes, and the critical role of genes in the evolution of viral pathogenicity.

Karyoptype has long been used as a representative taxonomic character, although the karyotypes of closely related species often differ [5]. To resolve the questionable taxonomic structure of the Calomyscus genus [6], Romanenko et al. [7] analyzed karyotype plasticity in 14 specimens of the mouse-like hamsters collected in various Iranian locations through comparative cytogenetics approaches. In this paper, the authors provided a detailed description of the karyotype and concluded that it cannot be used as an unambiguous indicator of Calumyscus species rank. This paper confirms the entangled relationship between the evolutionary histories of living organisms and their phenotypic outcomes.

The evolution of genes can also impact sexual behavior and thus reproduction [8]. These are complex phenotypic traits that require combined methodological approaches

Citation: Viggiano, L.; Marsano, R.M. The Stability and Evolution of Genes and Genomes. *Genes* **2023**, *14*, 1747. https://doi.org/10.3390/genes14091747

Received: 5 July 2023
Accepted: 13 July 2023
Published: 31 August 2023

to be dissected. Through a combination of transcriptomics, real time qPCR, and phylogenetics, Nieberding and collaborators provided a large-scale investigation of the genetic pathways underlying sex pheromone communication in the butterfly *Bicyclus anynana* [9]. Furthermore, in their review, Nieberding and collaborators [10] provide an interesting link between the genetic basis of behavioral variation and the evolution of social learning oviposition-related behavior. This paper could initially sound off topic; however, we have decided to include it because of its relevance in conjoining behavioral genetics and the evolution of complex traits.

Studying the relationship between gene evolution and environmental adaptation is crucial to understand how extant living organisms have originated. In a comparative study, Zhou et al. analyzed the amino acid substitution rate and natural selection of the OXPHOS genes in fig wasps [11] under the hypothesis that these genes have experienced adaptation to the compact, hypo-oxygenated, and dark environment of the syconia.

The regulation of the activity of transposable elements determines to what extent they can act as natural mutagens [12]. Two papers [13,14] showed that a *Drosophila melanogaster* ribosomal protein is able to bind TEs through its special histone-like domains and advanced the intriguing hypothesis that this interaction could reflect a regulation of the activity of transposable elements, especially in the heterochromatin.

Heterochromatin is a major structural feature of the eukaryotic genome stability, a specialized type of chromatin that contains a complex and still poorly understood genomic compartment, extremely enriched in repeats and transposable elements [15] in which expressed genes are rare but not completely absent [16,17]. Heterochromatin is also a hallmark of telomeres and centromeres, two important loci that stabilize chromosomes and ensure proper segregation. The effect of heterochromatic domains on gene expression is still poorly understood. *D. melanogaster* offers a precious model system for studying how stable and compact heterochromatic blocks influence resident genes and artificially inserted euchromatic coding sequences. Messina et al. [18] investigated on the epigenetic silencing of P-element reporter genes induced by transcriptionally active domains of constitutive heterochromatin in Drosophila melanogaster. The description of such a paradoxical phenomenon further entangles our current knowledge on the intricate mechanisms at the basis of heterochromatin structure, function, and stabilization.

Telomeres are specialized structures of the eukaryotic chromosome that allow for the stabilization of terminal genetic information on the chromosome [19]. Chromosome-ends shortening is a natural phenomenon which occurs in differentiated cells and is at the basis of the senescence process [20]. The genetic destabilization induced by telomere shortening in specific tissues or cell types can be potentially used as a biomarker of certain diseases. Zimnitskaya et al. surveyed the scientific literature in the field and suggested that the telomere length in leucocytes can be a promising marker of coronary heart disease [21].

An important aspect in gene evolution is how genetic code is used. The codon usage bias reflects a species-specific use of the expressed tRNA set which impacts the translation of all mRNAs [22]. In his review, Matsuo makes the point on the codon usage bias of histone-coding genes in Drosophila species and proposes a model to explain the GC-richness at the third position of codons in these genes [23].

Eukaryotic cells are internally compartmentalized. Mitochondria are semiautonomous organelles, functionally and genetically inter-dependent from the nucleus [24]. This form of symbiosis results in the destabilization of the mitochondrial genome (mtDNA) when mutations hit a subset of nuclear genes. Gilea and collaborators have reviewed the current scientific literature on the use of the budding yeast *S. cerevisiae* in the study of the mutations of nuclear genes associated with mtDNA instability [25]. This review also highlights the importance of using model organisms as tools to study genome stability.

In conclusion, the papers collected in this Special Issue cover various aspects of how genetic information evolves and is stabilized and emphasize the importance of further studies that combine different methodological approaches in order to provide a complete picture of the dynamics underlying the complex process of genome stabilization and its evolution.

Author Contributions: L.V. and R.M.M. equally contributed to the writing of this Editorial. All authors have read and agreed to the published version of the manuscript.

Acknowledgments: We are grateful to all the Authors and reviewers who provided contributions to this Special Issue.

Conflicts of Interest: The authors declare no conflict of interest.

References

1. Tocheva, E.I.; Ortega, D.R.; Jensen, G.J. Sporulation, bacterial cell envelopes and the origin of life. *Nat. Rev. Microbiol.* **2016**, *14*, 535–542. [CrossRef] [PubMed]
2. Leonard, R.R.; Sauvage, E.; Lupo, V.; Perrin, A.; Sirjacobs, D.; Charlier, P.; Kerff, F.; Baurain, D. Was the Last Bacterial Common Ancestor a Monoderm after All? *Genes* **2022**, *13*, 376. [CrossRef] [PubMed]
3. Nasir, A.; Romero-Severson, E.; Claverie, J.-M. Investigating the Concept and Origin of Viruses. *Trends Microbiol.* **2020**, *28*, 959–967. [CrossRef] [PubMed]
4. Pavesi, A. Origin, Evolution and Stability of Overlapping Genes in Viruses: A Systematic Review. *Genes* **2021**, *12*, 809. [CrossRef] [PubMed]
5. Robinson, T.J.; King, M. Species Evolution: The Role of Chromosome Change. *Syst. Biol.* **1995**, *44*, 578–580. [CrossRef]
6. Shahabi, S.; Aliabadian, M.; Darvish, J.; Kilpatrick, C.W. Molecular phylogeny of brush-tailed mice of the genus *Calomyscus* (Rodentia: Calomyscidae) inferred from mitochondrial DNA sequences (Cox1 gene). *Mammalia* **2013**, *77*, 425–431. [CrossRef]
7. Romanenko, S.A.; Malikov, V.G.; Mahmoudi, A.; Golenishchev, F.N.; Lemskaya, N.A.; Pereira, J.C.; Trifonov, V.A.; Serdyukova, N.A.; Ferguson-Smith, M.A.; Aliabadian, M.; et al. New Data on Comparative Cytogenetics of the Mouse-Like Hamsters (*Calomyscus Thomas*, 1905) from Iran and Turkmenistan. *Genes* **2021**, *12*, 964. [CrossRef]
8. Anholt, R.R.H.; O'Grady, P.; Wolfner, M.F.; Harbison, S.T. Evolution of Reproductive Behavior. *Genetics* **2020**, *214*, 49–73. [CrossRef]
9. Nieberding, C.M.; Beldade, P.; Baumle, V.; San Martin, G.; Arun, A.; Lognay, G.; Montagne, N.; Bastin-Heline, L.; Jacquin-Joly, E.; Noirot, C.; et al. Mosaic Evolution of Molecular Pathways for Sex Pheromone Communication in a Butterfly. *Genes* **2022**, *13*, 1372. [CrossRef]
10. Nieberding, C.M.; Marcantonio, M.; Voda, R.; Enriquez, T.; Visser, B. The Evolutionary Relevance of Social Learning and Transmission in Non-Social Arthropods with a Focus on Oviposition-Related Behaviors. *Genes* **2021**, *12*, 1466. [CrossRef]
11. Zhou, Y.; Huang, D.; Xin, Z.; Xiao, J. Evolution of Oxidative Phosphorylation (OXPHOS) Genes Reflecting the Evolutionary and Life Histories of Fig Wasps (Hymenoptera, Chalcidoidea). *Genes* **2020**, *11*, 1353. [CrossRef]
12. Bourque, G.; Burns, K.H.; Gehring, M.; Gorbunova, V.; Seluanov, A.; Hammell, M.; Imbeault, M.; Izsvák, Z.; Levin, H.L.; Macfarlan, T.S.; et al. Ten things you should know about transposable elements. *Genome Biol.* **2018**, *19*, 199. [CrossRef]
13. Minervini, C.F.; Berloco, M.F.; Marsano, R.M.; Viggiano, L. The Ribosomal Protein RpL22 Interacts In Vitro with 5′-UTR Sequences Found in Some Drosophila melanogaster Transposons. *Genes* **2022**, *13*, 305. [CrossRef]
14. Berloco, M.F.; Minervini, C.F.; Moschetti, R.; Palazzo, A.; Viggiano, L.; Marsano, R.M. Evidence of the Physical Interaction between Rpl22 and the Transposable Element Doc5, a Heterochromatic Transposon of *Drosophila melanogaster*. *Genes* **2021**, *12*, 1997. [CrossRef]
15. Marsano, R.M.; Dimitri, P. Constitutive Heterochromatin in Eukaryotic Genomes: A Mine of Transposable Elements. *Cells* **2022**, *11*, 761. [CrossRef]
16. Marsano, R.M.; Giordano, E.; Messina, G.; Dimitri, P. A New Portrait of Constitutive Heterochromatin: Lessons from Drosophila melanogaster. *Trends Genet.* **2019**, *35*, 615–631. [CrossRef]
17. Messina, G.; Prozzillo, Y.; Bizzochi, G.; Marsano, R.M.; Dimitri, P. The Green Valley of *Drosophila melanogaster* Constitutive Heterochromatin: Protein-Coding Genes Involved in Cell Division Control. *Cells* **2022**, *11*, 3058. [CrossRef]
18. Messina, G.; Celauro, E.; Marsano, R.M.; Prozzillo, Y.; Dimitri, P. Epigenetic Silencing of P-Element Reporter Genes Induced by Transcriptionally Active Domains of Constitutive Heterochromatin in *Drosophila melanogaster*. *Genes* **2022**, *14*, 12. [CrossRef]
19. Shay, J.W.; Wright, W.E. Telomeres and telomerase: Three decades of progress. *Nat. Rev. Genet.* **2019**, *20*, 299–309. [CrossRef]
20. Rossiello, F.; Jurk, D.; Passos, J.F.; d'Adda di Fagagna, F. Telomere dysfunction in ageing and age-related diseases. *Nat. Cell Biol.* **2022**, *24*, 135–147. [CrossRef]
21. Zimnitskaya, O.V.; Petrova, M.M.; Lareva, N.V.; Cherniaeva, M.S.; Al-Zamil, M.; Ivanova, A.E.; Shnayder, N.A. Leukocyte Telomere Length as a Molecular Biomarker of Coronary Heart Disease. *Genes* **2022**, *13*, 1234. [CrossRef] [PubMed]
22. Iriarte, A.; Lamolle, G.; Musto, H. Codon Usage Bias: An Endless Tale. *J. Mol. Evol.* **2021**, *89*, 589–593. [CrossRef] [PubMed]
23. Matsuo, Y. The Adenine/Thymine Deleterious Selection Model for GC Content Evolution at the Third Codon Position of the Histone Genes in *Drosophila*. *Genes* **2021**, *12*, 721. [CrossRef]

24. Lane, N.; Martin, W. The energetics of genome complexity. *Nature* **2010**, *467*, 929–934. [CrossRef] [PubMed]
25. Gilea, A.I.; Ceccatelli Berti, C.; Magistrati, M.; di Punzio, G.; Goffrini, P.; Baruffini, E.; Dallabona, C. *Saccharomyces cerevisiae* as a Tool for Studying Mutations in Nuclear Genes Involved in Diseases Caused by Mitochondrial DNA Instability. *Genes* **2021**, *12*, 1866. [CrossRef]

genes
MDPI

Article

Was the Last Bacterial Common Ancestor a Monoderm after All?

Raphaël R. Léonard [1,2], Eric Sauvage [1], Valérian Lupo [1,2], Amandine Perrin [3,4], Damien Sirjacobs [2], Paulette Charlier [1], Frédéric Kerff [1,*] and Denis Baurain [2,*]

1 InBioS–Centre d'Ingénierie des Protéines, Université de Liège, 4000 Liege, Belgium; rleonard@doct.uliege.be (R.R.L.); ericsauvage8@gmail.com (E.S.); valerian.lupo@doct.uliege.be (V.L.); paulette.charlier@uliege.be (P.C.)
2 InBioS–PhytoSYSTEMS, Unit of Eukaryotic Phylogenomics, Université de Liège, 4000 Liege, Belgium; d.sirjacobs@uliege.be
3 University Lille, CNRS, Centrale Lille, UMR 9189 CRIStAL, F-59000 Lille, France; amandine.perrin@pasteur.fr
4 Hub de Bioinformatique et Biostatistique-Département Biologie Computationnelle, Institut Pasteur, 75015 Paris, France
* Correspondence: fkerff@uliege.be (F.K.); denis.baurain@uliege.be (D.B.)

Abstract: The very nature of the last bacterial common ancestor (LBCA), in particular the characteristics of its cell wall, is a critical issue to understand the evolution of life on earth. Although knowledge of the relationships between bacterial phyla has made progress with the advent of phylogenomics, many questions remain, including on the appearance or disappearance of the outer membrane of diderm bacteria (also called Gram-negative bacteria). The phylogenetic transition between monoderm (Gram-positive bacteria) and diderm bacteria, and the associated peptidoglycan expansion or reduction, requires clarification. Herein, using a phylogenomic tree of cultivated and characterized bacteria as an evolutionary framework and a literature review of their cell-wall characteristics, we used Bayesian ancestral state reconstruction to infer the cell-wall architecture of the LBCA. With the same phylogenomic tree, we further revisited the evolution of the division and cell-wall synthesis (*dcw*) gene cluster using homology- and model-based methods. Finally, extensive similarity searches were carried out to determine the phylogenetic distribution of the genes involved with the biosynthesis of the outer membrane in diderm bacteria. Quite unexpectedly, our analyses suggest that all cultivated and characterized bacteria might have evolved from a common ancestor with a monoderm cell-wall architecture. If true, this would indicate that the appearance of the outer membrane was not a unique event and that selective forces have led to the repeated adoption of such an architecture. Due to the lack of phenotypic information, our methodology cannot be applied to all extant bacteria. Consequently, our conclusion might change once enough information is made available to allow the use of an even more diverse organism selection.

Keywords: bacterial evolution; cell-wall; outer membrane (OM); Bayesian inference (BI); phylogenomics; comparative genomics; ancestral traits

Citation: Léonard, R.R.; Sauvage, E.; Lupo, V.; Perrin, A.; Sirjacobs, D.; Charlier, P.; Kerff, F.; Baurain, D. Was the Last Bacterial Common Ancestor a Monoderm after All? *Genes* **2022**, *13*, 376. https://doi.org/10.3390/genes13020376

Academic Editors: Daisuke Kageyama and Renè Massimiliano Marsano

Received: 31 December 2021
Accepted: 15 February 2022
Published: 18 February 2022

1. Introduction

Cell-wall architecture has always been an important morphological character for bacterial classification [1]. Two main types of cell wall exist: the monoderm and the diderm architectures. While monoderm bacteria are generally surrounded by a thick peptidoglycan (and are positive to Gram coloration), in diderm bacteria, a thin peptidoglycan layer is sandwiched between the cytoplasmic membrane and the outer membrane (and are negative to Gram coloration) [2,3]. However, cell-wall features are insufficient to yield a classification that would correlate with phylogenetic trees based on molecular data [4]. Hence, distantly related phyla may have apparently identical cell walls (e.g., Negativicutes and Proteobacteria), whereas closely related phyla or families may present variations in their peptidoglycan thickness or composition, and even in the number of surrounding membranes (e.g., Negativicutes and Halanaerobiales compared to other Firmicutes) [5]. Nonetheless, the evolution of the bacterial cell wall should be addressed considering the phylogeny of the domain. The number of membranes (one or two) that surround a bacterial cell, their lipid composition and the thickness of the peptidoglycan layer are undoubtedly major characteristics of the bacterial cell wall, and these features often come into consideration when discussing the evolution of the bacterial domain. Hence, transition from one to two lipid membranes (or the opposite) has attracted much attention. Disappearance of the outer membrane going from "diderm" to "monoderm" architecture has been proposed by Cavalier-Smith [6,7] but evolution from monoderm to diderm bacteria is usually favoured by other evolutionary biologists [8–11]. It has been suggested that the endosymbiosis between an "actinobacterium" and a "clostridium" could be the starting point for the onset of double-membrane bacteria [12], but how exactly this symbiosis could have further evolved to form a diderm bacterium still is to be detailed. An attractive hypothesis accounting for the emergence of the outer membrane is its evolution from a forespore of a spore-former "firmicute". Based on 3D electron cryotomographic images of spore formation in the diderm firmicute *Acetonema longum*, Tocheva et al. showed that the inner membrane (IM) of the mother cell is inverted to become the outer membrane of the forespore and ultimately of the germinating cell [13], leading to the assumption that the outer membrane of diderm bacteria could have evolved from monoderms via sporulation [11,13–15]. In contrast, some studies of the evolution of the cell-wall architecture in the phylum Firmicutes interpreted the double membrane found in Halanaerobiales and Negativicutes (two classes of Firmicutes) as a reminiscence of the double membrane in the Firmicutes ancestor, and thus concluded that the outer membrane was lost multiple times in this phylum [16,17]. This interpretation further opens the possibility that the last bacterial common ancestor (LBCA) was a bona fide diderm bacterium.

Cell division in bacteria involves a series of proteins that fulfil many functions as diverse as cytoplasmic membrane invagination, DNA transfer control, peptidoglycan synthesis and daughter cell separation. They assemble into a dynamical complex that overpasses the cytoplasmic membrane and has components in both the cytoplasm and the periplasm. A small number of these proteins are essential and conserved in the genome of almost all bacteria [18]. Several of these proteins of cell division are generally clustered together with proteins involved in peptidoglycan synthesis in a single locus on the genome, the *dcw* (division and cell-wall synthesis) cluster [18]. This cluster is found in many bacteria and its composition and gene order are generally well conserved [19,20]. It has also been shown to be one of the most stable gene clusters (the cluster itself and the gene synteny within the cluster are conserved in a broad taxonomic range of genomes) [18], on par with the ribosomal clusters [21,22]. The longest version of the *dcw* cluster includes 17 genes and encompasses genes coding for proteins responsible for peptidoglycan precursors synthesis (DdlB, MurA, MurB, MurC, MurD, MurE, MurF, MurG, MraY), proteins integrated in the divisome (FtsA, FtsI, FtsL, FtsQ, FtsW, FtsZ), and proteins involved in regulation via DNA binding or RNA methylation (MraW, MraZ). The *E. coli dcw* cluster includes 15 genes, starting with *mraZ* and ending with *ftsZ*, but misses the *murA* and *murB* genes [23]. Many phyla, orders, classes, or families are apparently characterized by the lack of specific genes in the cluster, the absence of *ftsA* and *ftsZ* in Chlamydiae and Planctomycetes being well-known examples [24]. These observations suggest that the organization of the *dcw* cluster holds clues to bacterial evolution. Thus, its detailed study might complement sequence-based phylogenomic approaches, including in terms of rooting of the bacterial tree. For example, the integration of a gene in a specific position within the cluster probably happened only once in the history of the bacterial domain, whereas gene loss and genomic reorganization events, on the contrary, are expected to have been more frequent. Likewise, the phylogenetic distribution of the genes involved in the biosynthesis of the outer membrane in diderm bacteria might provide useful information about their evolutionary status, ancestral or derived, with respect to the bacterial domain as a whole [5,17,25].

In this work, we built a Bayesian phylogenomic tree of the bacterial domain using a supermatrix of 117 single-copy orthologous genes sampled from 85 species representative of the bacterial diversity and for which a descriptive literature exists. We then researched the cell-wall architectures for these species and used the tree to reconstruct the evolution of two cell-wall traits, the number of membranes and the presence and thickness of the peptidoglycan layer, again with Bayesian inference. Moreover, we compared the composition and gene order of the *dcw* cluster in our 85 representative species and used a new variant of a homology-based method to map the organization of the *dcw* cluster on the evolution of the bacterial domain. Contrary to our expectations based on recent literature and educated guesses, our Bayesian analyses inferred that the LBCA was a monoderm bacterium with a thick peptidoglycan. This reconstruction implies that the outer membrane of diderm bacteria appeared more than once, a hypothesis that is indeed supported by differences in the genetic machinery involved in its biosynthesis across the various diderm lineages, as shown by our extensive similarity searches. Our results also show that the LBCA already had a complete *dcw* cluster and that its organization does not correlate with cell-wall architecture.

2. Materials and Methods

2.1. Dataset Assembly

2.1.1. Data Download

The initial dataset of prokaryotic genomes and proteomes was downloaded from Ensembl Bacteria release 20 [26] using wget. This dataset had 8848 Bacteria and 238 Archaea represented.

2.1.2. Genome Dereplication and Selection

We first reduced the number of genomes based on genomic signatures [27] to regroup similar genomes into genome clusters with a prerelease version of our new software ToRQuEMaDA [28]. Briefly, for five different k-mer sizes (from 2 to 6-nt), we computed the frequency of each word in each genome using the program compseq from the EMBOSS software package [29]. The complete lineage of every genome was recovered from the NCBI Taxonomy database [30] using the program fetch-tax.pl from the Bio::MUST::Core distribution (D. Baurain, https://metacpan.org/dist/Bio-MUST-Core, accessed on 16 February 2022). Each signature file was further analysed in R [31] to cluster genomes into a predefined number of groups (300, 600, 900, 1200, 1500 and 2100) using various distance metrics (i.e., Euclidean, Pearson and Hamming) and clustering algorithms (i.e., k-means, ascending and descending hierarchical clusterings). To choose the best combination of methods and parameters, the available taxonomic information was used to evaluate the quality of the clustering. Briefly, we computed how many different taxa of each rank (phylum, class, order, family, genus, species) were found in each individual cluster or each set of clusters and chose the combination that best separated the higher-level taxa (phylum, class, order, family) while merging the lower-level taxa (genus, species) [28]. This led us to settle on the following set of methods and parameters: 6-nt k-mer, 900 clusters, Pearson distance and ascending hierarchical clustering algorithm. Then, we selected a single representative for each cluster, based on the quality of genome annotations, as evaluated by the number of gene names devoid of uninformative words like "hypothetical", "putative", "unknown" etc [28]. After including a few other well-characterized genomes (e.g., *Streptomyces coelicolor* A3(2), *Escherichia coli* O127:H6 str. E2348/69, *Staphylococcus aureus* subsp. *aureus* MRSA252), we ended up with a list of 903 genomes: 822 Bacteria and 81 Archaea.

2.1.3. Identification of Orthologous Groups

For every protein sequence of every one of these 903 genomes, we launched an all-versus-all BLAST-like similarity search using USEARCH v7.0.959 [32] with the following parameters (evalue = 1×10^{-5}; accel = 1; threads = 64). Then, we used OrthoMCL v2.0.3 [33] to cluster protein sequences into orthologous groups based on USEARCH reports, using an e-value cut-off of 1×10^{-5}, a similarity cut-off of 50% and an inflation parameter of 1.5. The total number of proteins for the 903 genomes was 2,467,263, and these were partitioned into

124,422 orthologous groups, whereas 326,269 sequences were considered as "singletons" by OrthoMCL (i.e., without homologues).

2.1.4. Database Creation

Gene metadata (organism, genomic coordinates, strand, putative function) for every protein was extracted from the definition lines of the Ensembl FASTA files and stored into a custom designed MySQL (Oracle Corporation) relational database (see Figure S16), along with orthology relationships, based on our protein sequence clustering.

2.2. Evolution of the Bacterial Domain

2.2.1. Supermatrix Assembly

To build a robust tree of the bacterial domain, we manually chose a subset of 85 genomes (out of the 903 genomes initially selected), trying to maximize the number of classes. Then, using classify-mcl-out.pl [34], we selected all orthologous groups of proteins featuring at least one representative of eight major bacterial phyla (Firmicutes, Chloroflexi, Actinobacteria, Deinococcus-Thermus, Proteobacteria, Spirochaetes, Planctomycetes and Bacteroidetes) and in which at most 10% of the selected genomes had more than one gene copy. This left us with a list of 176 broadly conserved and (mostly) single-copy genes. The final dataset was further reduced to 117 orthologous groups to ensure a maximum of 14 missing species in each individual orthologous group (Table S1). The corresponding orthologous groups were aligned with MAFFT v7.127b [35] using default parameters. The protein sequence alignments were then filtered with Gblocks v0.91b [36] using a set of "medium stringency" parameters (as predefined in Bio::MUST::Core) and concatenated with SCaFoS v1.30k [37]. Finally, the resulting concatenation was further filtered for sites >50% missing character states, yielding a supermatrix of 85 species × 19,959 unambiguously aligned amino-acid (AA) positions (4.29% missing character states). A preliminary (more diverse) supermatrix was also created in the process, including 101 species and 19,959 unambiguously aligned AA positions (4.72% missing states).

2.2.2. Phylogenomic Analyses

For Bayesian inference (BI), we used PhyloBayes MPI v1.5 [38] to produce six replicate Markov Chain Monte–Carlo (MCMC) chains of 50,000 cycles, with one tree sampled every 10 cycles, using the CAT+GTR+Γ model of sequence evolution [39–41]. Constant sites were deleted with the -dc option. Convergence was assessed using the program tracecomp from the PhyloBayes software package. Two consensus trees (along with their posterior probabilities) were extracted after a burn-in of 10,000 cycles: one over the six chains (A to F) and another over the two most congruent chains (A and C; maxdiff = 0.130; meandiff = 0.001), both with the -c option of bpcomp set to 0.01. Cross-validation tests to decide the best-fit model (CAT+GTR+Γ) were carried out using PhyloBayes v3.3f [42], as suggested in PhyloBayes manual (p. 38). For our preliminary tree, we ran two chains of 50,000 cycles, with one tree sampled every 10 cycles, under the simpler CAT+Γ model. The consensus tree was extracted after a burn-in of 5000 cycles (maxdiff = 0.580; meandiff = 0.011). All trees (including those described below) were formatted semi-automatically using the scripts format-tree.pl, export-itol.pl and import-itol.pl (also from Bio::MUST::Core) and iTOL v6 [43].

2.2.3. Congruence Tests

Congruence tests were performed on the 85-species supermatrix genes with Phylo-MCOA v1.4 [44], then Maximum Likelihood (ML) reconstruction with RAxML v8.1.17 [45] was used under the model PROTGAMMALGF (LG+F+Γ) to compare the topologies obtained with and without the "cell-by-cell outliers" (i.e., specific species in specific genes whose position is not concordant with their position in the other gene trees) found by Phylo-MCOA.

2.3. *Evolution of the Cell-Wall*

2.3.1. Cell-Wall Architecture of Extant Organisms

For each one of the 85 bacterial species, a dedicated survey of the literature was conducted (Table S2). When no information about the cell-wall architecture was available at the species level, we searched at a higher taxonomic level, sometimes up to the phylum. Based on the collected data, we summarized the cell-wall architecture using two different traits: the number of membranes and the presence and thickness of the peptidoglycan layer (Table S3). For the membrane trait, we used the following binary coding: 0 for one membrane and 1 for two membranes, whereas for the peptidoglycan trait, we used three different states: 0 for no peptidoglycan, 1 for a thin peptidoglycan and 2 for a thick peptidoglycan. Cell-wall trait analyses were then performed using BayesTraits V3 [46–48]. For *Parachlamydia acanthamoebae*, no clue about peptidoglycan thickness was found, so this trait was coded as "12", following the suggestion in BayesTraits manual (p. 9).

2.3.2. Correlation between Cell-Wall Traits

Correlation between cell-wall traits was tested by comparing the discrete independent and discrete dependent models using Bayes Factors (BF), as described in BayesTraits manual (p. 13). We applied the steppingstone sampler, using 100 stones with 10,000 iterations per stone. As this procedure only allows for the comparison of two binary traits, and as our peptidoglycan trait had three possible states, we had to combine two different states into a single state. Three different combinations were tested to check the robustness of the correlation. For case A, the absence of peptidoglycan was coded as 0 and the presence of peptidoglycan (either thin or thick) as 1. For case B, both the absence of peptidoglycan and the thin peptidoglycan were coded as 0, while the thick peptidoglycan was coded as 1. For case C, both the absence of peptidoglycan and the thick peptidoglycan were coded as 0, while the thin peptidoglycan was coded as 1. Because *P. acanthamoebae* is a Chlamydiae, which belong to the diderm-LPS group, its undocumented peptidoglycan layer (see above) was considered as thin when recoding the peptidoglycan trait.

2.3.3. Ancestral State Reconstruction of Cell-Wall Traits

For ancestral state reconstruction, the two traits were considered separately. We used the Bayesian phylogenomic tree rooted on Terrabacteria as an input tree, and further checked the robustness of our inferences to five alternative roots, all within Terrabacteria. Branch lengths were scaled to have a mean of 0.1, as suggested in BayesTraits manual (p. 10). Five different MultiState models were tested: prior exponential of 10 (model "E"), hyperprior exponential 0 to 10 (model "H1"), hyperprior exponential 0 to 100 (model "H2"), reverse-jump hyperprior exponential 0 to 10 (model "R1"), and reverse-jump hyperprior exponential 0 to 100 (model "R2"). Reversible-jump models had the opportunity to forbid some transitions (rate = 0) and/or to equate distinct rates. Ten MCMC chains were run for each combination of trait/root/model for 1,100,000 cycles, with one sample saved every 1000 cycles, and burnin set at 100,000 cycles. State probabilities and transition rates were summarized as means of the 10 × 10,000 samples. To investigate the sensitivity of the Bayesian inference of a monoderm LBCA to priors, one more analysis (biased on purpose towards reversion from diderm to monoderm state) was re-run as 100 MCMC chains with q01 and q10 exponential hyperpriors set to 0 to 1 for and 1 to 10, respectively.

2.3.4. Comparison of the Selected Models

Building on the steppingstones sampler files produced by the BayesTraits ancestral state reconstruction, we compared the fit of our five models (in a systematic pairwise fashion) to both the membrane and the peptidoglycan data (used for the ancestral state reconstruction) using Bayes Factors. We selected the steppingstones files from the runs with the tree rooted on the Terrabacteria. As above, the steppingstone sampler used 100 stones with 10,000 iterations per stone.

2.4. *Evolution of the dcw Cluster*

2.4.1. Synteny Analyses of Extant Genomes

To study the gene order of the *dcw* cluster across our 903 genomes, we developed a custom R script. This interactive interface allowed us to select any subset of genomes and to focus on any region of the bacterial chromosome chosen as the reference genome for the comparison. To maximize the robustness of these analyses, the data (genomic coordinates, orthology relationships, functions) needed for the visualization are fetched in real-time from the relational database. Examples of graphical outputs produced by this program (limited to the 85 final organisms) are shown in "synteny_85_dcw.pdf" available in the folder ProCARs. The orthologous groups corresponding to the genes of the *dcw* cluster were identified by a combination of homology searches using reference protein sequences as queries and our R interface for visual confirmation of synteny conservation. In most cases but the poorly conserved *ftsL* and *ftsQ*, a single orthologous group was found for each gene. For *ftsL* and, to a much lesser extent, *ftsQ*, several orthologous groups had to be merged, based on the presence of an unidentified gene sequence at their respective expected location, i.e., between *mraW* and *ftsI* for *ftsL*, and just before *ftsA* for *ftsQ*. Moreover, HMM profiles (pHMM) [49,50] (see also below) were built from unambiguous reference sequences to ensure proper identification of *ftsL* and *ftsQ* genes in genomes with a fragmented *dcw* cluster. Overall, *ftsL* and *ftsQ* were spread over 36 and 24 orthologous groups (many having only 2–3 sequences), respectively, whereas *mraW*, *mraZ* and *ftsA* were spread over 2, 3 and 4 orthologous groups, respectively.

2.4.2. Ancestral Gene Order Reconstruction

To reconstruct the evolution of the *dcw* cluster, we used the program ProCARs [51], modified to prevent gene inversions in the cluster (by enabling the -p option). ProCARs input files were built semi-automatically from the relational database, focusing on the 85 bacterial species of our phylogenomic analyses and informed by synteny analyses of extant genomes. Briefly, genes too far from other genes were encoded as lying on different "chromosomes" by introducing artificial telomeres. When several "orthologous" genes were available in a given genome for a specific gene, we first tried to select the gene copy lying on the artificial "chromosome" with the highest count of other *dcw* genes. If this failed due to ties, we turned to the gene copy located on the main DNA molecule (genuine chromosome or largest scaffold in the genome assembly); otherwise, as a last resort, we selected the gene copy in the same orientation as the *dcw* genes found on the genuine chromosome or largest scaffold. Finally, when two gene copies were in tandem, we considered them as a single (duplicated) gene for the purpose of the ancestral reconstruction.

2.4.3. Phylogenetic Analyses

For the single-gene analyses of the *dcw* cluster in the 85 genomes of interest, we used the 17 identified orthologous groups (possibly merged; see above) to produce trees according to two different approaches: (1) by ML using RAxML v8.1.17 under the PROTGAMMALGF (LG+F+Γ) model and (2) by BI using PhyloBayes v3.3f under the model GTR+C60+Γ, with two MCMC chains run for 10,000 cycles, with burnin of 5000 cycles and sampling every 10 cycles. Convergence was assessed as above (gene maxdiff's ranging between 0.208 and 1.000 and meandiff's between 0.013 and 0.062), with the -c option of bpcomp set to 0.25, which turned unresolved nodes to multifurcations. Then, a concatenation of 15 of the 17 genes of the *dcw* cluster was built using SCaFoS v1.30k, leaving out *ftsL* and *ftsQ* due to their poor conservation (see above). For these 15 genes, additional steps were carried out to ensure the orthology of the concatenated sequences. Briefly, we used our ProCARs input to select only the genes belonging to the *dcw* cluster (or sub-cluster) in each genome. Orthologues not supported by synteny evidence were removed from the alignments using prune-ali.pl (also from Bio::MUST::Core) before concatenation. We further filtered out sites with $\geq$50% missing character states, thereby yielding a sparser supermatrix of 85 species $\times$ 4571 AAs (8.47% missing character states). PhyloBayes MPI v1.4 was used to run two chains under the CAT+Γ model for 50,000 cycles. We chose a burnin of 10,000 cycles and kept only one sample every 10 cycles of the remaining 40,000 cycles. We selected both chains to compute the tree (maxdiff = 0.284; meandiff = 0.007), with the -c option of bpcomp set to 0.25. All trees were formatted as above.

2.5. Evolution of the Genes Related to the Outer Membrane

2.5.1. Homology Searches in Complete Proteomes

For our broader study of the taxonomic distribution of 16 genes involved in synthesis and in maintaining the integrity of the outer membrane across the 903 selected genomes (including previously discarded organisms like Thermotogae), we did not rely on synteny as those were not part of a single cluster in any organism. Instead, we searched for the orthologous groups containing unambiguous reference sequences for these genes. For each set of orthologous groups potentially corresponding to a gene of interest (merging from one to nine orthologous groups per gene), we computed an alignment over all sequences with MAFFT v7.453 (using the accurate LINSI strategy) and checked by eye if it was globally satisfactory or not, possibly after cleaning up a few divergent sequences. If the alignment was good enough, we built an HMM profile from it to search the complete proteomes of our 903 genomes using HMMER [49,50]. Then, based on the E-value, length, pHMM profile coverage, copy number and taxonomy of the HMMER hits, we selected the probably orthologous proteins using the visual software Ompa-Pa (A.R. Bertrand and D. Baurain; available at https://metacpan.org/dist/Bio-MUST-Apps-OmpaPa, accessed on 16 February 2022). In contrast, when the alignment of all sequences was too poor, we focused on the original orthologous group containing the *E. coli* sequence and tried to build a profile by adding up to 6 (for *lolB* and *lptC*) of the additional orthologous groups using an iterative strategy as implemented in the software Two-Scalp (A.R. Bertrand and D. Baurain; available at https://metacpan.org/dist/Bio-MUST-Apps-TwoScalp, 16 February 2022). Then, we followed the same route as if the pHMM had been computed from a "good-enough" alignment. For the specific case of the *bamA* gene, we first collected 28 orthologous groups containing proteins annotated as BamA, Omp85 and/or TspB, then we used InterProScan v.5.48-83.0 with default parameters and disabled use of the precalculated match lookup [52] to determine the number of POTRA domains [53] in the 1425 individual sequences. Two curated alignments based on preliminary ML trees (see below) were built: one from the five orthologous groups where the sequences mostly had 4 or 5 POTRA domains (Table S4), which we considered as the orthologues of the genuine BamA protein of true diderms-LPS, and one from five orthologous groups having 2 or 3 POTRA domains, which included the BamA "4" sequences of Cyanobacteria, as well as related proteins (i.e., BamA-like/Lipo/TamA) [54]. By "curation", we mean elimination of incomplete and/or divergent individual sequences but without discarding representatives of scarcer groups. Finally, these two alignments were used to build two pHMM profiles and perform HMMER searches as described above.

2.5.2. Taxonomic and Phylogenetic Analyses

For each gene of the 16 genes, we retrieved the list of genomes containing the (probably) orthologous proteins and tabulated the corresponding organisms at the phylum level. From these numbers, we tried to identify recurring patterns of gene distribution. For two genes, *tolA* and *ybgF*, the taxonomic distribution was discordant with respect to other genes (when present) in the atypical diderms group. In each case, only one of the expected phyla of the atypical diderms group had at least a copy, and this phylum was represented by a noticeably lower number of sequences compared to other genes present in the atypical diderms group (when they had copies of the gene). To decide if these discordances were due to genome contamination or very recent gene transfers, we aligned the sequences with MAFFT v7.453 (LINSI) and computed two phylogenetic trees using RAxML v8.1.17 under the PROTGAMMALGF (LG+F+Γ) model. Trees were also produced for the 14 other genes associated with the outer membrane following the same method. All trees were formatted as above, with unresolved nodes (BP < 25%) turned to multifurcations.

3. Results

3.1. A Robust Tree of the Bacterial Domain

To serve as the base for evolutionary analysis of the cell-wall architecture and reconstruction of the ancestral gene order in the *dcw* cluster, we needed a tree of Bacteria. With the growing availability of fully sequenced genomes, phylogenomics has developed as a discipline using the tools of phylogenetics but applied to tens to hundreds, or even thousands, sequences of broadly conserved genes [55]. Phylogenomic trees can either be inferred from supermatrices of concatenated genes [56] or through combination of single-gene trees into supertrees [57]. Hence, the phylogenomic tree shown in Figure 1 was computed by Bayesian inference based on a dense (4.29% missing character states) supermatrix of 117 single-copy orthologous genes (see Materials and Methods) sampled from 85 representative bacterial genomes with PhyloBayes MPI under the site-heterogeneous CAT+GTR+Γ model (CATegories + Generalised Time-Reversible + Gamma) of sequence evolution [38–41]. Congruence analyses were run on the 117 individual genes using Phylo-MCOA [44] and did not reveal incongruent genes or species, beyond 62 individual sequences, which might have experienced gene transfer and/or fast evolution. Once discarded, the overall results did not change, as demonstrated by comparing two control trees (i.e., before and after outlier removal) inferred with RAxML under the LG+F+Γ model (see Figures S1 and S2). Regarding model selection, cross-validation analyses on four different models confirmed that CAT+GTR+Γ had the best fit to our dataset, followed by CAT+Γ, then GTR+Γ and finally LG+Γ (Table S5).

Our unrooted tree is in good agreement with most recent concatenating phylogenomic studies aimed at resolving bacterial evolution [58–68]. In particular, we robustly recovered a bipartition of the bacterial lineages composing the Terrabacteria and the "Hydrobacteria" (=Gracilicutes sensu [69]). Within these "megaphyla" first defined by Hedges and Battistuzzi [58], resolution was weaker, as reflected in the lower posterior probabilities at medium phylogenetic depth, whereas phyla and known superphyla (e.g., FBC, for Fibrobacteres-Bacteroidetes-Chlorobi, and PVC, for Planctomycetes-Verrucomicrobia-Chlamydia) were always clearly resolved. In the Terrabacteria, relationships between member lineages slightly varied from run to run (we ran a total of six independent chains, Figure S3), while in the Hydrobacteria (e.g., FBC, PVC, Proteobacteria), Epsilonproteobacteria were occasionally separated from other groups of Proteobacteria (Figures S4 and S5). Some additional phyla initially present in our dataset (i.e., Synergistetes, Fusobacteria and Aquificae) were excluded from the tree shown in Figure 1 because they were difficult to robustly position (e.g., due to the chimerical nature of the Aquificae) without bringing more cell-wall architecture diversity (see also [70–72]). Likewise, we further discarded the Thermotogae, which are also chimeras [70], even though their toga might be akin to a modified outer membrane [73,74] (see Figure S6 for a preliminary 101-species tree including all these lineages). Such uncertainties are not uncommon in bacterial phylogenomics and are the result of a combination of weak phylogenetic signal, widespread lateral gene transfer and systematic error (e.g., long-branch attraction artifacts) [72,75–82].

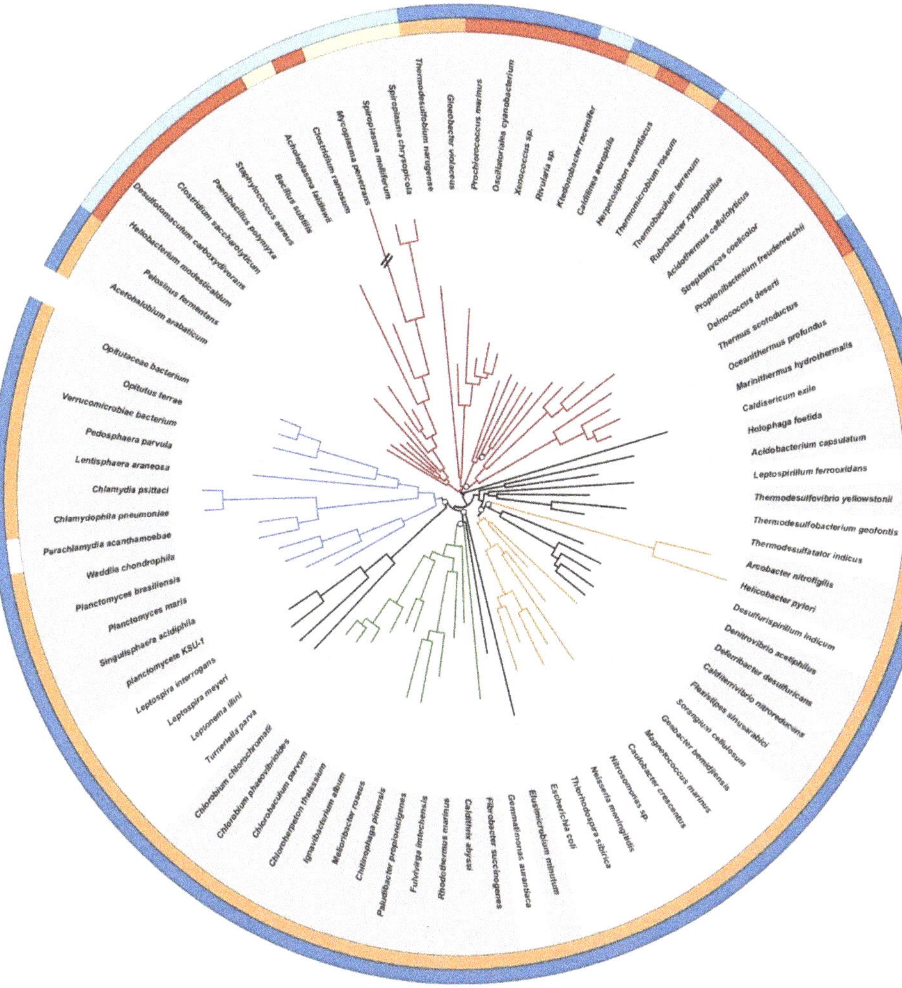

Figure 1. Phylogenomic tree of the bacterial domain based on a supermatrix concatenating 117 single-copy orthologous genes chosen for their broad conservation across Bacteria. The tree was rooted on Terrabacteria. The supermatrix had 85 species and 19,959 unambiguously aligned amino-acid positions (<5% missing character states). The tree was inferred from amino-acid sequences using PhyloBayes MPI and the CAT+GTR+Γ model of sequence evolution. Open symbols at the nodes are posterior probabilities (PP), and nodes without a symbol correspond to maximum statistical support for phylogenetic inference (posterior probabilities of 1.0; averaged over two MCMC chains). The length of the branch marked with "//" has been reduced by 50% for the sake of clarity. Colour key is red = Terrabacteria, orange = Proteobacteria, green = FBC superphylum, blue = PVC superphylum. Outer circles stand for the status of the peptidoglycan (PG) and of the outer membrane in the organisms, according to our literature survey: red = thick PG, orange = thin PG, yellow = no PG, dark blue = diderm, light blue = monoderm, white = no information. Alternating white and grey backgrounds highlight the alternance between differentially coloured groups or phyla.

Rooting the different domains of Life is not an easy issue [82]. In Figure 1, we chose to set the root of Bacteria between Terrabacteria and Hydrobacteria/Gracilicutes, following studies having included Archaea as an outgroup [25,41]. Remarkably, this basal split mirrors cell-wall architecture differences. In the first group, Firmicutes, Tenericutes, Actinobacteria, and presumably Chloroflexi (see below), are mostly monoderm bacteria. Together with the atypical diderms, i.e., Deinococcus-Thermus, Cyanobacteria, Synergistetes and Thermotogae, they compose Terrabacteria [58]. On the other hand, the remaining lineages are diderms mostly featuring lipopolysaccharides (LPS) and correspond to Hydrobacteria/Gracilicutes; these will be called "true diderms-LPS" in this study. Over time, several positions for the bacterial root have been proposed (Table S6). In the following, because our Bayesian analyses required a rooted tree, we tested several of them, yet excluding roots lying within the true diderms-LPS, which are likely monophyletic (see below). Beyond the root of Figure 1, we thus explored the effect of setting the bacterial root within Terrabacteria on our inferences.

3.2. Evolution of the Cell-Wall Architecture

To study the evolution of the cell-wall architecture, we carried out a thorough literature survey on all the bacteria kept in our tree (Tables S3 and S4). For each organism, we collected the number of membranes, the presence and thickness of the peptidoglycan layer and, if relevant, the type of spore, as there exists evidence of potential functional connection between sporulation and cell-wall remodelling processes [13,14]. However, preliminary analyses showed that the spore trait was difficult to encode reliably in terms of homologous states. Therefore, it was eventually discarded, whereas the two traits linked to the cell wall itself were analysed using BayesTraits under the MultiState model.

Based on this survey (Tables S3 and S4), most bacterial phyla have two membranes (diderm architecture) and a thin peptidoglycan layer. For example, Proteobacteria, Nitrospirae, Acidobacteria, Bacteroidetes and Chlorobi fall into this category and correspond to true diderms-LPS lineages. For the organisms belonging to the PVC superphylum, this architecture might be slightly different [83]. Actinobacteria are essentially monoderms with a thick peptidoglycan, whereas Firmicutes and Chloroflexi both have monoderm and diderm representatives. Firmicutes include Bacilli and Clostridia, two groups of endospore formers. Clostridia and Bacilli correspond to two well-defined classes, sharing many traits though being also very distinct. All Bacilli and most Clostridia are monoderms with a thick peptidoglycan, but some Clostridia [84] (Halanaerobiales and Thermoanaerobacteriales) and the Negativicutes have two membranes (some with lipopolysaccharides in the outer membrane) and a relatively thin peptidoglycan layer [16,85,86]. Regarding the status of the Chloroflexi cell-wall architecture, it is still controversial [68,87,88]. Beside these canonical diderm and monoderm phyla, respectively corresponding to classical Gram- and Gram+ bacteria, there exist a series of organisms with atypical cell-wall architectures. Hence, Deinococcus-Thermus and Cyanobacteria are diderm bacteria with an outer membrane, but their cell walls differ from those of the true diderms-LPS by having a thick peptidoglycan instead of a thin layer (Table S2).

Consequently, the number of membranes observed in the extant organisms is either one (state 0) or two (i.e., there is an outer membrane, state 1; Table S3). The evolutionary analysis of this trait suggests a LBCA surrounded by only one membrane. This inference is robust to five model variants (E, H1, H2, R1 and R2; see Materials and Methods) and six different positions for the bacterial root (P(0) = 94.2% to 98.2%; Figure S7). Due to the robustness of our results to alternative rootings, we will only present those obtained with a root located between Terrabacteria and true diderms-LPS (as in Figure 1). In accordance with the inference of a monoderm LBCA, the posterior transition rates indicate that it is easier to gain (q01) an outer membrane (range of the five model's mean = 2.288–2.495, Table 1) than losing (q10) an existing one (range = 0.008–0.132). If we try to alter the H1/H2 model hyperpriors to promote the loss (q10 = 1–10) at the expense of the gain (q01 = 0–1), the LBCA remains inferred as a monoderm in 67.1% of the cases (mean P(0)), whereas it is inferred as a diderm in 32.9% of the cases (mean P(1)) (Table 1). Concerning the rates, the inferred loss rate remains weak (mean q10 = 0.000–0.187; Table 1), while the distribution of the gain rate (q01) becomes bimodal, with a mode at 0.2 and another at 1.8 (Figure S8A) and remain low for the loss rate (q10) (Figure S8B). Consequently, under this extreme parameterization, we distinguish two main configurations for the pair of rates (Figure S8C) and the monoderm probability P(0) (Figure S8D).

Table 1. Overview of BayesTraits results. qij design posterior transition rates, whereas P(i) correspond to posterior ancestral state probabilities. For the membrane (MBN) trait, state 0 = one MBN and state 1 = two MBN, while for the peptidoglycan (PG) trait, state 0 = no PG, state 1 = thin PG and state 2 = thick PG. "H biased" is the model where the hyperprior has been purposely biased to favour a diderm LBCA (see Materials and Methods for details).

Node	Trait	Statistic	E	H1	H2	R1	R2	H Biased
LBCA	MBN	mean q01	2.495	2.352	2.477	2.288	2.411	1.431
LBCA	MBN	mean q10	0.132	0.113	0.121	0.012	0.008	0.210
LBCA	MBN	mean P(0)	94.951	94.204	95.375	97.134	98.161	67.092
LBCA	PG	mean P(0)	22.068	4.022	38.604	0.397	0.594	N/A
LBCA	PG	mean P(2)	76.497	94.622	60.147	99.535	99.358	N/A
LBCA	PG	mean q01	4.626	1.634	7.317	0.798	0.827	N/A
LBCA	PG	mean q02	6.935	2.020	20.967	0.953	1.041	N/A
LBCA	PG	mean q10	0.166	0.102	0.187	0.000	0.000	N/A
LBCA	PG	mean q12	0.128	0.109	0.118	0.001	0.000	N/A
LBCA	PG	mean q20	2.088	0.937	4.941	1.347	1.413	N/A
LBCA	PG	mean q21	1.890	2.165	1.600	1.398	1.419	N/A
Firmicutes	PG	mean P(0)	17.631	3.936	30.120	0.611	0.738	N/A
Firmicutes	PG	mean P(2)	81.891	95.648	69.435	99.378	99.237	N/A

In the 85 extant organisms considered in our study, the peptidoglycan layer is either absent (state 0), present and thin (state 1) or present and thick (state 2; Table S3). The LBCA is inferred with a thick peptidoglycan. While this result is robust to alternative positions of the root, some models (E and H2) let the possibility open (22.0–38.6%, Table 1) for the LBCA having been devoid of peptidoglycan (Figure S9). Moreover, the posterior rates are highly heterogeneous, depending on the transition considered, and present a sensitivity to the model used (mean range = 0.000–20.967; Figure S10 and Table 1). Based on the values of the rates, the thin peptidoglycan state (state 1), once acquired, is unlikely to change towards another state, whereas the other two states (states 0 and 2) can exchange freely or change towards the thin peptidoglycan state (Figure S10 and Table 1).

In a second step, we used BayesTraits to reconstruct the state of the characters for the Last Common Ancestor (LCA) of every one of the 15 bacterial phyla included in our study, as well as the LCA of several larger groups (e.g., PVC, Terrabacteria), still based on the Terrabacteria root (Figure 2). As expected, the LCA of the true diderms-LPS bacteria is inferred as a diderm organism featuring a thin peptidoglycan layer, whereas the Terrabacteria LCA is reconstructed as a monoderm with thick peptidoglycan. The results obtained for the larger groups are homogeneous across the different models (Figure S11). For Firmicutes, which is the only phylum with some architectural diversity in our dataset, two of the five models (E and H2) do not completely settle on an LCA monoderm with a thick peptidoglycan, and instead do not dismiss an LCA without peptidoglycan (17.6% and 30.1%, respectively; Table 1). Finally, a comparison of the fit of the five models using Bayes Factors (Table 2) showed that model R1 was the best, followed by models R2, H1, E, and finally H2. Therefore, the two models that do not fully agree with the others about the peptidoglycan trait are also those that are deemed less fit by Bayes Factors (E and H2).

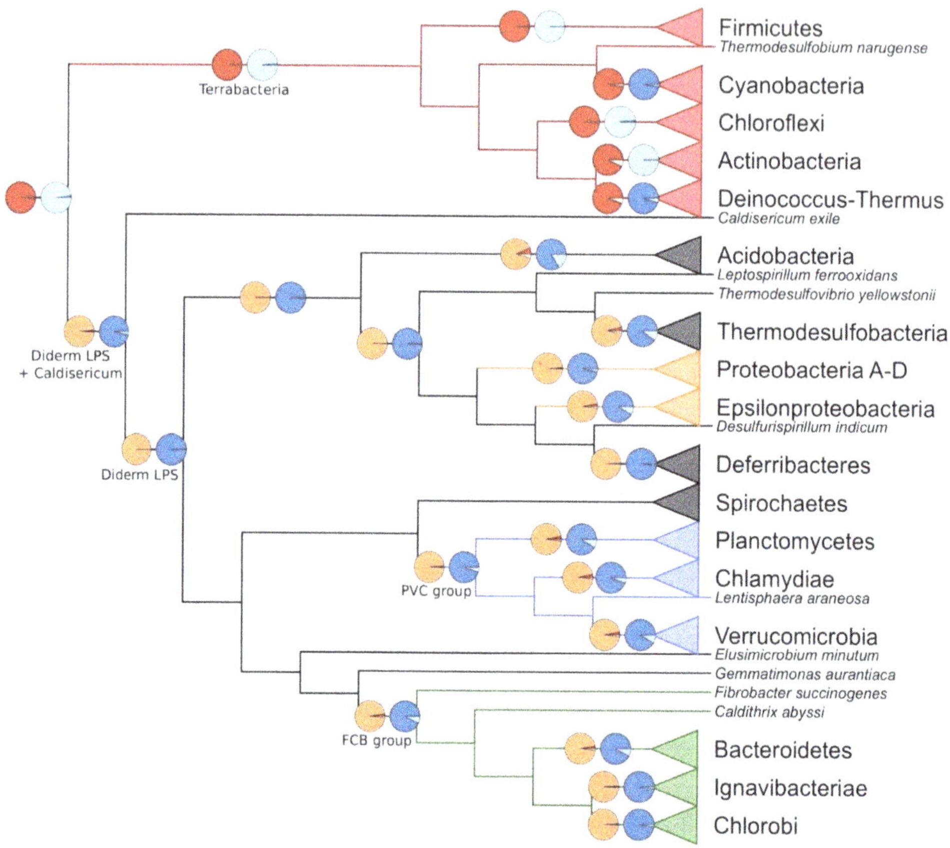

Figure 2. Cladogram derived from the tree of Figure 1 featuring the cell-wall architecture inferred for selected last common ancestors among Bacteria. Colour key is red = Terrabacteria, orange = Proteobacteria, green = FBC superphylum, blue = PVC superphylum Branches ending with a triangle represent collapsed groups (for details, see Figure 1 or Table S3). The pie chart sectors correspond to the posterior probabilities of the model reverse-jump hyperprior exponential 0 to 100 (R2). Colour key is red = thick PG, orange = thin PG, yellow = no PG, dark blue = diderm, light blue = monoderm.

Table 2. Pairwise comparisons of BayesTraits model fit using Bayes Factors (BF). BF > 2 are interpreted as positive evidence, $5 \leq$ BF < 10 as strong evidence and BF > 20 as very strong evidence in favour of the more complex model [89].

Complex	Simple	MBN	PG
R1	H2	7.41	22.86
	E	5.95	17.47
	H1	2.69	8.38
	R2	2.42	1.91
R2	H2	4.99	20.95
	E	3.53	15.56
	H1	0.27	6.47
H1	H2	4.71	14.47
	E	3.25	9.09
E	H2	1.46	5.39

Hitherto, the two cell-wall traits were analysed separately, owing to the limitations of the MultiState model used. However, from a biological point of view, their evolution might be correlated. To account for this possibility, we conducted the BayesTraits procedure to estimate the correlation between two traits, which revealed that the peptidoglycan and the membrane characters are indeed linked. The actual strength of the correlation depended on the scheme used to recode the three-state peptidoglycan trait into a binary character, which was needed to estimate the correlation with the membrane trait (see Materials and Methods). When the coding scheme rewarded the mere presence of the peptidoglycan layer, whatever its thickness, the correlation was supported by strong evidence (log Bayes Factor for case A = 9.0), while it raised to very strong evidence when the scheme emphasized either a thick peptidoglycan (case B = 27.6) or a thin peptidoglycan (case C = 37.8). These differences in correlation can easily be explained. In case A, almost all organisms of our study without peptidoglycan are also deprived of the outer membrane (see *Parachlamydia acanthamoebae* in Figure 1), whereas organisms with a peptidoglycan layer often have an outer membrane. In case B, all organisms without peptidoglycan or with a thin peptidoglycan layer are put in the same category. In our study, all organisms with a thin peptidoglycan layer have an outer membrane, and they are more numerous than the organisms without a peptidoglycan layer. In case C, the organisms with a thin peptidoglycan layer have their own category and, in our study, all these organisms also feature an outer membrane.

3.3. *Evolution of the Gene Order within the dcw Cluster*

Initially, we studied the organization of the *dcw* cluster in extant organisms based on the output of a custom visualization software showing orthologous gene groups in their syntenic context (see Materials and Methods for details and "synteny_85_dcw.pdf" available in the folder ProCARs from our Figshare, for the status of the *dcw* cluster in the 85 bacteria of our phylogenomic tree). This approach led us to identify the orthologous groups for the 17 genes of (the most complete form of) the *dcw* cluster. In Cyanobacteria, the nearly total absence of the *dcw* cluster is noteworthy: *mraZ* and *ftsA* are missing from all cyanobacterial genomes examined, and all other genes of the cluster are generally present but completely dispersed on almost as many loci as the number of genes, with some exceptions, the doublet *murC* and *murB* or the doublet *ftsQ* and *ftsZ* (see .xlsx file available in the folder ProCARs). The *murA* gene can be found in clusters or sub-clusters in several genomes. The complete form of the *dcw* cluster is only seen in a single order of Clostridia, the Halanaerobiales (more precisely, in *Acetohalobium arabaticum*). Halanaerobiales are robustly affiliated to Firmicutes yet branching at the root of the phylum [90]. However, *murA* is also present in sub-clusters in Cyanobacteria, Planctomycetes, Lentisphaerae and *Caldithrix abyssi*. Otherwise, if present in the genome, *murA* is usually outside of the *dcw* cluster. Beside this specific gene and particular phyla, several true diderms-LPS phyla are characterized by the loss of specific genes from the cluster (*ftsW* in Thermodesulfobacteria, *murB* and *ddlB* in the FBC superphylum, *ftsA* and *ftsZ* in Chlamydiae and Planctomycetes) (see .xlsx file available in the folder ProCARs).

Taking the rooted phylogenomic tree of Figure 1 as an evolutionary framework and the orthologous groups identified just above as input extant data, we used a new variant of a homology-based reconstruction method (ProCARs) [51] to retrace the evolution of the organization of the *dcw* cluster in our 85 representative organisms. Our reconstruction shows that both the LBCA and the LCA of the Terrabacteria group were organisms featuring a complete 17-gene *dcw* cluster. In contrast, the reconstructed cluster for the ancestor of the true diderms-LPS group included 16 genes, with the *murA* gene outside of the cluster (even if present in the genome). Detailed study revealed that the *murA* gene was also outside of the main cluster in every reconstructed ancestor among true diderms-LPS (Figure 3A). This gene is at best found on a small sub-cluster, and most of the time it exists as a singleton. An example of such a small sub-cluster reconstructed by ProCARs can be seen in the LCA of the FBC superphylum where *murA* and *murB* are in tandem. A parsimonious way to explain these observations would be that the *murA* gene has left the *dcw* gene cluster (but persisted in the genome) of the LCA of true diderms-LPS and the LCA of Actinobacteria, Deinococcus-Thermus and Chloroflexi (assuming these three phyla share a common ancestor). Alternatively, it was lost independently in the three latter phyla. Overall, the *dcw* cluster is conserved in almost all high-level ancestors down to the phyla (see Figure 3A for a summary and .xlsx file available in the folder ProCARs, for details). This conservation mostly takes the form of a single cluster (e.g., Proteobacteria LCA) or of a limited number of sub-clusters, with the synteny retained within individual sub-clusters (e.g., Chloroflexi LCA, Planctomycetes LCA). Thus, the *dcw* cluster appears as an ancient locus with mainly a history of gene loss or gene delocalization, but likely no gene gain since its establishment before the advent of the LBCA.

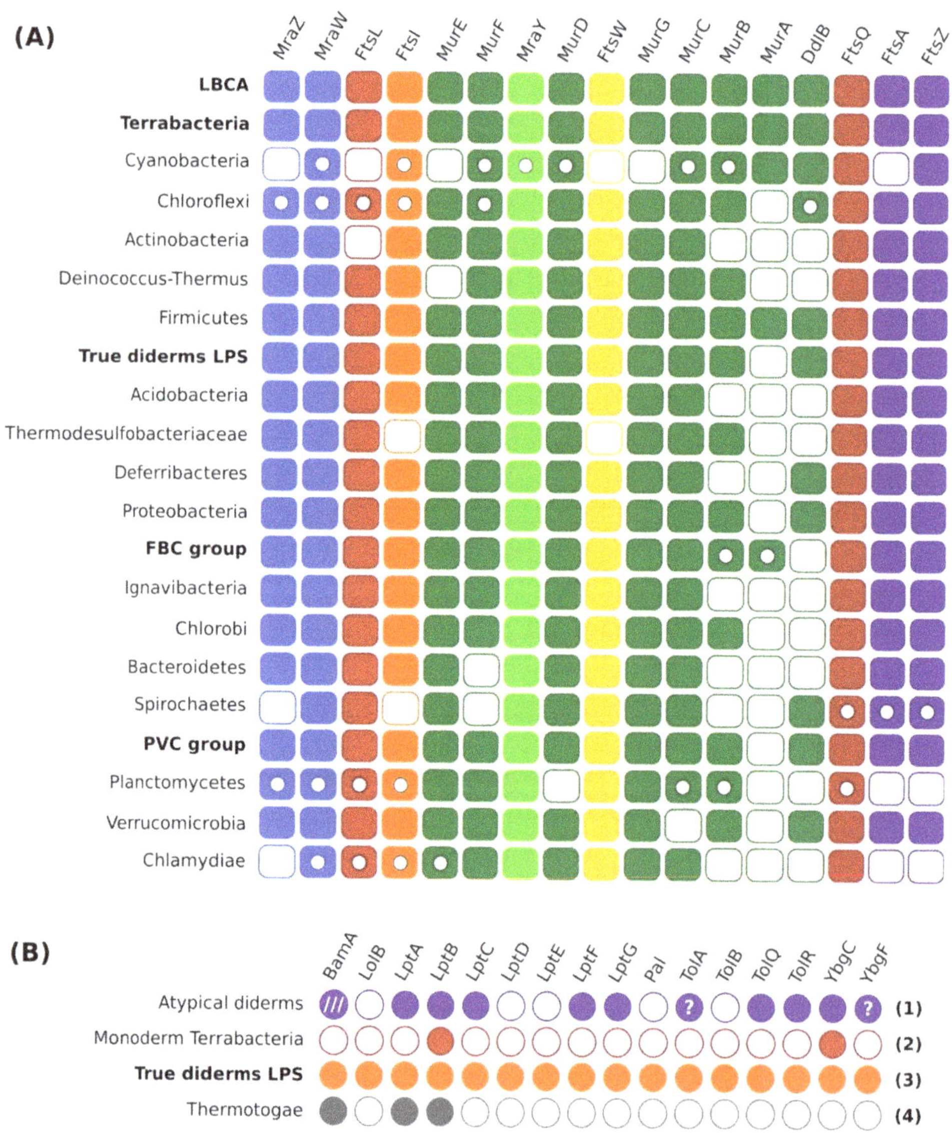

Figure 3. Overview of gene distribution and synteny analyses. (**A**) ProCARs results for *dcw* cluster

organization in selected LCA among Bacteria. Full rectangle = gene present and in the main cluster; empty circle in rectangle = gene present but in a sub-cluster; empty rectangle = gene present but outside of any cluster. Note that the reconstruction procedure prevents the complete lack of a gene in an ancestral genome. (**B**) Recurring distribution patterns at the phylum level for the proteins involved with the outer membrane. Full circle = gene present in the group; empty circle = gene absent in the group; "?" in a circle = potential presence of the gene in the group; /// = presence in a sub-group only (i.e., Deinococcus-Thermus). Numbers in bold are the pattern numbers. Names written in bold are the names of groups regrouping several phyla.

Phylogenetic trees for the 17 genes of the *dcw* cluster were computed from protein sequences, but these trees are not well resolved ("DCW_17_SG.pdf" available in the folder Trees). Known phyla can be supported by low to high bootstrap proportions (BP: 9–100%) and posterior probabilities (0.3–1.0), while the support is always too low to resolve the relationships between phyla, even though general trends, such as the bipartition between Terrabacteria and true diderms-LPS (Firmicutes–Chloroflexi–Actinobacteria–Deinococcus-Thermus vs. Proteobacteria–FBC–PVC), are observable in several single-gene trees. Moreover, trees inferred from genes often found outside of the *dcw* cluster (e.g., *murC*, *murB* and *ddlB*) are blurrier than those computed from genes kept in the cluster. Finally, the trees of the genes *ftsQ* and *ftsL*, for which the orthologous groups had to be manually reconstructed (see Materials and Methods) are particularly chaotic. In contrast, the *mraY* tree (Figure S12) is better supported (BP: 39–100%; posterior probabilities: 0.5–1.0) at the phylum level and is the most congruent with the tree resulting from the 117-gene supermatrix (Figure 1). When concatenated, the *dcw* genes (all but *ftsQ* and *ftsL*) recover a similar tree (Figure S13), notably featuring the Terrabacteria group, the FBC group and the true diderms-LPS, but with one exception: the PVC group is split in three, with the Planctomycetes and Verrucomicrobia on one side, the Chlamydia on the other side and the Lentisphaerae within the FBC group. This suggests that the *dcw* cluster mostly experienced a vertical evolution.

3.4. Evolution of the Genes Related to the Outer Membrane

According to our ancestral reconstruction of the cell wall, the LBCA had a single membrane around its cell, which implies that the atypical diderms lineages within Terrabacteria (Cyanobacteria, Deinococcus-Thermus and some Firmicutes, i.e., the Halanaerobiales and the Negativicutes) had to acquire their outer membrane independently and in distinct events from the event at the origin of true diderms-LPS. At face value, this inference might seem less parsimonious than hypothesizing a diderm LBCA and multiple independent outer membrane losses over the evolution of the bacterial domain, as suggested repeatedly [5,25,68]. To decide whether the outer membrane could indeed have evolved several times independently, we studied the taxonomic distribution of 16 genes involved in outer membrane synthesis and integrity: *bamA*, *lolB*, *lptA*, *lptB*, *lptC*, *lptD*, *lptE*, *lptF*, *lptG*, *pal*, *tolA*, *tolB*, *tolQ*, *tolR*, *ybgC*, *ybgF*. Briefly, BamA is the main protein of the Bam complex (to which the other Bam proteins attach to), which is responsible for the assembly of beta-barrel proteins in the outer membrane [91]. LolB is the only outer membrane-anchored protein of the Lol pathway, which delivers lipoproteins to the outer membrane [3]. The Lpt system (LptA to LptG) ensures the transport of the lipopolysaccharides from the cytoplasm to the outer membrane [92]. Finally, the Tol-Pal system (Pal, TolA, TolB, TolQ, TolR, YbgC, YbgF) is involved in the uptake of colicin, the uptake of filamentous bacteriophage DNA and the integrity of the outer membrane [93].

The distribution of these genes was examined across our first selection of 903 bacterial genomes (all genomes even the previously discarded ones) using curated Hidden Markov Model (HMM) profiles built from orthologous groups including *E. coli* reference sequences and complemented by phylogenetic analyses when orthology was doubtful (see Materials and Methods for details). These results were then summarized at the phylum level to identify recurring patterns of gene distribution (Figure 3B and "OM_genes_presence-hmms.csv" available in the folder Outer_membrane, for details), while single-gene trees inferred from the corresponding protein sequences are available ("LBCA_OM_16_SG.pdf" available in the folder Trees). Altogether, our study of the genes encoding the proteins BamA, LolB, the Lpt system and the Tol-Pal system revealed four different patterns of presence/absence in bacterial phyla with diderm organisms. These four gene distribution patterns correspond to: (1) "atypical diderms" (see references in Table S2), i.e., Cyanobacteria, Deinococcus-Thermus and diderm Firmicutes; (2) "monoderm Terrabacteria", i.e., Chloroflexi, of which some may be monoderms but all are devoid of lipopolysaccharides [68,87], Actinobacteria, and monoderm Firmicutes; (3) "true diderms with LPS" (TDL = typical Gram–bacteria); (4) Thermotogae, in which the outer membrane has been replaced by a toga made of structural proteins and polysaccharide hydrolases (xylanases) [73,74,94]. Below, we briefly comment on these gene distributions from a functional perspective.

First, according to our comprehensive homology searches, *bamA* is exclusive to true diderms-LPS, Deinococcus-Thermus and Thermotogae, even though the latter lack nearly all other outer membrane-related genes studied here. This result suggests a true diderms-LPS origin for Thermotogae, which are now considered as chimeras partly derived from (or at least related to) Aquificales [70,72,95]. This chimerical nature of Thermotoga is the reason we did not include them in our phylogenomic tree (see above). Regarding the presence of the *bamA* gene in the atypical diderms of the group Deinococcus-Thermus, it has already been reported [96] and this result appears less compatible with a monoderm LBCA. However, in other atypical diderms, we could not find a genuine BamA protein. Instead, Cyanobacteria and diderm Firmicutes feature proteins that have a quite different domain architecture (see BamA4 and BamA-like in Heinz et al., 2014 [54]) and for which the orthology (i.e., overall sequence similarity due to vertical descent only) with the typical BamA is at best dubious. Therefore, we currently disagree with the idea that BamA per se would be common outside true diderms-LPS [97]. Nonetheless, BamA, taken as a family regrouping the typical BamA, "BamA4" and "BamA-like" proteins, might indeed be an essential family (each sub-group sharing a similar function) to all diderm (i.e., featuring an outer membrane) but its members do not necessarily share a vertical transmission from a single ancestral protein. To verify this hypothesis would require a whole new study and is thus not expanded in the current article. Second, *lolB* is exclusive to Proteobacteria, a member of true diderms-LPS, whereas *lptB* (Lpt system) and *ybgC* (Tol-Pal system) are found in all (or almost every) bacterial phylum of our selection of 903 genomes (including Chloroflexi) and are thus not informative about the origins of the outer membrane. It is likely that these two genes have function(s) outside their respective system, functions that could be unrelated to the outer membrane. This has already been proposed for *ybgF*, which might be part of a protein network involved in phospholipid biosynthesis [98]. On the opposite, the LptB protein is known to assemble with LptF and LptG to form an ABC transporter for lipopolysaccharides [92,99], but the two corresponding genes are apparently lacking in Acidobacteria (true diderms-LPS), Tenericutes and Chloroflexi. Perhaps unexpectedly, this is also the case for Actinobacteria, these monoderm bacteria further sharing with Chloroflexi the same distribution pattern for the 16 genes involved with the outer membrane.

Beyond *lptB* and *ybgC*, the Lpt and Tol-Pal systems are found in both atypical diderms and true diderms-LPS but to a different extent. Indeed, both systems are present in atypical diderms, albeit only in a largely reduced form, whereas in true diderms-LPS, they range from a largely reduced form (e.g., Chlamydiae or Planctomycetes) to a (almost) complete form (e.g., Proteobacteria or Bacteroidetes), and this distribution is phylum-specific (Figure 3B). Hence, two genes from each system are only present in (most) true diderms-LPS genomes, *lptD* and *lptE* on one side, *pal* and *tolB* on the other side, whereas all four genes are never found in atypical diderms genomes. Regarding *tolA* and *ybgF*, they may or may not be exclusive to true diderms-LPS, depending on the biological reality of their scarce occurrence in some organisms belonging to atypical diderms (Firmicutes for *tolA* and Cyanobacteria for *ybgF*). Based on our trees of the corresponding proteins, the dubious sequences (denoted by "?" in Figure 3B and by stars in "OM_genes_presence-hmms.csv" available in the folder Outer_membrane) are sisters to Bacteroidetes (member of true diderms-LPS) in both cases, plus one case with a sequence sister to Moraxella in *tolA* tree (Figures S14 and S15, see also "LBCA_OM_16_SG.pdf" available in the folder Trees). Therefore, provided they are not the product of genome contamination [100], these genes are unlikely to have been vertically inherited.

From a functional point of view, the genes retained by atypical diderms for the Lpt system (*lptA*, *lptB*, *lptC*, *lptF* and *lptG*) are involved in the transport of the lipopolysaccharides from the cytoplasm to the outer membrane and thus are not directly associated to the outer membrane itself, contrarily to *lptD* and *lptE*, which form a complex at the outer membrane that may serve as the recognition site for the lipopolysaccharides [101]. Similarly, for the Tol-Pal system, atypical diderms genomes lack *pal* and *tolB*, two genes encoding proteins located in the periplasm and therefore directly associated to the outer membrane [102,103]. Overall, the Lpt and Tol-Pal systems in atypical diderms are thus restricted to components that might have a function in the absence of an outer membrane.

Remarkably, the genes of the Tol-Pal system are clustered in most genomes of Proteobacteria and Chlorobi, as well as in the lone genomes we studied within Fibrobacter and Gemmatimonadetes, and sporadically in those of Verrucomicrobia and Acidobacteria (available in the folder Outer_membrane sub-folder synteny_output). As all these lineages belong to the true diderms-LPS, we cannot exclude that the conservation of the Tol-Pal cluster appears patchier than it really is, owing to uneven levels of genome assembly. Regarding the genes of the Lpt system, they are not clustered in any of the genomes examined, except in Proteobacteria, where five of the seven genes are grouped on two loci (*lptFG* and *lptABC*) (available in the folder Outer_membrane sub-folder synteny_output). Nevertheless, as the synteny of the genes of both Lpt and Tol-Pal systems was only studied in the 85 genomes of our phylogenomic tree, we may have missed non-Proteobacterial genomes in which some of the *lpt* genes are indeed clustered, as reported in the recent study of Taib et al. [17].

4. Discussion

The nature of the LBCA is unknown, especially the architecture of its cell wall. The lack of reliably affiliated bacterial fossils outside Cyanobacteria [104] makes it elusive to decide the very nature of the LBCA. Nevertheless, phylogenomic inference leads to informative results, and our analysis of the cell-wall characteristics of extant bacteria, combined with ancestral state reconstruction and distribution of key genes, opens interesting possibilities: the LBCA might have been a monoderm bacterium featuring a complete 17-gene *dcw* cluster, two genes more than in the model *E. coli* cluster. This result was also supported by the recent study of [105], in which the authors found 146 protein families that formed a predicted core for the metabolic network of the LBCA. From these families, phylogenetic trees were produced and the divergence of the modern genomes from the root to the tips was analysed. It appears that the Clostridia (a class of Firmicutes) are the least diverged of the modern genomes and thus the first lineage to diverge from the predicted LBCA were similar to the modern Clostridia. Based on these results, the authors suggested that the LBCA could have been a monoderm bacteria.

As diderm bacteria are not monophyletic, whatever the root used for the bacterial domain, our reconstruction of a monoderm LBCA implies that the diderm character state has appeared several times, which goes against the principle of parsimony commonly invoked in such matters [68]. Indeed, acquiring an outer membrane is more than a simple mutation: it requires the acquisition of a whole new complex system. This makes the "monoderm-first" result counter-intuitive to the opposite of the alternative, widely held "diderm-first" hypothesis, in which the outer membrane is an ancestral feature having evolved only once in the LBCA and later lost in monoderms [5,17,25,68]. However, such an observation can be made in Archaea, where most of the studied organisms have a monoderm cell wall featuring a S-layer and/or pseudomurein, methanochondroitin and protein sheaths. In this context, some diderm Archaea have been reported in different distant phyla, like the Crenarchaeon *Ignicoccus hospitalis*, the Euryarchaeon ARMAN (Archaeal Richmond Mine Acidophilic Nanoorganisms) or the *Candidatus Altiarchaeum hamiconnexum* (SM1 Euryarchaeum) in the DPANN group [106]. Although it has not been proved that a monoderm cell wall is the general architecture in Archaea, the discovery of diderm Archaea within different phyla shows that acquisition of a second membrane has occurred multiple times during archaeal evolution. Moreover, our results are model-based, congruent across different roots and models and robust to a heavily biased hyperprior towards the diderm-first hypothesis. It contrasts with other recent studies, which do not rely on probabilistic models [5,68] and conclude to a diderm LBCA, based on qualitative considerations. That being said, the diderm-first view has also been supported in the recent work of Coleman et al. [25]. The latter study featured a reconciliation tree and infered the diderm state of the LBCA based on the genes involved in lipopolysaccharides synthesis and the flagellar subunits, notably PilQ, which is part of the Type IV pili. While the approach of Coleman and co-workers was also model-based, it differed from ours by first inferring the gene catalogue of the LBCA and then deducing its cell-wall architecture, whereas we directly infer the LBCA architecture and then studied the underlying gene distribution patterns to corroborate our inference. It is of note that the Type IV pili is also present in monoderm bacteria [107], thus its presence does not automatically entail the inference of a diderm LBCA.

Hence, following a bibliographic search for proteins with functions exclusive to diderms (without distinguishing between diderms with and without lipopolysaccharides), we identified 16 candidates: BamA, which is part of a complex assembling the proteins in the outer membrane [91], LolB, which is part of the proteins inserting the lipopolysaccharides in the outer membrane [3], the Lpt proteins, which serve as a transport chain from the inner, i.e., cytoplasmic [108], membrane (IM) to the outer membrane [92], and the Tol-Pal system, the exact function of which is still unknown but important to the integrity of the outer membrane [93]. Then, we studied the distribution of the 16 corresponding genes in 903 broadly sampled bacterial genomes. Four recurring patterns of outer membrane gene distribution were identified (Figure 3B): (1) atypical diderms (Deinococcus-Thermus, Cyanobacteria and diderm Firmicutes), (2) monoderm Terrabacteria (Actinobacteria, Chloroflexi and monoderm Firmicutes), (3) true diderms-LPS, and (4) Thermotogae. Thermotogae have chimerical genomes [70] and are likely derived with respect to other bacteria; thus, their cell-wall architecture is of secondary origin. Therefore, we do not elaborate further on their case. For similar reasons, the atypical cell-wall of the Corynebacteriales (an order of the Actinobacteria phylum) is not considered in this work. Indeed, Corynebacteriales are positioned deeply within Actinobacteria [109], which again implies a secondary origin for their peculiar cell-wall architecture.

From these patterns, it appears that even monoderm Terrabacteria share some genes involved with the outer membrane despite their lack of an outer membrane. It implies that these genes provide at best circumstantial evidence concerning the presence or the absence of an outer membrane. Thus, solely relying on their detection to infer the presence of an outer membrane would be hazardous. In the study of Coleman et al. [25], the authors build upon two types of genes to justify their inference of a diderm LBCA: the genes involved with the lipopolysaccharides synthesis and the genes involved with the pili type IV. However, our results show that the mere presence of lipopolysaccharides genes is an unreliable feature to infer the presence of an outer membrane, given that even monoderm bacteria can carry some of them. Similarly, the study of [107] showed that the type IV pili is not exclusive to the diderm bacteria. Therefore, the inference of a diderm LBCA by Coleman et al. was based on genes that only provide ambiguous evidence for the outer membrane.

Pattern 2 shows that Chloroflexi share the same gene distribution as monoderm Terrabacteria, despite being mostly considered as diderms (3 out of 4 genomes) in our reconstruction of the cell wall. Currently, there is still debate on whether Chloroflexi are monoderm or diderm organisms, microscopical observations having been inconclusive so far but hinting at the presence of an outer membrane in some Chloroflexi [87,88]. The fact that they share the same outer membrane gene distribution pattern as monoderm Terrabacteria is a clue in favour of Chloroflexi having only one membrane too. In this case, our reconstruction of the LBCA's cell wall would have had a small bias towards the diderm state and, despite that unwarranted handicap, we still recovered the LBCA as a monoderm bacterium. In our opinion, this result can be taken as more evidence for a genuinely strong signal for a monoderm LBCA.

Patterns 1, 2 and 3 may be arranged following a gradual complexification, with pattern 2 being the simplest, pattern 1 the intermediate and pattern 3 the most complex. The study of the functions of the proteins characterizing the different patterns reveals that pattern 3 is the only one including proteins directly involved with the outer membrane (i.e., linked to the outer membrane), whereas pattern 1 only includes proteins indirectly involved with the outer membrane (i.e., linked to the IM or interacting with the IM or located in the cytoplasm) and pattern 2 only includes proteins indirectly involved with the outer membrane and located in the cytoplasm. Although we know (some of) the outer membrane pathways functioning in true diderms-LPS, for atypical diderms, we only identified the common parts between their pathways and the true diderms-LPS pathways. The rest of the true diderms-LPS pathways should have an equivalent in the atypical diderms pathways but our approach by candidate genes did not allow us to identify them. This hints at the possibility of a different evolution from a common base, as some of the functions performed by the genes present in pattern 3 (true diderms-LPS) but absent in pattern 1 (atypical diderms) should be carried out in one way or another (e.g., the maintenance of the outer membrane or the outer membrane invagination during cell division) [110]. In this case, the common base would be the partial Lpt and Tol-Pal systems, upon which at least two different systems for handling the outer membrane would have built in the true diderms-LPS and (all or some) atypical diderms. On the other hand, if the LBCA was a diderm, then extant monoderms would have been the result of several independent secondary simplifications. Consequently, the monoderms dispersed within the Terrabacteria group would share the same origin, a diderm ancestor, but would not necessarily end up with the same remaining genes after their respective simplification. Yet, they all display the same single pattern (pattern 1).

Assuming a monoderm LBCA, single-gene trees might suggest that some outer membrane genes found in atypical diderms (e.g., LptF and LptG) stem from horizontal transfer from true diderms-LPS, rather than through vertical inheritance from a diderm LBCA ancestor. However, because most of these trees are poorly resolved (despite good multiple sequence alignments), the evidence is weak at best. Based on a parsimony reasoning, the exclusivity of pattern 3 to true diderms-LPS and the fact that it is shared between all of them suggest, alongside their well-supported branch in our phylogenomic tree, the monophyly of the true diderms-LPS group. Indeed, if all current genomes of a group have the same subset of genes, the LCA of the group is likely to have had these genes (in a form or another). If correct, the bacterial root cannot lie within true diderms-LPS and as already mentioned, a root on (or within) Terrabacteria implies that the diderm cell-wall architecture appeared at least on two separate occasions. The latter inference is necessary to account for diderms other than true diderms-LPS in Firmicutes, Cyanobacteria, Chloroflexi and Deinococcus-Thermus, which then raises the issue of how the lipopolysaccharides are transported from the IM to the outer membrane for these atypical diderms nested within Terrabacteria. Indeed, they do not share the same Lpt system as true diderms-LPS as theirs is "reduced", so they must have developed another system grafted (or not) onto the "reduced" Lpt system.

Another clue that might confirm our reconstruction is that the rare organisms amongst the CPR (Candidate Phylum Radiation, also known as Patescibacteria [62,111]) to have been described to feature a monoderm cell-wall architecture [112]. In several trees including the CPR (with the Archaea used as the outgroup), these are the first to diverge from the other bacteria, while the remaining of those trees have the same structure as ours [64,65].

However, in [25,113], the CPR subtree is found within the Terrabacteria with strong support. Consequently, depending on the accepted topology, the CPR could either be another (small) clue for a monoderm LBCA (CPR at the base of the bacterial tree) or only for a monoderm ancestor for the Terrabacteria group (CPR within the Terrabacteria group). Nonetheless, as most CPR genomes still lack detailed reliable information about the cell-wall architecture of the corresponding organisms, there was no point adding them to our study for now.

When it comes to the reconstruction of the *dcw* cluster, the LBCA is inferred as featuring a complete 17-gene cluster. This complete cluster has probably been vertically transmitted since then and often subject to parallel reduction, either by escape of one or several genes from the cluster or by disappearance of those genes from the genome. As it is shared by both monoderm and diderm organisms, the *dcw* cluster does not give a clue about the issue of the number of membranes of the LBCA. However, it confirms that the LBCA had a cell wall with a peptidoglycan layer, even if it does not inform on its original thickness.

In true diderms-LPS and Terrabacteria, the *murA* gene is (almost) always absent from the main *dcw* cluster. In Firmicutes, which are at the base of Terrabacteria, this gene is nevertheless considered located within the cluster by our reconstruction, as this is the situation for five (out of nine) genomes from our selection of 85 representatives. The gene is also found in sub-clusters distributed relatively patchily across Cyanobacteria, Firmicutes, Epsilon-proteobacteria, Elusimicrobia, *Caldithrix abyssi*, *planctomycete KSU1*, and *Lentisphaera araneosa*. Both extant and reconstructed ancestors show that true diderms-LPS have excised their *murA* from the main cluster after diverging from Terrabacteria, whereas Terrabacteria kept it longer in the main cluster. However, *murA* is found located on sub-clusters in both groups.

For the moment, there is no scenario to explain the appearance of the outer membrane in the lineage leading to true diderms-LPS, but such a scenario exists for the appearance of diderms in Firmicutes: it is the failed endospore origin [11,13,15,114]. According to this hypothesis, an ancestral monoderm endospore former would have experienced a failed sporulation, thereby locking the endospore within the cell while never finishing the spore. With time, it would have become a diderm bacteria. Indeed, during sporulation, the prespore engulfed in the bacterial mother cell has two membranes. A thin layer of the mother peptidoglycan subsists between these membranes before the cortex is added around the prespore between this small layer and the outer membrane. Although not yet a diderm-LPS architecture, a cortex-less spore could be a starting point for the emergence of diderm bacteria in the specific case of Firmicutes. In 2016, Tocheva [14] amended the model by arguing that this founding event would have taken place in an ancestor not only to diderm Firmicutes but to all diderm bacteria. Regarding the origin of the outer membrane in atypical diderms other than Firmicutes, we have already mentioned that Chloroflexi might be monoderms, based on their shared pattern (pattern 2) with monoderm Terrabacteria. This leaves us with Cyanobacteria and Deinococcus-Thermus, along with the large true diderms-LPS group. Because pattern 3 looks like a complexification of pattern 1, the origin of didermia in true diderms-LPS might come from one of these atypical diderms phyla by horizontal gene transfer of outer membrane genes, followed by complexification in an ancestor of true diderms-LPS. Alternatively, true diderms-LPS ancestors might have transferred outer membrane genes to distinct ancestors of atypical diderms phyla, thus in the opposite direction. At this stage, this remains an open question because of the lack of resolution of the corresponding single-gene trees, which prevents any definitive answer. However, it is of note that the failed sporulation scenario is compatible with the inferences of [105].

5. Conclusions

Our results suggest that the LBCA might have been, against familiar parsimony reasoning, a monoderm bacteria with a thick peptidoglycan layer, which is also supported by the recent study of [105]. The reconstruction of the *dcw* cluster adds a strong hint towards an LBCA with a peptidoglycan layer but does not discriminate between a thick and a thin peptidoglycan layer. Concerning our study of the outer membrane genes, their distribution suggests that indeed a monoderm ancestor is possible, but the evidence is not decisive. Yet, further improving our results using the same methods would require a more accurate description of the cell-wall architecture of the extant organisms, notably the presence or absence of the lipopolysaccharides, an information which, in our experience, is often lacking. When available, it is concentrated in the older literature, when organisms were cultivated and characterized before being sequenced, in contrast to the numerous candidate bacterial phyla that populate recent phylogenomic trees [66,67]. Nevertheless, even older genomes do not guarantee an exploitable description, like *Rivularia* sp. (Table S2: 38). Moreover, we observe that some outer membrane genes involved with the precursors of lipopolysaccharides synthesis are also present in genomes of bacteria that does not have lipopolysaccharides on their outer membrane (or even an outer membrane), thus relying solely on the presence of specific genes to determine the presence or absence of lipopolysaccharides is not adequate.

One could argue that the current study does not concern the LBCA but the LCA of cultured (and characterized) Bacteria and we would not completely disagree as we ourselves see it as a proof of concept of the method. A follow-up would be interesting to carry out once accurate information for the cell wall of more phyla are available. In such a follow-up study, it could be interesting to add supplementary genomes such as the "rogue" lineages (e.g., Aquificae and Thermotogae), additional phyla of uncertain phylogenetic position (e.g., basal Terrabacteria), completely new genomes (e.g., CPR) or even an outgroup to root the tree (e.g., Archaea). Aquifex being "just another" group of diderms and Thermotogae being a chimera with a specific diderm architecture, their inclusion would only provide a limited amount of information compared to considering additional Terrabacteria genomes or representatives of the recently discovered CPR. Regarding the difficulty to place such lineages accurately in a phylogenomic tree, it could be overcome by adding genes that are not single copy but at the expense of more work to sort out the orthologous copies. The CPR group would be a particularly welcome addition, provided a useful description of their cell wall could be obtained. Concerning the addition of an outgroup, the question of how it will be used should be answered first: will it be included in the cell-wall reconstruction analyses or will it only be used to root the bacterial subtree. Indeed, if it is not used for reconstruction, any slow evolving fully sequenced Archaea would be usable. On the other hand, if we are interested in reconstructing their cell wall too, we would need to select them very carefully, just as we did for Bacteria. In this respect, the cell-wall diversity of Archaea is as complicated as the bacterial one, if not more, which would add another level of difficulty, and thus uncertainty, to the inferred results.

Supplementary Materials: The following are available online at https://www.mdpi.com/article/10.3390/genes13020376/s1. Figure S1: Unrooted phylogenomic tree of the bacterial domain based on a supermatrix concatenating 117 single-copy orthologous genes chosen for their broad conservation across Bacteria. Figure S2: Unrooted phylogenomic tree of the bacterial domain based on a supermatrix concatenating 117 single-copy orthologous genes chosen for their broad conservation across Bacteria. Figure S3: Evolution of the log likelihood of six PhyloBayes MCMC chains running under the CAT+GTR+Γ model of sequence evolution. Figure S4: Phylogenomic tree of the bacterial domain based on a supermatrix concatenating 117 single-copy orthologous genes chosen for their broad conservation across Bacteria. Figure S5: Trees inferred by the six individual MCMC chains running under the CAT+GTR+Γ model of sequence evolution. Figure S6: Phylogenomic tree of the bacterial domain based on a supermatrix concatenating 117 single-copy orthologous genes chosen for their broad conservation across Bacteria. Figure S7: Posterior probabilities for a monoderm LBCA according to five different models and six possible roots for the bacterial domain. Figure S8: Posterior transition rates and posterior probability of being monoderm for the model where the hyper-prior was purposely biased towards the "diderm-first" hypothesis. Figure S9: Posterior probabilities for a LBCA featuring a thick peptidoglycan (PG) layer according to the five different models and the six possible bacterial roots. Figure S10: Posterior transition rates for the peptidoglycan (PG) trait. Figure S11: Posterior probabilities for the peptidoglycan (PG) and membrane traits in the LCA of four bacterial groups. Figure S12: MraY tree inferred using RAxML under the LG+F+Γ model of sequence evolution. Figure S13: Phylogenomic tree based on a supermatrix of 85 species × 4571 unambiguously aligned amino-acid positions (8.47% missing character states) using 15 of the *dcw* cluster genes. Figure S14: Unrooted TolA tree inferred using RAxML under the LG+F+Γ model. Figure S15: Unrooted YbgF tree inferred using RAxML under the LG+F+Γ model. Figure S16: Schema of the MySQL database used by the synteny tool. Table S1: List of the 117 genes used for the phylogenomic tree of Figure 1. Table S2: List of references used to determine the cell-wall architecture for the 85 representative organisms of Figure 1. Table S3: Details of the data given to BayesTraits for the ancestral trait reconstruction. Table S4: Number of POTRA domains predicted by InterProScan in the majority of the sequences composing each orthologous group (OG) identified as a member of the Omp85/TpsB family. Table S5: Results of the cross-validation procedure comparing four different models of sequence evolution available in PhyloBayes MPI. Table S6: Possible roots for the bacterial domain reported in the phylogenomic literature since 2006.

Author Contributions: R.R.L. performed the experiments, analyzed the data, prepared figures and/or tables, authored or reviewed drafts of the paper, developed the dereplication tool and the synteny tool, and approved the final draft. E.S., F.K. and D.B. conceived and designed the experiments, analyzed the data, authored or reviewed drafts of the paper and approved the final draft. V.L. prepared figures, reviewed drafts of the paper and approved the final draft. A.P. developed a specific version of ProCARs for this study, reviewed drafts of the paper and approved the final draft. D.S. provided technical support for the high-performance computing cluster, reviewed drafts of the paper and approved the final draft. P.C. reviewed drafts of the paper and approved the final draft. All authors have read and agreed to the published version of the manuscript.

Funding: This work was supported in part by the Belgian Program on Interuniversity Poles of Attraction initiated by the Belgian State, Prime Minister's Office, Science Policy programming (IAP no. P6/19). It also benefited from computational resources made available on the Tier-1 supercomputer of the Fédération Wallonie-Bruxelles, infrastructure funded by the Walloon Region under the grant agreement n°1117545, and on the "*durandal*" grid computer, partially funded by two grants to DB (University of Liège "Crédit de démarrage 2012" SFRD-12/04; F.R.S.-FNRS "Crédit de recherche 2014" CDR J.0080.15). R.R.L. and V.L. are the recipients of FRIA (Fonds de la Recherche pour l'Industrie et l'Agriculture) fellowships (F.R.S.-FNRS, Brussels, Belgium).

Institutional Review Board Statement: Not applicable.

Informed Consent Statement: Not applicable.

Data Availability Statement: The datasets generated and analysed for this study can be found in the FigShare repository available here: https://doi.org/10.6084/m9.figshare.14932386.v2 (16 February 2022). Similarly, the database schema and corresponding table dump are available at: https://doi.org/10.6084/m9.figshare.17102651.v1 (16 February 2022).

Acknowledgments: F.K. is a research associate of the F.R.S.-FNRS, Belgium. We thank D. de Vienne for his advice on the interpretation of Phylo-MCOA output and Nguyen The Anh for preliminary trees of the genes in the *dcw* cluster.

Conflicts of Interest: The authors declare no conflict of interest.

References

1. Schleifer, K.H.; Kandler, O. Peptidoglycan Types of Bacterial Cell Walls and Their Taxonomic Implications. *Bacteriol. Rev.* **1972**, *36*, 407–477. [CrossRef] [PubMed]
2. Coico, R. Gram Staining. *Curr. Protoc. Microbiol.* **2006**, A-3C.
3. Silhavy, T.J.; Kahne, D.; Walker, S. The Bacterial Cell Envelope. *Cold Spring Harb. Perspect. Biol.* **2010**, *2*, a000414. [CrossRef] [PubMed]
4. Woese, C.R. Bacterial Evolution. *Microbiol. Rev.* **1987**, *51*, 221–271. [CrossRef]
5. Megrian, D.; Taib, N.; Witwinowski, J.; Beloin, C.; Gribaldo, S. One or Two Membranes? Diderm Firmicutes Challenge the Gram-Positive/Gram-Negative Divide. *Mol. Microbiol.* **2020**, *113*, 659–671. [CrossRef] [PubMed]
6. Cavalier-Smith, T. The Origin of Eukaryote and Archaebacterial Cells. *Ann. N. Y. Acad. Sci.* **1987**, *503*, 17–54. [CrossRef] [PubMed]
7. Cavalier-Smith, T. Deep Phylogeny, Ancestral Groups and the Four Ages of Life. *Philos. Trans. R. Soc. B Biol. Sci.* **2010**, *365*, 111–132. [CrossRef]
8. Koch, A.L. Were Gram-Positive Rods the First Bacteria? *TRENDS Microbiol.* **2003**, *11*, 166–170. [CrossRef]

9. Sutcliffe, I.C. A Phylum Level Perspective on Bacterial Cell Envelope Architecture. *Trends Microbiol.* **2010**, *18*, 464–470. [CrossRef]
10. Gupta, R.S. Origin of Diderm (Gram-Negative) Bacteria: Antibiotic Selection Pressure Rather than Endosymbiosis Likely Led to the Evolution of Bacterial Cells with Two Membranes. *Antonie Van Leeuwenhoek Int. J. Gen. Mol. Microbiol.* **2011**, *100*, 171–182. [CrossRef]
11. Errington, J. L-Form Bacteria, Cell Walls and the Origins of Life. *Open Biol.* **2013**, *3*, 120143. [CrossRef]
12. Lake, J.A. Evidence for an Early Prokaryotic Endosymbiosis. *Nature* **2009**, *460*, 967–971. [CrossRef] [PubMed]
13. Tocheva, E.I.; Matson, E.G.; Morris, D.M.; Moussavi, F.; Leadbetter, J.R.; Jensen, G.J. Peptidoglycan Remodeling and Conversion of an Inner Membrane into an Outer Membrane during Sporulation. *Cell* **2011**, *146*, 799–812. [CrossRef]
14. Tocheva, E.I.; Ortega, D.R.; Jensen, G.J. Sporulation, Bacterial Cell Envelopes and the Origin of Life. *Nat. Rev. Microbiol.* **2016**, *14*, 535–542. [CrossRef]
15. Vollmer, W. Bacterial Outer Membrane Evolution via Sporulation? *Nat. Chem. Biol.* **2011**, *8*, 14–18. [CrossRef]
16. Antunes, L.C.S.; Poppleton, D.; Klingl, A.; Criscuolo, A.; Dupuy, B.; Brochier-Armanet, C.; Beloin, C.; Gribaldo, S. Phylogenomic Analysis Supports the Ancestral Presence of LPS-Outer Membranes in the Firmicutes. *eLife* **2016**, *5*, 1–21. [CrossRef]
17. Taib, N.; Megrian, D.; Witwinowski, J.; Adam, P.; Poppleton, D.; Borrel, G.; Beloin, C.; Gribaldo, S. Genome-Wide Analysis of the Firmicutes Illuminates the Diderm/Monoderm Transition. *Nat. Ecol. Evol.* **2020**, *4*, 1661–1672. [CrossRef]
18. Mingorance, J.; Tamames, J. The Bacterial Dcw Gene Cluster: An Island in the Genome? In *Molecules in Time and Space;* Springer: Boston, MA, USA, 2004; pp. 249–271, ISBN 978-0-306-48579-4.
19. Tamames, J. Evolution of Gene Order Conservation in Prokaryotes. *Genome Biol.* **2001**, *2*, 1–11. [CrossRef]
20. Real, G.; Henriques, A.O. Localization of the Bacillus Subtilis MurB Gene within the Dcw Cluster Is Important for Growth and Sporulation. *J. Bacteriol.* **2006**, *188*, 1721–1732. [CrossRef] [PubMed]
21. Barloy-Hubler, F.; Lelaure, V.; Galibert, F. Ribosomal Protein Gene Cluster Analysis in Eubacterium Genomics: Homology between Sinorhizobium Meliloti Strain 1021 and Bacillus Subtilis. *Nucleic Acids Res.* **2001**, *29*, 2747–2756. [CrossRef] [PubMed]
22. Nikolaichik, Y.A.; Donachie, W.D. Conservation of Gene Order amongst Cell Wall and Cell Division Genes in Eubacteria, and Ribosomal Genes in Eubacteria and Eukaryotic Organelles. *Genetica* **2000**, *108*, 1–7. [CrossRef]
23. Eraso, J.M.; Markillie, L.M.; Mitchell, H.D.; Taylor, R.C.; Orr, G.; Margolin, W. The Highly Conserved MraZ Protein Is a Transcriptional Regulator in Escherichia Coli. *J. Bacteriol.* **2014**, *196*, 2053–2066. [CrossRef]
24. Pilhofer, M.; Rappl, K.; Eckl, C.; Bauer, A.P.; Ludwig, W.; Schleifer, K.H.; Petroni, G. Characterization and Evolution of Cell Division and Cell Wall Synthesis Genes in the Bacterial Phyla Verrucomicrobia, Lentisphaerae, Chlamydiae, and Planctomycetes and Phylogenetic Comparison with RRNA Genes. *J. Bacteriol.* **2008**, *190*, 3192–3202. [CrossRef]
25. Coleman, G.A.; Davín, A.A.; Mahendrarajah, T.A.; Szánthó, L.L.; Spang, A.; Hugenholtz, P.; Szöllősi, G.J.; Williams, T.A. A Rooted Phylogeny Resolves Early Bacterial Evolution. *Science* **2021**, *372*. [CrossRef] [PubMed]
26. Kersey, P.J.; Allen, J.E.; Christensen, M.; Davis, P.; Falin, L.J.; Grabmueller, C.; Hughes, D.S.T.; Humphrey, J.; Kerhornou, A.; Khobova, J.; et al. Ensembl Genomes 2013: Scaling up Access to Genome-Wide Data. *Nucleic Acids Res.* **2014**, *42*, D546–D552. [CrossRef] [PubMed]
27. Moreno-Hagelsieb, G.; Wang, Z.; Walsh, S.; ElSherbiny, A. Phylogenomic Clustering for Selecting Non-Redundant Genomes for Comparative Genomics. *Bioinformatics* **2013**, *29*, 947–949. [CrossRef] [PubMed]
28. Léonard, R.R.; Leleu, M.; Vlierberghe, M.V.; Cornet, L.; Kerff, F.; Baurain, D. ToRQuEMaDA: Tool for Retrieving Queried Eubacteria, Metadata and Dereplicating Assemblies. *PeerJ* **2021**, *9*, e11348. [CrossRef]
29. Rice, P.; Longden, I.; Bleasby, A. EMBOSS: The European Molecular Biology Open Software Suite. *Trends Genet.* **2000**, *16*, 276–277. [CrossRef]
30. Sayers, E.W.; Barrett, T.; Benson, D.A.; Bolton, E.; Bryant, S.H.; Canese, K.; Chetvernin, V.; Church, D.M.; DiCuccio, M.; Federhen, S.; et al. Database Resources of the National Center for Biotechnology Information. *Nucleic Acids Res.* **2011**, *39*, D38–D51. [CrossRef]
31. R Core Team, Rf. *R: A Language and Environment for Statistical Computing;* R Foundation for Statistical Computing: Vienna, Austria, 2013.
32. Edgar, R.C. Search and Clustering Orders of Magnitude Faster than BLAST. *Bioinformatics* **2010**, *26*, 2460–2461. [CrossRef]
33. Li, L.; Stoeckert, C.J.; Roos, D.S. OrthoMCL: Identification of Ortholog Groups for Eukaryotic Genomes. *Genome Res.* **2003**, *13*, 2178–2189. [CrossRef]
34. Van Vlierberghe, M.; Philippe, H.; Baurain, D. Broadly Sampled Orthologous Groups of Eukaryotic Proteins for the Phylogenetic Study of Plastid-Bearing Lineages. *BMC Res. Notes* **2021**, *14*, 143. [CrossRef] [PubMed]
35. Katoh, K.; Standley, D.M. MAFFT Multiple Sequence Alignment Software Version 7: Improvements in Performance and Usability. *Mol. Biol. Evol.* **2013**, *30*, 772–780. [CrossRef]
36. Castresana, J. Selection of Conserved Blocks from Multiple Alignments for Their Use in Phylogenetic Analysis. *Mol. Biol. Evol.* **2000**, *17*, 540–552. [CrossRef] [PubMed]
37. Roure, B.; Rodriguez-Ezpeleta, N.; Philippe, H. SCaFoS: A Tool for Selection, Concatenation and Fusion of Sequences for Phylogenomics. *BMC Evol. Biol.* **2007**, *7* Suppl. S1, S2. [CrossRef]
38. Lartillot, N.; Rodrigue, N.; Stubbs, D.; Richer, J. PhyloBayes MPI: Phylogenetic Reconstruction with Infinite Mixtures of Profiles in a Parallel Environment. *Syst. Biol.* **2013**, *62*, 611–615. [CrossRef]

39. Lartillot, N.; Philippe, H. A Bayesian Mixture Model for Across-Site Heterogeneities in the Amino-Acid Replacement Process. *Mol. Biol. Evol.* **2004**, *21*, 1095–1109. [CrossRef]
40. Lartillot, N.; Philippe, H. Computing Bayes Factors Using Thermodynamic Integration. *Syst. Biol.* **2006**, *55*, 195–207. [CrossRef]
41. Lartillot, N.; Brinkmann, H.; Philippe, H. Suppression of Long-Branch Attraction Artefacts in the Animal Phylogeny Using a Site-Heterogeneous Model. *BMC Evol. Biol.* **2007**, *7* Suppl. S1, S4. [CrossRef]
42. Lartillot, N.; Lepage, T.; Blanquart, S. PhyloBayes 3: A Bayesian Software Package for Phylogenetic Reconstruction and Molecular Dating. *Bioinformatics* **2009**, *25*, 2286–2288. [CrossRef] [PubMed]
43. Letunic, I.; Bork, P. Interactive Tree Of Life (ITOL) v5: An Online Tool for Phylogenetic Tree Display and Annotation. *Nucleic Acids Res.* **2021**, *49*, W293–W296. [CrossRef] [PubMed]
44. De Vienne, D.M.; Ollier, S.; Aguileta, G. Phylo-MCOA: A Fast and Efficient Method to Detect Outlier Genes and Species in Phylogenomics Using Multiple Co-Inertia Analysis. *Mol. Biol. Evol.* **2012**, *29*, 1587–1598. [CrossRef] [PubMed]
45. Stamatakis, A. RAxML Version 8: A Tool for Phylogenetic Analysis and Post-Analysis of Large Phylogenies. *Bioinformatics* **2014**, *30*, 1312–1313. [CrossRef] [PubMed]
46. Pagel, M.; Meade, A.; Barker, D. Bayesian Estimation of Ancestral Character States on Phylogenies. *Syst. Biol.* **2004**, *53*, 673–684. [CrossRef] [PubMed]
47. Pagel, M.; Meade, A. Bayesian Analysis of Correlated Evolution of Discrete Characters by Reversible-Jump Markov Chain Monte Carlo. *Am. Nat.* **2015**, *167*, 808–825. [CrossRef]
48. Meade, A.; Pagel, M. BayesTraits V3. 0.1. 2017. Available online: http://www.evolution.rdg.ac.uk/BayesTraitsV3.0.1/BayesTraitsV3.0.1.html(accessed on 9 February 2022).
49. Eddy, S.R. Accelerated Profile HMM Searches. *PLoS Comput. Biol.* **2011**, *7*, e1002195. [CrossRef]
50. Mistry, J.; Finn, R.D.; Eddy, S.R.; Bateman, A.; Punta, M. Challenges in Homology Search: HMMER3 and Convergent Evolution of Coiled-Coil Regions. *Nucleic Acids Res.* **2013**, *41*, e121-e121. [CrossRef]
51. Perrin, A.; Varré, J.-S.; Blanquart, S.; Ouangraoua, A. ProCARs: Progressive Reconstruction of Ancestral Gene Orders. *BMC Genomics* **2015**, *16 Suppl 5*, S6. [CrossRef]
52. Jones, P.; Binns, D.; Chang, H.-Y.; Fraser, M.; Li, W.; McAnulla, C.; McWilliam, H.; Maslen, J.; Mitchell, A.; Nuka, G.; et al. InterProScan 5: Genome-Scale Protein Function Classification. *Bioinformatics* **2014**, *30*, 1236–1240. [CrossRef]
53. Sánchez-Pulido, L.; Devos, D.; Genevrois, S.; Vicente, M.; Valencia, A. POTRA: A Conserved Domain in the FtsQ Family and a Class of β-Barrel Outer Membrane Proteins. *Trends Biochem. Sci.* **2003**, *28*, 523–526. [CrossRef] [PubMed]
54. Heinz, E.; Lithgow, T. A Comprehensive Analysis of the Omp85/TpsB Protein Superfamily Structural Diversity, Taxonomic Occurrence, and Evolution. *Front. Microbiol.* **2014**, *5*, 370. [CrossRef]
55. Delsuc, F.; Brinkmann, H.; Philippe, H. Phylogenomics and the Reconstruction of the Tree of Life. *Nat. Rev. Genet.* **2005**, *6*, 361–375. [CrossRef]
56. Philippe, H.; de Vienne, D.M.; Ranwez, V.; Roure, B.; Baurain, D.; Delsuc, F. Pitfalls in Supermatrix Phylogenomics. *Eur. J. Taxon.* **2017**, *283*, 1–25. [CrossRef]
57. Liu, L.; Anderson, C.; Pearl, D.; Edwards, S.V. Modern Phylogenomics: Building Phylogenetic Trees Using the Multispecies Coalescent Model. In *Evolutionary Genomics: Statistical and Computational Methods*; Anisimova, M., Ed.; Humana: New York, NY, USA, 2019; pp. 211–239.
58. Battistuzzi, F.U.; Hedges, S.B. A Major Clade of Prokaryotes with Ancient Adaptations to Life on Land. *Mol. Biol. Evol.* **2009**, *26*, 335–343. [CrossRef]
59. Wu, D.; Hugenholtz, P.; Mavromatis, K.; Pukall, R.; Dalin, E.; Ivanova, N.N.; Kunin, V.; Goodwin, L.; Wu, M.; Tindall, B.J.; et al. A Phylogeny-Driven Genomic Encyclopaedia of Bacteria and Archaea. *Nature* **2009**, *462*, 1056–1060. [CrossRef] [PubMed]
60. Yutin, N.; Puigbo, P.; Koonin, E.V.; Wolf, Y.I. Phylogenomics of Prokaryotic Ribosomal Proteins. *Curr. Sci.* **2012**, *101*, 1435–1439. [CrossRef] [PubMed]
61. Lasek-nesselquist, E.; Gogarten, J.P. The Effects of Model Choice and Mitigating Bias on the Ribosomal Tree of Life. *Mol. Phylogenet. Evol.* **2013**, *69*, 17–38. [CrossRef]
62. Rinke, C.; Schwientek, P.; Sczyrba, A.; Ivanova, N.N.; Anderson, I.J.; Cheng, J.-F.; Darling, A.E.; Malfatti, S.; Swan, B.K.; Gies, E.A.; et al. Insights into the Phylogeny and Coding Potential of Microbial Dark Matter. *Nature* **2013**, *499*, 431–437. [CrossRef] [PubMed]
63. Raymann, K.; Brochier-Armanet, C.; Gribaldo, S. The Two-Domain Tree of Life Is Linked to a New Root for the Archaea. *Proc. Natl. Acad. Sci. USA* **2015**, *112*, 6670–6675. [CrossRef] [PubMed]
64. Hug, L.A.; Baker, B.J.; Anantharaman, K.; Brown, C.T.; Probst, A.J.; Castelle, C.J.; Butterfield, C.N.; Hernsdorf, A.W.; Amano, Y.; Kotaro, I.; et al. A New View of the Tree of Life. *Nat. Microbiol.* **2016**, *1*, 16048. [CrossRef]
65. Castelle, C.J.; Banfield, J.F. Major New Microbial Groups Expand Diversity and Alter Our Understanding of the Tree of Life. *Cell* **2018**, *172*, 1181–1197. [CrossRef]
66. Parks, D.H.; Chuvochina, M.; Waite, D.W.; Rinke, C.; Skarshewski, A.; Chaumeil, P.-A.; Hugenholtz, P. A Standardized Bacterial Taxonomy Based on Genome Phylogeny Substantially Revises the Tree of Life. *Nat. Biotechnol.* **2018**, *36*, 996–1004. [CrossRef]
67. Zhu, Q.; Mai, U.; Pfeiffer, W.; Janssen, S.; Asnicar, F.; Sanders, J.G.; Belda-ferre, P.; Al-ghalith, G.A.; Kopylova, E.; Mcdonald, D.; et al. Phylogenomics of 10,575 Genomes Reveals Evolutionary Proximity between Domains Bacteria and Archaea. *Nat. Commun.* **2019**, *10*, 1–14. [CrossRef]

68. Cavalier-Smith, T.; Ema, E.; Chao, Y. Multidomain Ribosomal Protein Trees and the Planctobacterial Origin of Neomura (Eukaryotes, Archaebacteria). *Protoplasma* **2020**, 1–133. [CrossRef] [PubMed]
69. Cavalier-Smith, T. Rooting the Tree of Life by Transition Analyses. *Biol. Direct* **2006**, *1*, 19. [CrossRef]
70. Zhaxybayeva, O.; Swithers, K.S.; Lapierre, P.; Fournier, G.P.; Bickhart, D.M.; DeBoy, R.T.; Nelson, K.E.; Nesbø, C.L.; Doolittle, W.F.; Gogarten, J.P.; et al. On the Chimeric Nature, Thermophilic Origin, and Phylogenetic Placement of the Thermotogales. *Proc. Natl. Acad. Sci. USA* **2009**, *106*, 5865–5870. [CrossRef]
71. Bhandari, V.; Naushad, H.S.; Gupta, R.S. Protein Based Molecular Markers Provide Reliable Means to Understand Prokaryotic Phylogeny and Support Darwinian Mode of Evolution. *Front. Cell. Infect. Microbiol.* **2012**, *2*, 1–14. [CrossRef]
72. Eveleigh, R.J.M.; Meehan, C.J.; Archibald, J.M.; Beiko, R.G. Being Aquifex Aeolicus: Untangling a Hyperthermophile's Checkered Past. *Genome Biol. Evol.* **2013**, *5*, 2478–2497. [CrossRef] [PubMed]
73. Rachel, R.; Wildhaber, I.; Stetter, K.O.; Baumeister, W. The Structure of the Surface Protein of Thermotoga Maritima. In *Crystalline Bacterial Cell Surface Layers*; Springer: Berlin/Heidelberg, Germany, 1988; pp. 83–86.
74. Rachel, R.; Engel, A.M.; Huber, R.; Stetter, K.-O.; Baumeister, W. A Porin-Type Protein Is the Main Constituent of the Cell Envelope of the Ancestral Eubacterium Thermotoga Maritima. *FEBS Lett.* **1990**, *262*, 64–68. [CrossRef]
75. Bapteste, E.; Boucher, Y.; Leigh, J.; Doolittle, W.F. Phylogenetic Reconstruction and Lateral Gene Transfer. *Trends Microbiol.* **2004**, *12*, 406–411. [CrossRef]
76. Mira, A.; Pushker, R.; Legault, B.A.; Moreira, D.; Rodríguez-Valera, F. Evolutionary Relationships of Fusobacterium Nucleatum Based on Phylogenetic Analysis and Comparative Genomics. *BMC Evol. Biol.* **2004**, *4*, 50. [CrossRef]
77. Beiko, R.G.; Harlow, T.J.; Ragan, M. a Highways of Gene Sharing in Prokaryotes. *Proc. Natl. Acad. Sci. USA* **2005**, *102*, 14332–14337. [CrossRef]
78. Koonin, E.V. Orthologs, Paralogs, and Evolutionary Genomics. *Annu. Rev. Genet.* **2005**, *39*, 309–338. [CrossRef]
79. Koonin, E.V. Horizontal Gene Transfer: Essentiality and Evolvability in Prokaryotes, and Roles in Evolutionary Transitions. *F1000Research* **2016**, *5*, F1000 Faculty Rev-1805. [CrossRef]
80. Boussau, B.; Guéguen, L.; Gouy, M. Accounting for Horizontal Gene Transfers Explains Conflicting Hypotheses Regarding the Position of Aquificales in the Phylogeny of Bacteria. *BMC Evol. Biol.* **2008**, *8*, 1–18. [CrossRef]
81. Philippe, H.; Brinkmann, H.; Lavrov, D.V.; Littlewood, D.T.J.; Manuel, M.; Wörheide, G.; Baurain, D. Resolving Difficult Phylogenetic Questions: Why More Sequences Are Not Enough. *PLoS Biol.* **2011**, *9*. [CrossRef] [PubMed]
82. Gouy, R.; Baurain, D.; Philippe, H. Rooting the Tree of Life: The Phylogenetic Jury Is Still Out. *Philos. Trans. R. Soc. B Biol. Sci.* **2015**, *370*, 20140329. [CrossRef] [PubMed]
83. Rivas-Marín, E.; Canosa, I.; Devos, D.P. Evolutionary Cell Biology of Division Mode in the Bacterial Planctomycetes-Verrucomicrobia- Chlamydiae Superphylum. *Front. Microbiol.* **2016**, *7*. [CrossRef] [PubMed]
84. Cruz-Morales, P.; Orellana, C.A.; Moutafis, G.; Moonen, G.; Rincon, G.; Nielsen, L.K.; Marcellin, E. Revisiting the Evolution and Taxonomy of Clostridia, a Phylogenomic Update. *Genome Biol. Evol.* **2019**, *11*, 2035–2044. [CrossRef]
85. Mavromatis, K.; Ivanova, N.; Anderson, I.; Lykidis, A.; Hooper, S.D.; Sun, H.; Kunin, V.; Lapidus, A.; Hugenholtz, P.; Patel, B.; et al. Genome Analysis of the Anaerobic Thermohalophilic Bacterium Halothermothrix Orenii. *PLoS ONE* **2009**, *4*, e4192. [CrossRef]
86. Kivistö, A.T.; Karp, M.T. Halophilic Anaerobic Fermentative Bacteria. *J. Biotechnol.* **2011**, *152*, 114–124. [CrossRef] [PubMed]
87. Sutcliffe, I.C. Cell Envelope Architecture in the Chloroflexi: A Shifting Frontline in a Phylogenetic Turf War. *Environ. Microbiol.* **2011**, *13*, 279–282. [CrossRef] [PubMed]
88. Gaisin, V.A.; Kooger, R.; Grouzdev, D.S.; Gorlenko, V.M.; Pilhofer, M. Cryo-Electron Tomography Reveals the Complex Ultrastructural Organization of Multicellular Filamentous Chloroflexota (Chloroflexi) Bacteria. *Front. Microbiol.* **2020**, *11*, 1373. [CrossRef] [PubMed]
89. Gilks, W.R.; Richardson, S.; Spiegelhalter, D. *Markov Chain Monte Carlo in Practice*; CRC Press: Boca Raton, FL, USA, 1995; ISBN 978-1-4822-1497-0.
90. Yutin, N.; Galperin, M.Y. A Genomic Update on Clostridial Phylogeny: Gram-Negative Spore Formers and Other Misplaced Clostridia. *Environ. Microbiol.* **2013**, *15*, 2631–2641. [CrossRef] [PubMed]
91. Hagan, C.L.; Silhavy, T.J.; Kahne, D. β-Barrel Membrane Protein Assembly by the Bam Complex. *Annu. Rev. Biochem.* **2011**, *80*, 189–210. [CrossRef]
92. Bowyer, A.; Baardsnes, J.; Ajamian, E.; Zhang, L.; Cygler, M. Characterization of Interactions between LPS Transport Proteins of the Lpt System. *Biochem. Biophys. Res. Commun.* **2011**, *404*, 1093–1098. [CrossRef]
93. Walburger, A.; Lazdunski, C.; Corda, Y. The Tol/Pal System Function Requires an Interaction between the C-Terminal Domain of TolA and the N-Terminal Domain of TolB. *Mol. Microbiol.* **2002**, *44*, 695–708. [CrossRef]
94. Ranjit, C.; Noll, K.M. Distension of the Toga of Thermotoga Maritima Involves Continued Growth of the Outer Envelope as Cells Enter the Stationary Phase. *FEMS Microbiol. Lett.* **2016**, 363.
95. Bernard, G.; Chan, C.X.; Ragan, M.A. Alignment-Free Microbial Phylogenomics under Scenarios of Sequence Divergence, Genome Rearrangement and Lateral Genetic Transfer. *Sci. Rep.* **2016**, *6*, 1–12. [CrossRef]
96. Yu, J.; Lu, L. BamA Is a Pivotal Protein in Cell Envelope Synthesis and Cell Division in Deinococcus Radiodurans. *Biochim. Biophys. Acta BBA-Biomembr.* **2019**, *1861*, 1365–1374. [CrossRef]
97. Voulhoux, R.; Bos, M.P.; Geurtsen, J.; Mols, M.; Tommassen, J. Role of a Highly Conserved Bacterial Protein in Outer Membrane Protein Assembly. *Science* **2003**, *299*, 262–265. [CrossRef]

98. Gully, D.; Bouveret, E. A Protein Network for Phospholipid Synthesis Uncovered by a Variant of the Tandem Affinity Purification Method in Escherichia Coli. *Proteomics* **2006**, *6*, 282–293. [CrossRef]
99. Narita, S.; Tokuda, H. Biochemical Characterization of an ABC Transporter LptBFGC Complex Required for the Outer Membrane Sorting of Lipopolysaccharides. *FEBS Lett.* **2009**, *583*, 2160–2164. [CrossRef] [PubMed]
100. Cornet, L.; Meunier, L.; Van Vlierberghe, M.; Léonard, R.R.; Durieu, B.; Lara, Y.; Misztak, A.; Sirjacobs, D.; Javaux, E.J.; Philippe, H.; et al. Consensus Assessment of the Contamination Level of Publicly Available Cyanobacterial Genomes. *PLoS ONE* **2018**, *13*, e0200323. [CrossRef]
101. Wang, Z.; Xiang, Q.; Zhu, X.; Dong, H.; He, C.; Wang, H.; Zhang, Y.; Wang, W.; Dong, C. Structural and Functional Studies of Conserved Nucleotide-Binding Protein LptB in Lipopolysaccharide Transport. *Biochem. Biophys. Res. Commun.* **2014**, *452*, 443–449. [CrossRef] [PubMed]
102. Rigal, A.; Bouveret, E.; Lloubes, R.; Lazdunski, C.; Benedetti, H. The TolB Protein Interacts with the Porins of Escherichia Coli. *J. Bacteriol.* **1997**, *179*, 7274–7279. [CrossRef]
103. Ray, M.-C.; Germon, P.; Vianney, A.; Portalier, R.; Lazzaroni, J.C. Identification by Genetic Suppression OfEscherichia Coli TolB Residues Important for TolB-Pal Interaction. *J. Bacteriol.* **2000**, *182*, 821–824. [CrossRef] [PubMed]
104. Demoulin, C.F.; Lara, Y.J.; Cornet, L.; François, C.; Baurain, D.; Wilmotte, A.; Javaux, E.J. Cyanobacteria Evolution: Insight from the Fossil Record. *Free Radic. Biol. Med.* **2019**, *140*, 206–223. [CrossRef]
105. Xavier, J.C.; Gerhards, R.E.; Wimmer, J.L.; Brueckner, J.; Tria, F.D.; Martin, W.F. The Metabolic Network of the Last Bacterial Common Ancestor. *Commun. Biol.* **2021**, *4*, 1–10. [CrossRef]
106. Kuhn, A. The Bacterial Cell Wall and Membrane—A Treasure Chest for Antibiotic Targets. In *Bacterial Cell Walls and Membranes*; Kuhn, A., Ed.; Subcellular Biochemistry; Springer International Publishing: Cham, Switzerland, 2019; pp. 1–5. ISBN 978-3-030-18768-2.
107. Melville, S.; Craig, L. Type IV Pili in Gram-Positive Bacteria. *Microbiol. Mol. Biol. Rev.* **2013**, *77*, 323–341. [CrossRef]
108. Baurain, D.; Wilmotte, A.; Frère, J.-M. Gram-Negative Bacteria:" Inner" vs." Cytoplasmic" or" Plasma Membrane": A Question of Clarity Rather than Vocabulary. *J. Microb. Biochem. Technol.* **2016**, *8*, 325–326. [CrossRef]
109. Verma, M.; Lal, D.; Kaur, J.; Saxena, A.; Kaur, J.; Anand, S.; Lal, R. Phylogenetic Analyses of Phylum Actinobacteria Based on Whole Genome Sequences. *Res. Microbiol.* **2013**, *164*, 718–728. [CrossRef]
110. Yakhnina, A.A.; Bernhardt, T.G. The Tol-Pal System Is Required for Peptidoglycan-Cleaving Enzymes to Complete Bacterial Cell Division. *Proc. Natl. Acad. Sci. USA* **2020**, *117*, 6777–6783. [CrossRef]
111. Parks, D.H.; Rinke, C.; Chuvochina, M.; Chaumeil, P.; Woodcroft, B.J.; Evans, P.N.; Hugenholtz, P.; Tyson, G.W. Recovery of Nearly 8,000 Metagenome-Assembled Genomes Substantially Expands the Tree of Life. *Nat. Microbiol.* **2017**, *2*. [CrossRef]
112. Luef, B.; Frischkorn, K.R.; Wrighton, K.C.; Holman, H.-Y.N.; Birarda, G.; Thomas, B.C.; Singh, A.; Williams, K.H.; Siegerist, C.E.; Tringe, S.G.; et al. Diverse Uncultivated Ultra-Small Bacterial Cells in Groundwater. *Nat. Commun.* **2015**, *6*, 1–8. [CrossRef] [PubMed]
113. Martinez-Gutierrez, C.A.; Aylward, F.O. Phylogenetic Signal, Congruence, and Uncertainty across Bacteria and Archaea. *Mol. Biol. Evol.* **2021**, *38*, 5514–5527. [CrossRef] [PubMed]
114. Dawes, I.W. Sporulation in Evolution. *Mol. Cell. Asp. Microb. Evol. Camb. Univ. Press N. Y.* **1981**, 85–130.

genes MDPI

Review

Origin, Evolution and Stability of Overlapping Genes in Viruses: A Systematic Review

Angelo Pavesi

Department of Chemistry, Life Sciences and Environmental Sustainability, University of Parma, Parco Area delle Scienze 23/A, I-43124 Parma, Italy; angelo.pavesi@unipr.it; Tel.: +39-0521905647

Abstract: During their long evolutionary history viruses generated many proteins *de novo* by a mechanism called "overprinting". Overprinting is a process in which critical nucleotide substitutions in a pre-existing gene can induce the expression of a novel protein by translation of an alternative open reading frame (ORF). Overlapping genes represent an intriguing example of adaptive conflict, because they simultaneously encode two proteins whose freedom to change is constrained by each other. However, overlapping genes are also a source of genetic novelties, as the constraints under which alternative ORFs evolve can give rise to proteins with unusual sequence properties, most importantly the potential for novel functions. Starting with the discovery of overlapping genes in phages infecting *Escherichia coli*, this review covers a range of studies dealing with detection of overlapping genes in small eukaryotic viruses (genomic length below 30 kb) and recognition of their critical role in the evolution of pathogenicity. Origin of overlapping genes, what factors favor their birth and retention, and how they manage their inherent adaptive conflict are extensively reviewed. Special attention is paid to the assembly of overlapping genes into *ad hoc* databases, suitable for future studies, and to the development of statistical methods for exploring viral genome sequences in search of undiscovered overlaps.

Keywords: asymmetric evolution; codon usage; *de novo* protein creation; modular evolution; multivariate statistics; negative selection: phylogenetic distribution; positive selection; prediction methods; sequence-composition features; symmetric evolution; virus evolution

Citation: Pavesi, A. Origin, Evolution and Stability of Overlapping Genes in Viruses: A Systematic Review. *Genes* **2021**, *12*, 809. https://doi.org/10.3390/genes12060809

Academic Editors: Luigi Viggiano and Renè Massimiliano Marsano

Received: 6 May 2021
Accepted: 24 May 2021
Published: 26 May 2021

1. Introduction

Modification of existing genes, such as duplication followed by functional divergence, fusion (two adjacent genes fuse into a single gene), fission (a single gene splits into two genes), exon shuffling (rearrangement of protein modules), or horizontal gene transfer (gene exchange between unrelated species), is a common mechanism by which new genes arose during the evolution of living organisms [1–4]. However, genes can also originate *de novo* by taking place in non-coding regions, such as intergenic regions or introns [5,6].

During their long evolutionary history viruses generated many proteins *de novo* by a mechanism called "overprinting" [7]. Overprinting is a process in which critical nucleotide substitutions in a pre-existing gene can induce the expression of a novel protein by translation of an alternative open reading frame (ORF), while preserving the function of pre-existing gene [8]. It is thought that most overlapping genes evolve by this mechanism, and that consequently each overlap contains one ancestral frame and one originated *de novo* [9]. It is also believed that overprinting is a source of genetic novelties, because the *de novo* proteins, unlike the ancestral ones, usually lack any remote homologs in databases [10].

Most of new genes originated by overprinting are expressed by the sense strand. They are classified as same-strand, or parallel, overlapping genes because of transcription from the same strand of DNA. They are usually denoted as +1 overlapping genes, when the *de novo* frame is shifted one nucleotide 3′ with respect to the ancestral frame (Figure 1A), or as +2 overlapping genes when the *de novo* frame is shifted two nucleotides

3′ (Figure 1B). According to genetic code, 71.6% of substitutions in the third codon position are synonymous, compared to only 0 and 4.6% of substitutions in the second and first codon positions respectively. In overlapping genes, therefore, a nucleotide substitution that is synonymous in one frame is highly likely to be non-synonymous in the alternative frame.

(A)

	Nucleotide codon position
Upstream overlapping reading frame	.. **1** 2 *3* **1** 2 *3* **1** 2 *3*..
Downstream overlapping reading frame (+1 overlap)	.. **3** 1 *2* **3** 1 *2* **3** 1 *2*..

(B)

	Nucleotide codon position
Upstream overlapping reading frame	.. **1** 2 *3* **1** 2 *3* **1** 2 *3*..
Downstream overlapping reading frame (+2 overlap)	.. **2** 3 *1* **2** 3 *1* **2** 3 *1*..

Figure 1. Orientation of same-strand overlapping genes. (**A**) Overlapping gene with the downstream frame shifted one nucleotide 3′ with respect to the upstream frame. There are 3 types of codon position (cp): cp13 (bold character), in which the first codon position of the upstream frame overlaps the third codon position of the downstream frame; cp21 (underlined character), in which the second codon position of the upstream frame overlaps the first codon position of the downstream frame; cp32 (italic character), in which the third codon position of the upstream frame overlaps the second codon position of the downstream frame. (**B**) Overlapping gene with the downstream frame shifted two nucleotides 3′ with respect to the upstream frame. There are 3 types of codon position (cp): cp12 (bold character), in which the first codon position of the upstream frame overlaps the second codon position of the downstream frame; cp23 (underlined character), in which the second codon position of the upstream frame overlaps the third codon position of the downstream frame; cp31 (italic character), in which the third codon position of the upstream frame overlaps the first codon position of the downstream frame. According to the genetic code, a nucleotide substitution at first codon position causes an amino acid change in 95.4% of cases, at second position in 100% of cases, and at third position in 28.4% of cases.

One of the reasons why overlapping genes have long attracted attention of researchers is that they represent an intriguing example of adaptive conflict. Indeed, they simultaneously encode two proteins whose freedom to change is constrained by each other, which would be expected to severely reduce the ability of the virus to adapt. On the other hand, the unusual constraints under which alternative ORFs evolve can give rise to proteins with unusual sequence properties, most importantly the potential for novel structural folds and mechanisms of action.

This review deals with the origin of overlapping genes, what factors favor their birth and retention, how they influence the evolution of viral genome, and how they manage their inherent adaptive conflict. The review is focused on overlapping genes from small viruses (genomic length below 30 kb), in which both members of the pair are known to be expressed during infection. Special attention is paid to the genealogy of the overlap, that is inferring which frame is ancestral and which one is de novo. Special attention is paid to the assembly of overlapping genes into *ad hoc* databases, suitable for future studies, and to the development of statistical methods for exploring viral genome sequences in search of undiscovered overlapping coding regions.

2. Discovery of Overlapping Genes and Evolutionary Implications

Overlapping genes, also called "dual-coding genes", were first discovered by Barrell et al. [11] in the genome of ΦX174, a small single-stranded DNA virus (5386 nt) that infects *Escherichia coli*. Analysis of the fully sequenced genome revealed that it contains, thanks to overprinting, 15% more coding ability than a co-linear relationship between nucleotide and protein sequences would suggest [12].

Genome sequence analysis of ΦX174 showed that there are two types of overlaps: in "internal overlaps" one overlapping gene is contained entirely within the other gene (e.g., gene E is nested within gene D) whereas "terminal overlaps" involve only the 3′ terminal region of one gene and the 5′ start region of another (e.g., the 3′ end of gene A overlaps the 5′ end of gene K) [12]. The strength of selection pressure acting on the gene overlap was estimated by a mathematical model, which pointed out a strong reduction of amino acid changes (at most 40 or 50%) in the overlapping genes B and D of ΦX174 [13].

By sequence analysis of the paired overlapping genes D and E, Fiddes and Godson proposed a simple method to predict the genealogy of the overlap [14]. Genealogy means to recognize which frame is ancestral and which frame is *de novo*. The authors first found that the genome of ΦX174 is rich in T nucleotides (31%) and these tend to occur at third codon position. They then found that in the region of D overlapping E (279 nt) the high incidence of T-ending codons is a feature of the frame D (39%) rather than the frame E (14%). Based on this finding, D was predicted as the ancestral gene and E as the *de novo* gene.

In addition, displacement of the high T content from the third codon position in frame D to the second codon position in frame E yields a high incidence of codons that specify leucine, one of the most hydrophobic amino acids, in frame E. The high content of leucine in protein E is mainly localized within a transmembrane domain, which induces lysis of the cell host *Escherichia coli* [15] by inhibiting biosynthesis of cell wall [16]. This finding suggests that de novo protein creation can be a significant factor in the evolution of pathogenicity.

The Fiddes's method to predict the genealogy of the overlap [14] was improved by means of a correlation analysis of the codon usage [17]. It was based on the assumption that the ancestral gene, which has co-evolved with the other viral genes over a long period of time, has a distribution of synonymous codons closer to that of the viral genome than the *de novo* gene. The codon-usage correlation analysis of ΦX174 demonstrated that E and K are *de novo* overlapping proteins and that the C-terminal region of protein A is a *de novo* overlapping extension [17].

When applied to ΦX174, α3 and G4 (the three evolutionary clades of the genus *Microvirus*, family *Microviridae*), the codon-usage correlation analysis predicted a gradual increase in the genome information content due to overprinting [18]. It predicted an ancestral genome having only single-coding genes, whose coding capacity increased over time due to the birth of novel overlapping coding regions (Figure 2). This fine evolutionary process led to the present genome, which contains two *de novo* overlapping genes (K and E) and two *de novo* overlapping extensions of genes A and C.

As said in introduction, an intriguing paradox of overlapping genes is that the biological information in the encoded proteins is strongly interdependent, yet each of the two proteins has evolved to its own well-defined function. Sander and Schultz [19] developed a mathematical model and applied it to the overlapping proteins A and B of ΦX174. The model postulated that the paradox can be explained by assuming sufficiently large degeneracy of the information content of amino acid sequences with respect to function.

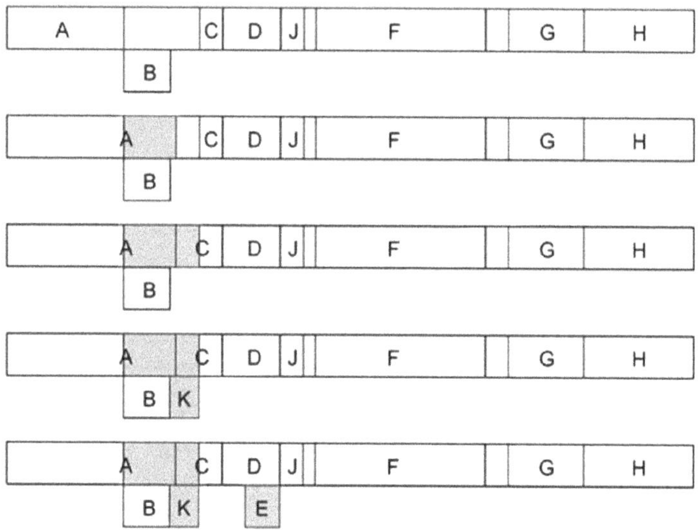

Figure 2. Increase in the genome information content during the evolution of microviruses (family *Microviridae*). The nomenclature of genes, from A to J, is that originally proposed in ΦX174 [11–13]. Empty boxes indicate ancestral pre-existing genes, while grey boxes indicate the new genes (or gene regions) that originated by overprinting. Figure reproduced from [18] with the permission of the Microbiology Society.

3. *De Novo* Overlapping Genes Show a Restricted Phylogenetic Distribution and Encode Accessory Proteins

In 1992, Keese and Gibbs published a seminal paper [9] in which the birth of new genes by overprinting was described as a continuous, and significant, evolutionary process. They proposed a new method to predict the genealogy of overlapping genes. It is based on the assumption that the protein with the most restricted phylogenetic distribution is encoded by the *de novo* frame, while that with the widest distribution is encoded by the ancestral frame.

As an example to explain the phylogenetic method, the genome of tymoviruses contains a large dual-coding region in which the 5′ one-third of replicase, encoding a methyltransferase domain, overlaps an ORF that encodes a movement protein necessary for viral spread [20]. While the methyltransferase domain has a wide phylogenetic distribution, including the closely related sister groups of potexviruses and closteroviruses or outgroups such as tricornaviruses and furoviruses, the movement protein is unique to tymoviruses (Figure 3).

Based on this finding, Keese and Gibbs inferred that replicase is the ancestral gene and that the overlapping ORF arose later, de novo, after the evolutionary divergence between tymoviruses and potexviruses. It is unlikely, indeed, that this ORF was present earlier but was subsequently lost in all virus groups with the exception of tymoviruses. It follows that the genome region of potexviruses homologous to the gene overlap unique to tymoviruses should have sequence-composition features typical of a "pre-overlapping" coding region.

Using the phylogenetic method, Rancurel et al. [21] were able to recognize the ancestral and the *de novo* frame for 17 pairs of overlapping genes, covering a wide evolutionary range of RNA viruses. Almost all de novo frames resulted to encode accessory proteins, rather than proteins central to viral replication or to the structure of capsid. "Accessory" does not mean that they are dispensable in vivo, because most novel proteins play an important role in viral pathogenicity or spread. Indeed, six de novo proteins promote a systemic diffusion of infection in plants [20,22–25], for example by binding viral RNA and forming protective ribonucleoprotein complexes [26]. Two de novo proteins contribute to evade or counteract the innate host defense, acting as inhibitor of interferon response [27] or suppressor of RNA silencing [28].

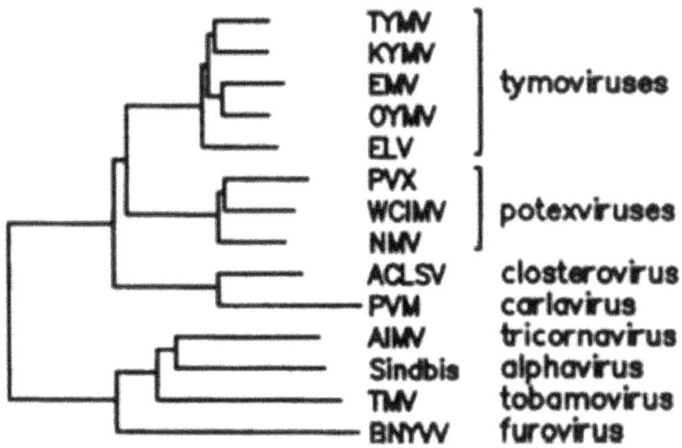

Figure 3. Dendrogram of the methyltransferase-like domain of replicase from Turnip yellow mosaic virus (TYMV), Kennedya yellow mosaic tymovirus (KYMV), Eggplant mosaic tymovirus (EMV), Ononis yellow mosaic tymovirus (OYMV), Erysimum latent tymovirus (ELV), Potato X potexvirus (PVX), White clover mosaic potexvirus (WClMV), Narcissus mosaic potexvirus (NMV), Apple chlorotic leaf spot closterovirus (ACLSV), Potato M carlavirus (PVM), Alfalfa mosaic alfamovirus (AIMV), Tobacco mosaic tobamovirus (TMV), and Beet necrotic yellow vein furovirus (BNYVV). The overlapping ORF encoding a movement protein (entirely nested within replicase) is a genetic novelty unique to tymoviruses. Figure reproduced from [9] with the permission of the authors.

The same study [21] demonstrated that most de novo proteins have a sequence composition globally biased toward disorder-promoting amino acids and that overlapping proteins are predicted to contain significantly more structural disorder than non-overlapping proteins (the term disorder applies to proteins which lack a stable secondary and tertiary structure, at least in the absence of a binding partner) (Figure 4). Based on the notion that disordered proteins are generally subjected to less structural constraint than ordered ones [29], Rancurel et al. proposed that presence of disorder in one or both overlapping proteins could relieve the evolutionary constraints imposed by the overlap.

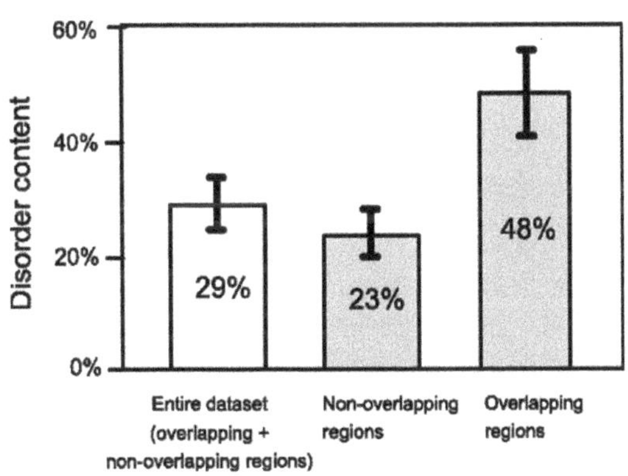

Figure 4. Predicted disorder content of proteins encoded by overlapping genes. The error bars correspond to a 95% confidence interval. Figure reproduced from [21] with the permission of the American Society of Microbiology.

This feature was further investigated by Willis and Masel [30], who analyzed a dataset of 92 overlapping genes spanning 33 viral families, 47 of them with a predicted ancestral and *de novo* frame. In accordance to [21], the authors found that the mean predicted value of the intrinsic structural disorder (ISD) in overlapping proteins is significantly higher than that in non-overlapping proteins. In addition, they found that the *de novo* proteins have a higher ISD than the ancestral ones, but this feature is specific to overlapping genes with a *de novo* frame shifted two nucleotides 3′ (+2 overlap) with respect to the ancestral frame.

The Willis study also demonstrated that the majority of overlapping genes (75%) shows a de novo frame shifted one nucleotide 3′ (+1 overlap) with respect to the ancestral frame. This feature was stronger for internal overlaps, in which one gene is completely contained within its overlapping partner, and was not found for terminal overlaps, in which the 3′ end of the upstream gene overlaps with the 5′ end of the downstream member of the pair. The prevalence of +1 gene births, despite the advantage of higher ISD in +2 gene births, was explained by the mutation bias. By sequence analysis of a control set of non-overlapping genes, Willis and Masel found that +1 frameshifts are evolutionary advantaged, because they yield significantly more ATG start codons (1 per 27 codons) than +2 frameshifts (1 per 111) and slightly fewer termination codons (1 per 14 codons) than +2 frameshifts (1 per 11).

4. Advanced Evolutionary Studies and Creation of a Curated Dataset of Overlapping Genes with Known Expression

As reported in the previous paragraph, identifying which frame of a gene overlap is ancestral and which one is de novo can be done by assessing their phylogenetic distribution (the frame phylogenetically most restricted is assumed to be the de novo one). This approach is simple and reliable but is not applicable to cases where the two frames have an identical phylogenetic distribution.

To overcome this drawback, Pavesi et al. [31] developed a new method to identify the de novo proteins. Like the previous ones [14,17,18], the method relied on the codon usage but was statistically more robust (the method assumes that the novel frame has a codon usage significantly less related to that of viral genome than the ancestral frame). It used as benchmark a reference dataset of 27 overlapping genes whose genealogy was predicted using the phylogenetic criterion. For each overlap, the method calculated: (i) the correlation coefficient (r_1) between the codon usage of the ancestral frame and that of the viral genome; (ii) the correlation coefficient (r_2) between the codon usage of the novel frame and that of the viral genome. Using the t-Hotelling test, the method evaluated the significance of the difference between r_1 and r_2, and predicted the genealogy of the overlap only in the case of r_2 significantly lower (and not simply lower) than r_1.

The method was applied to seven cases of overlap in which both frames have the same phylogenetic distribution, making the phylogenetic criterion not applicable. It demonstrated that the codon usage of overlapping frames was significantly different (or very close to significance) in only three cases: the overlap Tax protein/Rex protein of *Deltaretrovirus* and the overlap replicase/protein B2 of *Alphanodavirus* and *Betanodavirus*. Indeed, Tax and replicases had a codon usage significantly closer to that of the viral genome than the alternative frames, suggesting that they are the ancestral frames. Therefore, the *de novo* frames are those encoding the Rex protein, a post-transcription regulatory factor [32], and the protein B2, a suppressor of RNA silencing [33]. In the four other overlaps, both frames had a comparable codon usage, preventing prediction of genealogy.

The discrepancy between overlapping genes in which the novel frame has a codon usage significantly different from that of the ancestral frame and overlapping genes in which there is no significant difference was investigated by Sabath et al. [34]. They analyzed the evolution of 12 viral genes that arose de novo by overprinting and estimated their relative ages. They found that young de novo genes have a different codon usage from the rest of the genome and that evolve rapidly, under positive or weak purifying selection. In contrast, older de novo genes have a codon usage that is similar to the rest of the genome. They evolve slowly and are under strong purifying selection. Therefore, de novo genes can evolve very rapidly shortly after their origin. As they age, they tend to experience increasingly severe selective constraints, and their codon usage tends to approach that of the ancestral gene from which they originate [34].

To provide a benchmark for systematic studies, Pavesi et al. [35] assembled a high-quality dataset of 80 overlapping genes experimentally proven. They were selected from small or medium-sized eukaryotic viruses with a genome shorter than 30 kb, including single-stranded and double-stranded DNA viruses and single-stranded and double-stranded RNA viruses. The authors found that the overall nucleotide and amino acid composition of overlapping genes is significantly different from that of non-overlapping genes for several composition features. In particular, the proteins they encode show an enrichment in amino acids with high codon degeneracy (the 6-fold degenerate amino acids L, R, and S) and a depletion in amino acids with low codon degeneracy (the 2- and 1-fold degenerate amino acids C, D, E, F, H, K, N, Q, Y, M, and W), a feature that could have been selected because it mitigates the constraints under which the two frames evolve. Using a multivariate statistical method, that is the principal component analysis [36], the study demonstrated that the vast majority of overlapping genes (75 out of 80) follow a similar composition bias, despite their heterogeneity in length and function [35].

A valuable feature of the dataset is that it contains detailed biological information for each pair of overlapping genes (type of experimental evidence for expression, mechanism of translation, function of the two gene products, phenotypic effects upon mutation, and bibliography). By examining this information, Pavesi et al. [35] identified 11 overlaps in which the two encoded proteins take part in the same pathway and interact directly each other. This interaction is critical for viral assembly [37], viral replication [38], relocation of viral genome from nucleus to cytoplasm [39], and viral entry in the host cell [40].

The same study [35] pointed out that the most common mechanisms to express overlapping genes occur at the level of translation. Indeed, more than two thirds of overlapping genes with a known or suspected mechanism of expression (54 out of 71 cases) are expressed by translational processes, such as the use of an alternative start codon [41], ribosomal frameshifting [42], and internal ribosome entry site [43]. The remaining third of overlapping genes is expressed by transcriptional mechanisms, such as the use of sub-genomic RNAs [44] and transcriptional slippage [45].

5. Symmetric and Asymmetric Evolution in Viral Overlapping Genes

As first proposed by Miyata and Yasunaga [13], we would expect, in principle, that overlapping genes evolve under strong constraints, because a single nucleotide substitution can simultaneously impair two proteins (e.g., codon position 12 in Figure 1B). An example of "constrained evolution" is that observed in hepatitis B virus (HBV), a small double-stranded DNA virus (3.2 kb) with a high content of overlapping genes. Mizokami et al. [46] found that the mean number of synonymous nucleotide substitutions per site in the five overlapping coding regions of HBV is significantly lower (0.234) than that in non-overlapping regions (0.508).

However, dual-coding genes can also show a less constrained pattern of change, as a consequence of a high rate of non-synonymous substitution in one frame (positive adaptive selection) with concurrent dominance of synonymous substitution in the other (negative purifying selection). In simian immunodeficiency virus, Hughes et al. [47] found that the region of protein Tat under strongest positive selection is encoded by a frame which overlaps, for a length of 150 nt, the frame encoding protein Vpr. Another case is the overlapping gene protein p19/protein p22 (549 nt) of tombusviruses. Allison et al. [48] demonstrated that p19, a suppressor of the host RNA interference mechanism in response to viral infection [49], is under positive selection, whereas p22, a membrane-bound protein essential for cell-to-cell movement of virus [50], is under purifying selection.

These studies suggest that the evolution of overlapping genes can be summarized in accordance to two different models. The first claims that the two proteins encoded by the overlap can evolve under similar selection pressures. In the case of strong selection against amino acid change, both proteins (or protein regions) are highly conserved. For example, comparative analysis of 27 strains of HBV showed that the RNase domain of polymerase and the N-terminal half of protein X have both a percentage of conserved amino acids higher than 90% [46]. In the case of weak selection against amino acid change, both proteins can vary considerably. For example, the same study [46] showed that the spacer domain of polymerase and the pre-S1 region of surface protein show a percentage of conserved amino acids of 30 and 40%, respectively. This model was named "symmetric evolution", because the number of amino acid substitutions of one protein is expected to be not significantly different from that of the other [51]. It corresponds to the "shared model" described by Fernandes et al. [52].

The other model claims that the two proteins encoded by the overlap can evolve under significantly different selection pressures. Support for this model, which implies positive selection on one frame and negative selection on the other, was provided by a number of studies. In addition to those mentioned previously [47,48], they concern the overlapping gene P/C of Sendai virus [53], the overlapping genes ORF0/ORF1 and ORF3/ORF4 of potato leafroll virus [54], and the overlapping gene VP1/VP2 of human parvovirus B19 [55]. Interestingly, an accordance to this model was also found in the overlapping gene p16INK4a/p19ARF of mammals [56]. This model was named "asymmetric evolution", because the number of amino acid substitutions of one protein is expected to be significant different from that of the other [51]. It corresponds to the "segregated model" described by Fernandes et al. [52].

As most individual overlapping genes examined in [35] have at least one homolog, I assembled a dataset of 75 pairs of homologous overlaps and analyzed it to determine which of the two evolutionary models is the prevailing one [51]. The study demonstrated that half of overlaps (38 out of 75) evolve in accordance with the asymmetric model. A clear example was the overlapping gene of apple stem grooving virus (ASGV) that encodes a movement protein and a linker-region connecting the RdRp (RNA-dependent RNA polymerase) domain to the coat-protein domain. In detail, the percent amino acid diversity between the linker-region of ASGV and the homolog from citrus tatter leaf virus (39%; 125 differences and 195 identities) resulted to be ten-fold higher than that between the movement protein and the homolog (4%; 13 differences and 307 identities).

The same study [51] pointed out that in all overlapping genes evolving asymmetrically and with known genealogy (23 cases) the most variable protein is that encoded by the de novo frame. Despite the small number of cases, this finding suggests that de novo proteins are the preferred target of selection. As shown in Table 1, most of de novo proteins (14 out of 23) are known to play a role in viral pathogenicity: six act as suppressor of interferon response, four as suppressor of RNA silencing, two as suppressor of interferon response and apoptosis factor, one as apoptosis factor, and one has the ability to selectively degrade the host RNA-polymerase II transcripts. Very interesting is the notion that two *de novo* proteins are known to exert functions that are not virus-specific. They are the apoptin of *Chicken anemia virus*, which induces cell death in a broad range of human tumour cell lines but not in normal cells [57,58], and the protein X of Borna disease virus, which shows protective properties against neurodegeneration in vitro and in vivo [59,60].

Symmetric evolution (similar selection pressures on the two proteins) was found in the remaining 37 overlaps of the dataset [51]. A strong selection against amino acid change was found in the overlapping gene protein 3a/protein 3b of human severe acute respiratory syndrome-related coronavirus (SARS-CoV): the amino acid diversity between protein 3a of human SARS-CoV and the homolog from bat SARS-CoV was rather low (5.3%), as well as that between protein 3b and the homolog (8.8%). A weak selection against amino acid change was found in the overlapping gene of spinach latent virus (SLV) encoding the zinc-finger domain of polymerase and protein 2b: the amino acid diversity between the zinc-finger domain of SLV and the homolog from elm mottle virus was high (47%), as well as that between protein 2b and the homolog (44%).

Table 1. List of 14 overlapping genes evolving asymmetrically and with a known function of the *de novo* protein.

Virus Species and Genome Ac. Number	Overlapping Gene (Protein Products)	Predicted *De Novo* Protein (Prediction Criterion)	Most Variable Protein (Length of Overlap)	Function [Bibliographic Reference]
Theiler's murine encephalomyelitis virus (NC_001366)	Polyprotein region encoding the leader and VP4 capsid proteins/protein L*	Protein L* (phylogeny and codon usage)	Protein L* (156 aa)	Suppressor of interferon response [61]
Hepatitis C virus (NC_004102)	Polyprotein region encoding the core protein/protein F	Protein F (codon usage)	Protein F (151 aa)	Suppressor of interferon response [62]
Puumala virus (NC_005224)	Nucleocapsid protein/non-structural protein NSs	Non-structural protein NSs (codon usage)	Non-structural protein NSs (90 aa)	Suppressor of interferon response [63]
Infectious pancreatic necrosis virus (NC_001915)	Protein VP5/polyprotein region encoding the N-half of capsid protein VP2	Protein VP5 (phylogeny and codon usage)	Protein VP5 (131 aa)	Suppressor of interferon response [64]
Borna disease virus (NC_001607)	Protein X/phosphoprotein (P)	Protein X (codon usage)	Protein X (71 aa)	Suppressor of interferon response [65]
Infectious salmon anemia virus (NC_006497)	Protein p6/protein p7	Protein p6 (codon usage)	Protein p6 (183 aa)	Suppressor of interferon response [66]
Apple chlorotic leaf spot virus (NC_001409)	Protein p50/capsid protein	Protein p50 (phylogeny)	Protein p50 (105 aa)	Suppressor of RNA silencing [23]
Tomato bushy stunt virus (NC_001554)	Protein p19/protein p22	Protein p19 (phylogeny)	Protein p19 (172 aa)	Suppressor of RNA silencing [67]
Turnip yellow mosaic virus (NC_004063)	Protein p69/replicase (methyltransferase domain and downstream region)	Protein p69 (phylogeny and codon usage)	Protein p69 (626 aa)	Suppressor of RNA silencing [68]
East African cassava mosaic virus (NC_004674)	Protein AC1/protein AC4	Protein AC4 (phylogeny)	Protein AC4 (77 aa)	Suppressor of RNA silencing [69]
Murine norovirus (NC_008311)	Capsid protein VP1/virulence factor VF1	Virulence factor VF1 (phylogeny and codon usage)	Virulence factor VF1 (213 aa)	Suppressor of interferon response and apoptosis factor [70]
Influenza A virus (NC_002021)	Subunit PB1 of RdRp/protein PB1-F2	Protein PB1-F2 (phylogeny and codon usage)	Protein PB1-F2 (87 aa)	Suppressor of interferon response and apoptosis factor [71,72]
Chicken anemia virus (NC_001427)	Capsid protein VP4/apoptin	Apoptin (phylogeny)	Apoptin (119 aa)	Apoptosis factor [73]
Influenza A virus (NC_002022)	Subunit PA of RdRp/protein PA-X	Protein PA-X (codon usage)	Protein PA-X (61 aa)	Degradation of the host RNA-polymerase II transcripts [74]

6. Overlapping Genes Show a Peculiar Pattern of Nucleotide and Amino Acid Composition

Overlapping genes represent an unusual pattern of the genetic language [75,76], as two, or exceptionally three, reading frames may lie inside a single nucleotide sequence. The first attempts to detect composition features peculiar to the overlap were carried out using the information theory indices [77]. They are D_1, the divergence from a random nucleotide composition, and D_2, the divergence from a random dinucleotide composition [78,79]. The assumption is that the smallness of D_1, which implies a frequency of each nucleotide around 25%, represents the richness of vocabulary, while the largeness of D_2 represents the clarity of grammatical rules, that is the constraints against a random dinucleotide composition [80]. Thus, information theory predicts that dual-coding genes should have a lower D_1 value and a higher D_2 value when compared to single-coding genes, as hallmarks of a greater information content.

However, comparative analysis of overlapping and non-overlapping genes in the genome of three microviruses (ΦX174, α3 and G4), two avian hepadnaviruses, three strains of HIV-1, two plant luteoviruses, and two plant tymoviruses showed that the pattern predicted by information theory is valid for the first three groups of viruses, but weak for luteoviruses and inconsistent for tymoviruses [17].

In the following years, comparative analyses of overlapping and non-overlapping genes were limited to individual virus species, such as *Infectious bursal disease virus* [81], to virus families such as *Papillomaviridae* [82], or to a small dataset of RNA viruses [21]. Only recently, it was possible to perform a wide-scale analysis using the curated dataset assembled in [35]. It contains, indeed, not only the nucleotide sequence of 80 overlapping genes but also that of the entire complement of non-overlapping genes in the virus genome.

Pavesi et al. [35] found that overlapping genes differ significantly from non-overlapping genes for 20 composition features (Figure 5). Some of them are clearly linked. For example, the enrichment in C of overlapping genes is linked to that in dinucleotide CC, codons CCC and CCG, and proline. The depletion in A and T of overlapping genes is linked to that in amino acids with a low codon degeneracy, because they are encoded by codons rich in A and T. Depletion in T, A, and TA of overlapping genes reduces the probability of occurrence of stop codons (TGA, TAG and TAA) and thereby increases that of occurrence of long overlapping frames.

The dataset in [35] was also a valuable start point to assemble a much larger one [83]. For each overlapping gene, it included all the homologs gathered from the NCBI Viral Genome Database [84]. The size of the sample increased from 80 to 319 overlaps, coming from 244 virus species (the number of virus species is lower than that of overlaps because some viruses contain more than one overlap). Consider for example the overlapping gene replicase/movement protein of tymoviruses. The dataset in [35] contains only the overlap of turnip yellow mosaic virus (TYMV), the dataset in [51] contains the overlap of TYMV and the homolog of watercress white vein virus (nucleotide diversity of 28%), while the dataset in [83] contains as many as 20 homologous overlaps, covering a nucleotide diversity from 28 to 50%.

By comparative analysis of overlapping and non-overlapping genes (319 overlaps and 244 non-overlaps), I detected a total of 37 significantly different composition features [83]. Principal component analysis, aimed to evaluate whether the observed differences were homogeneously distributed in individual overlapping genes, revealed the presence of only four outliers (Figure 6). This finding confirmed that overlapping genes follow a common pattern of composition bias, despite their different length and function.

With the aim to distinguish overlapping from non-overlapping genes with the best accuracy, I compared the sample set of 319 overlaps to the control-set of 244 non-overlaps using multivariate statistics [83]. The methods were the Fisher's linear discriminant analysis (LDA) [85,86] and the partial least squares-discriminant analysis (PLS-DA) [87,88].

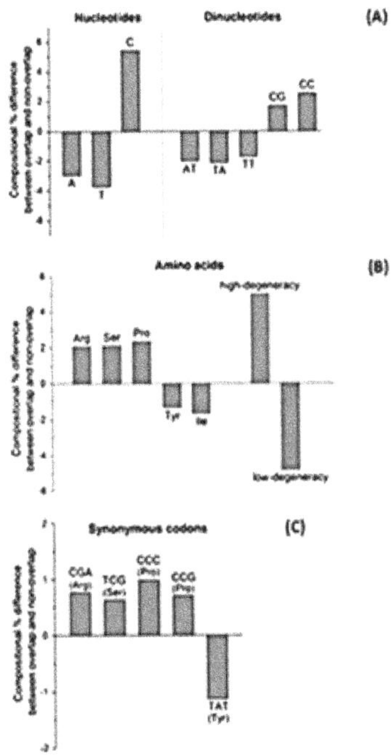

Figure 5. Difference between the pooled sets of overlapping and non-overlapping genes for the 20 most critical composition features. (**A**) Nucleotides and dinucleotides. (**B**) Amino acids and amino acids grouped in accordance to codon degeneracy. (**C**) Synonymous codons. The figure, made by A. Vianelli, was reproduced from [35].

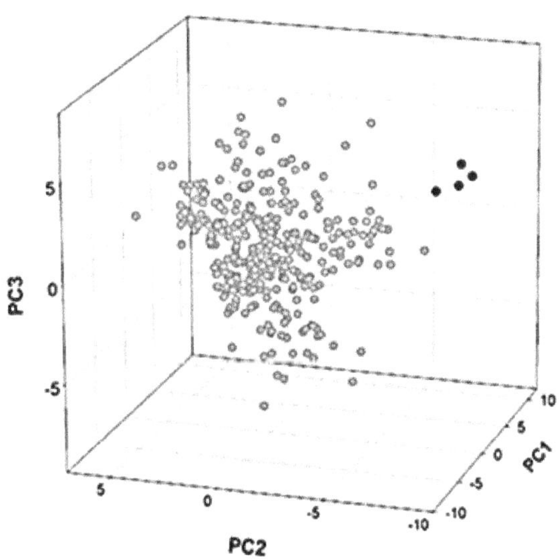

Figure 6. Principal component analysis (PCA) of a sample set of 319 overlapping genes. The three-dimensional map was obtained using the first (PC1), second (PC2), and third (PC3) principal component. Black circles indicate the 4 homologs of the overlapping gene polymerase/protein X of Hepatitis B virus. They were classified as outlier because of a highly atypical sequence composition. Figure reproduced from [83] with the permission of Elsevier.

The best performance of LDA was given by a linear function of 21 coefficients, corresponding to 21 significantly different composition features between overlap and non-overlap (two from nucleotides, four from dinucleotides, eight from amino acids, and seven from synonymous codons). As shown in Figure 7, the strong discriminant power of the function is highlighted by the different distribution of the LDA score in overlapping genes (grey columns) compared to that in non-overlapping genes (black columns).

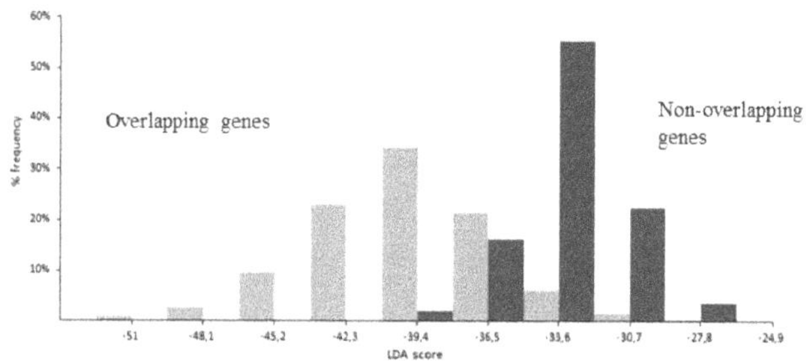

Figure 7. Histogram of the distribution of LDA score in overlapping genes (grey columns) and in non-overlapping genes (black columns). With a discriminant score of −35.31, a high percentage (96.5%) of overlapping genes were correctly classified as overlap (score below −35.31) and a high percentage (97.1%) of non-overlapping genes were correctly classified as non-overlap (score above −35.31). Figure was reproduced from [83] with the permission of Elsevier.

The best performance of PLS-DA was given by a linear regression function of 23 regression coefficients, corresponding to 23 significantly different composition features between overlap and non-overlap (one from nucleotides, six from dinucleotides, seven from amino acids, and nine from synonymous codons). The strong discriminant power of the function is evident in Figure 8, which shows the distribution of the PLS-DA score in overlapping (grey columns) and non-overlapping genes (black columns).

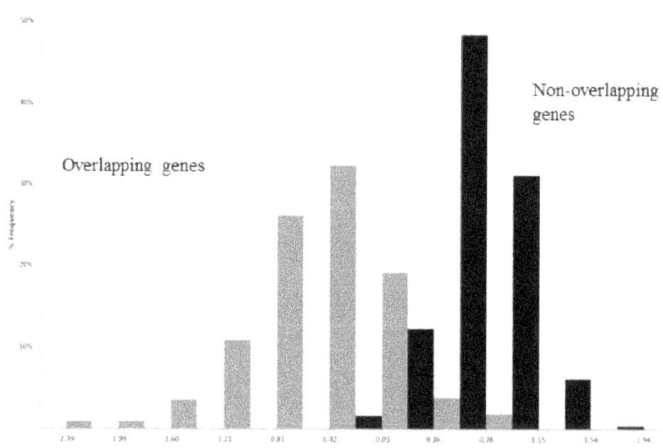

Figure 8. Histogram of the distribution of PLS-DA score in overlapping genes (grey columns) and in non-overlapping genes (black columns). With a discriminant score of 0, a high percentage of overlapping genes (94.9%) were correctly classified as overlap (score below 0) and a high percentage (98.4) of non-overlapping genes were correctly classified as non-overlap (score above 0). Figure reproduced from [83] with the permission of Elsevier.

7. Birth of Overlapping Genes in Viruses: Gene Compression or Gene Novelty?

The abundance of overlapping genes in viruses [89] was explained by two, not mutually exclusive, theories. The gene-compression theory states that the gene overlap is a valuable strategy to maximize the coding ability of small genomes [13,17,90–92], as a consequence of biophysical constraints on the size of the capsid structure [93] or of a high mutation rate such that occurring in RNA viruses [94]. As most mutations are deleterious, the high mutation rate will limit the genome size, and thus new genes must come from overprinting [95]. The gene-novelty theory claims that the birth of novel proteins by overprinting is driven by selection pressures providing the virus with a fitness advantage that lead to their fixation [9,21,96].

Using as benchmark the dataset of overlapping genes assembled in [83], I could determine which of the two theories is the most plausible one. Using the phylogenetic and codon-usage criteria, I first predicted the genealogy of 46 overlapping genes. By extending the inferred genealogy to the homologs, I then obtained a dataset of 194 overlapping genes with a known ancestral and de novo frame: 126 overlaps with a +1 de novo frame and 68 overlaps with a +2 de novo frame. Analysis of amino acid and synonymous codon composition revealed that the +1 and +2 de novo frames differ significantly from the respective ancestral frames for 25 and 23 composition features, respectively [83].

On the basis of these differences in composition, the linear discriminant analysis clearly separated the ancestral frames from the +1 de novo frames (Figure 9A), as well as the ancestral frames from the +2 de novo frames (Figure 9B). When compared to the respective ancestral proteins, the +1 de novo proteins were found enriched in hydrophobic residues and depleted in acidic residues, while the +2 de novo proteins were found enriched in basic residues and cysteine and depleted in hydrophobic residues [83]. Although one theory does not entirely exclude the other, the different amino acid composition of de novo proteins vs. the ancestral ones should better support gene-novelty than gene-compression.

In the same study [83], I examined the 244 virus species in the dataset to determine whether there is a negative relationship between the length of their genomes and that of their overlapping genes, a feature in accordance to the gene-compression theory. Using the Spearman rank correlation coefficient, I found a significant negative correlation of −0.31, too weak however for supporting the gene-compression theory. A similar study demonstrated that gene overlap is not a significant factor in the compression of viral genomes [96].

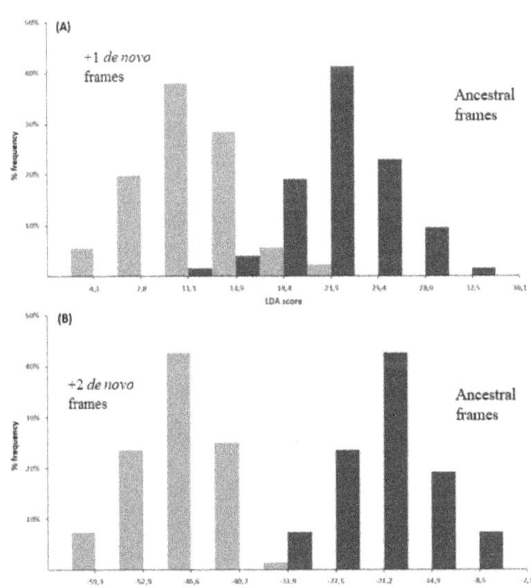

Figure 9. (**A**) Histogram of the distribution of the LDA score in 126 ancestral frames (black columns)

and in the respective +1 de novo frames (grey columns). With a discriminant score of 17.20, a high percentage (96.8%) of ancestral frames were correctly classified as ancestral (score above 17.20) and a high percentage (97.6%) of +1 de novo frames were correctly classified as de novo (score below 17.20). (**B**) Histogram of the distribution of the LDA score in 68 ancestral frames (black columns) and in the respective +2 de novo frames (grey columns). With a discriminant score of −34.98, all ancestral frames and all +2 de novo frames were correctly classified as ancestral and de novo, respectively. Figure reproduced from [83] with the permission of Elsevier.

8. Modular Evolution in Overlapping Genes: The Case of Hepatitis B Virus

The theory of modular evolution for viruses predicts that various coding sequences are used as functional modules during recombination events [97]. This is thought to speed up virus evolution by utilizing various combinations of functional modules to gain novel genes [98,99]. However, viruses can also evolve through a mechanism in which the gain of novel modules depends on overprinting. Two studies showed that modular evolution played a critical role in the genesis of the overlapping gene polymerase/surface protein of hepadnaviruses [100,101].

Hepatitis B virus (HBV), a member of the familiy *Hepadnaviridae*, is a DNA reverse-transcribing virus with a circular genome of 3.2 kb. About 50% of the genome contains overlapping coding regions, due to the large overlap between the gene for polymerase (P) and the genes for capsid (C), X, and surface (S) proteins (Figure 10).

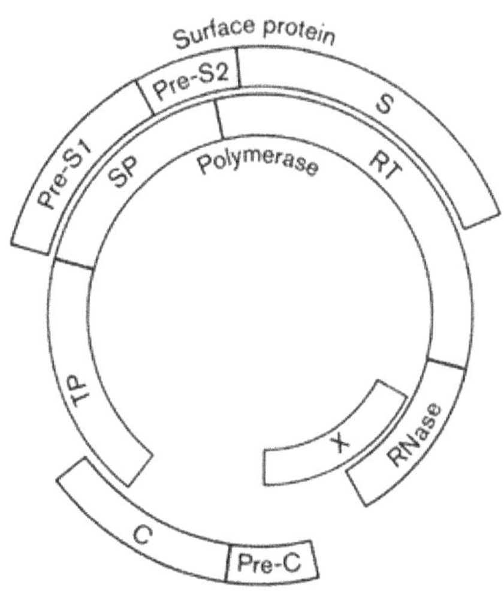

Figure 10. Map of the genome of HBV with overlapping and non-overlapping coding regions. Pre-S1, Pre-S2, and S are the domains of surface protein. TP, SP, RT, and RNase are the domains of polymerase. TP, terminal protein domain; SP, spacer domain; RT, reverse transcriptase domain; RNase, ribonuclease domain; C, capsid. Figure reproduced from [100] with the permission of the Microbiology Society.

Several studies were carried out to investigate the role of gene overlap in the evolution of HBV [46,102–105]. The genetic diversity of the overlapping proteins P and S was also related to virus survival in response to antiviral drugs [106], to virus escape from neutralizing antibodies [107], and to the clinical significance of mutations induced by selection [108].

Using the phylogenetic method, the genealogy of the overlap between the RNase domain of polymerase and the N-terminal half of protein X was clearly elucidated. The method predicts that protein X arose de novo, because of its presence in *Orthohepadnavirus* but not in the sister genus *Avihepadnavirus* [21,109]. In contrast, the genealogy of the overlap between the surface protein and the spacer (SP) and reverse-transcriptase (RT) domains of polymerase was difficult to predict. In this case, the phylogenetic criterion was not applicable because the homologs of both frames show an identical phylogenetic distribution, making possible only the codon-usage approach.

By a sliding-window analysis of the codon usage along the entire overlapping coding region (1200 nt), I found that the overlap P/S can be subdivided into two regions, each with its own pattern of codon usage [100]. By predicting the ancestral and the de novo frame in each region, I hypothesized a primordial genome with a short gene S placed between the gene encoding the terminal protein (TP) and the gene encoding the RT and RNase domains of polymerase (Figure 11A). A first increase in coding density was due to the birth, within gene S, of a de novo frame encoding the spacer (SP) domain (Figure 11B). Acting as linker, it led to creation of a multi-domain polymerase (TP, SP, RT, and RNase domains).

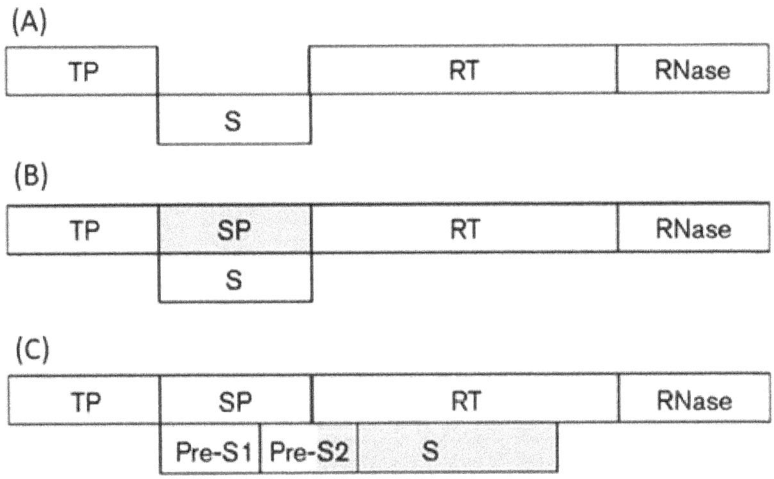

Figure 11. Modular evolution in the genesis of the overlapping gene polymerase/surface protein of hepadnaviruses. (**A**) Putative primordial genome of HBV. (**B**) Birth of a novel frame encoding the SP domain of polymerase (shaded box). (**C**) Birth of a novel frame encoding the C-terminal region of the Pre-S2 domain and the S domain of surface protein (shaded box). Figure reproduced from [100] with the permission of the Microbiology Society.

A further increase in coding density was due to a long overlapping extension of gene S. In addition to a full-length Pre-S2 domain, it led to a de novo creation of the S domain of surface protein (Figure 11C). As a result, this overlapping extension generated a surface gene consisting of three in-phase ORFs, whose co-translation yields the large surface protein. Taken together, these evolutionary inferences suggest that the overlapping gene polymerase/surface protein attained its present complexity through modular evolution [100].

The hypothesis that the Pre-S/S ORF is an innovation unique to the hepadnaviral lineage was confirmed by Lauber et al. [101]. In addition, they dated the de novo emergence of Pre-S/S about 400 million years ago. This date corresponds to the inferred separation time between hepadnaviruses (enveloped viruses with a surface-protein gene) and nackednaviruses (non-enveloped fish viruses lacking a surface-protein gene). Both studies [100,101] pointed out that overprinting is a source not only of de novo accessory proteins with regulatory function [20–28], but also of de novo essential structural proteins, such as the large surface protein of hepadnaviruses.

9. Estimation of Selection Intensities in Overlapping Genes by *ad hoc* Methods

The strength of selection pressure in protein-coding genes is usually inferred by comparing the number of non-synonymous nucleotide substitutions per site (d_n) with that of synonymous nucleotide substitutions per site (d_s), with $d_n/d_s > 1$ indicative of positive selection and $d_n/d_s < 1$ of negative selection [110,111]. Extending this standard approach to overlapping genes is inappropriate, because a nucleotide substitution that is synonymous in one frame is highly likely to be non-synonymous in the alternative frame. It follows that the constraints against synonymous substitutions in a frame significantly lowers its d_s value, causing an artifactual increase of d_n/d_s and a wrong inference of positive selection if $d_n/d_s > 1$.

To overcome this problem, several researchers have developed methods for correctly estimating the strength of selection intensities in overlapping genes. The maximum-likelihood model by Hein and Støvlbæk [112] was an extension of the notion of degeneracy class of a site [111] to that of a combination of two degeneracy classes (one for each frame to which a site belongs). De Groot et al. integrated this model into a statistical alignment framework and estimated selection in the overlapping genes of HBV and HIV-2 [113]. McCauley et al. developed a Hidden Markov Model (HMM) capable of accounting for varying levels of selection along the viral genome, including those acting on overlapping ORFs [114]. When applied to a multiple alignment of HIV-2 sequences, HMM was able to make truly statistically significant statements about the nature of selection on dual-coding regions. The Markov-chain Monte Carlo model by Pedersen and Jensen [115] incorporated the constraints imposed by both of the overlapping genetic codes in an exact manner. This model, indeed, included parameters representing the degrees of selection constraints operating in the different frames.

Sabath et al. proposed a non-stationary method, similar to that of Pedersen and Jensen but with the advantage to avoid the need for computationally-expensive procedure [116]. The method was tested on the overlapping genes PB1-F2 and NS1 of influenza A virus, because they were previously reported to exhibit values of d_n/d_s remarkably higher than 1 (9.4 for PB1-F2 and 1.9 for NS1) and thus indicative of strong positive selection [117,118]. The method demonstrated that PB1-F2 and NS1 appear to be under weak negative selection, because of a d_n/d_s value of 0.50 and 0.70 respectively. Therefore, the previous estimates of selection on PB1-F2 and NS1 were wrong, because they were calculated ignoring the interdependence with the respective overlapping frames PB1 and NS2. A limitation of the Sabath's method is that it restricts the analysis to homologous overlapping genes in which the two encoded proteins have both an amino acid diversity smaller than 50% or greater than 5%.

The method developed by Wei and Zhang [119] was an extension of the standard method for protein-coding genes originally proposed by Nei and Gojobori [111]. The method first classifies each site in the reference overlapping gene into four categories (NN, NS, SN, and SS, where N stands for non-synonymous and S for synonymous), depending on the impacts of potential mutations on the two overlapping ORFs (ORF1 and ORF2). The method then classifies all nucleotide differences between the reference overlapping gene and its homolog into four categories (NN, NS, SN, and SS) and counts their numbers (M_{NN}, M_{NS}, M_{SN}, and M_{SS}, respectively). Finally, the method estimates the strength of natural selection acting on ORF1 by $\omega_1 = d_{NN}/d_{SN}$ and that acting on ORF2 by $\omega_2 = d_{NN}/d_{NS}$.

10. Computational Methods to Predict Overlapping Genes in Viruses

To identify overlapping genes by sequence analysis, several groups have developed methods that detect the atypical pattern of nucleotide substitution induced by the overlap. Firth and Brown developed a method called Maximum-Likelihood Overlapping Gene Detector (MLOGD), which was designed to detect the mutation signature of overlapping coding sequences in pairwise alignments of two sequences, under a double-coding model [120]. The same authors presented an improved version of MLOGD, whose ability to estimate the magnitude of constraints on the gene overlap yielded a sensitivity of 90% in the detection of known overlapping genes [121].

A further improvement was provided by the computational tool Synplot2 [122]. It analyzed alignments of protein-coding virus sequences to identify regions where there is a statistically significant reduction in the degree of variability at synonymous sites, a characteristic signature of overlapping functional elements such an overlapping gene or a conserved RNA structure. The same approach was followed by Sealfon et al., who developed a phylogenetic codon-model based method (FRESCo, that is Finding Regions of Excess Synonymous Constraints) for detecting virus regions with a significantly reduced synonymous variability [123]. When applied to a multiple alignment of over 2000 whole-genome sequences of HBV, FRESCo detected strong synonymous constraint elements within known regions of overlapping function (overlapping ORFs or regulatory elements).

By modifying the method in [119], Nelson et al. [124] developed a computational tool named OLGenie, where OLG means OverLapping Gene. It estimated signs of strong purifying (negative) selection in aligned sequences as hallmark of functional overlapping genes. Assessment with simulations and controls from viral genomes (58 OLGs and 176 non-OLGs) demonstrated low false-positive rates and good ability in differentiating true OLGs from non-OLGs.

Although powerful, these computational methods are necessarily constrained by the requirement for multiple sequences of sufficient diversity to reliably detect overlapping genes. Therefore, these methods are not applicable in the case of a single nucleotide sequence or sequences with a low nucleotide diversity. To overcome this drawback, Schlub et al. developed a statistical method that relies on only a single gene, or genome, nucleotide sequence [125]. The method detects candidate overlapping genes in viruses by selecting overlapping ORFs that are significantly longer than expected by chance. It consists of a codon-permutation test and a synonymous-mutation test. The limit of the method is that the sensitivity was high (90% for codon-permutation test and 95% for synonymous-mutation test) for overlapping genes longer than 300 nt, but rather low for those longer than 100 nt (65% for codon-permutation test and 71% for synonymous-mutation test).

Another prediction method that relies on single nucleotide sequences was the combined use of linear discriminant analysis (LDA) and partial least squares-discriminant analysis (PLS-DA) [83]. Taken individually, LDA correctly classified 96.5% of overlapping genes and 97.1% of non-overlapping genes (Figure 7) and PLS-DA 94.9% of overlapping genes and 98.4% of non-overlapping genes (Figure 8). The performance of the combined use of LDA and PLS-DA is summarized in Figure 12. Grey circles in part A indicate the overlaps correctly classified by both methods (94.2% of the total). Black circles in part C indicate the non-overlaps correctly classified by both methods (97.1% of the total). Application of the method to the genome sequence of SARS-CoV-2 (isolate Wuhan-Hu-1), the etiological agent of current pandemic [126], led to detection of two new potential overlapping ORFs (asterisks in part A of the figure).

Another method analyzing single, or closely related, genome sequences was GOFIX [127]. It detects overlapping ORFs on the basis of a significant enrichment in the X motif (a set of 20 codons over-represented in viral genes).

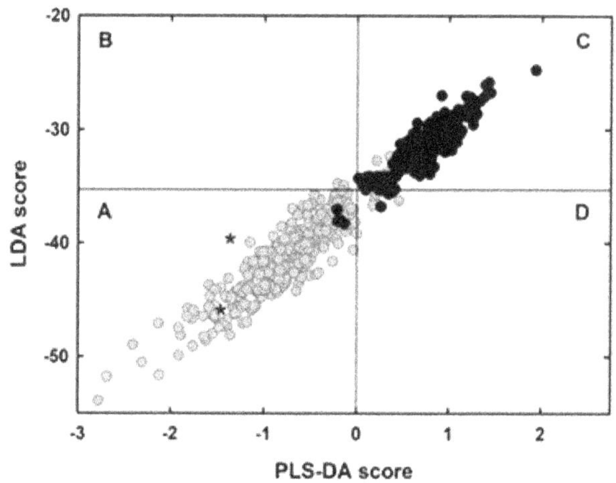

Figure 12. Map of overlapping genes (grey circles) and non-overlapping genes (black circles), in which the PLS-DA score is plotted against the respective LDA score. Grey circles in part (**A**) indicate overlaps correctly classified by both methods (94.2% of the total). Black circles in part C indicate non-overlaps correctly classified by both methods (97.1% of the total). Gray circles in part (**B–D**) indicate overlaps misclassified by one or both methods (5.8% of the total). Black circles in part (**A**) and (**D**) indicate non-overlaps misclassified by one or both methods (2.9% of the total). Asterisks in part (**A**) indicate two new potential overlapping genes detected in the genome of SARS-CoV-2 (isolate Wuhan-Hu-1). Figure reproduced from [83] with the permission of Elsevier.

11. Brief Note on the Presence of Overlapping Genes in Prokaryotes and Eukaryotes

Although the present review is focused on viral overlapping genes, it is important to note that experimental and computational reports suggest that the birth of new genes by overprinting is not confined to viruses. It is a much wider phenomenon than previously thought, both in prokaryotic [128,129] and eukaryotic genomes [130–135]. Thus, the expression of two proteins from the same mRNA has changed the traditional view that a mature eukaryotic mRNA is a mono-cistronic molecule with a single translated ORF [136,137]. Interestingly, it has also been found that some human cancer-specific antigens, silent in normal tissues, are translated from alternative open reading frames (AltORFs) [138–142]. These neoantigens are promising targets for the development of anti-tumour immunotherapies with a potentially broader coverage of patients [143].

12. Brief Note on the Presence of Anti-Sense Overlapping Genes in Viruses

Overlapping genes can be classified broadly into two types: (1) same-strand overlapping genes, which are transcribed from the same strand of DNA (also known as sense-overlap); (2) different-strand overlapping genes, which are transcribed from two opposite strands of DNA (also known as anti-sense overlap).

As the great majority of known overlapping genes are of same-strand type, they were the primary focus this review. However, I would briefly report two cases of anti-sense overlap experimentally validated. The first was found in the pX region of Human T-lymphotropic virus 1 (HTLV-1). The sense strand encodes p30, a protein playing a role in viral replication, host immunity, and cellular proliferation [144]. The anti-sense strand encodes HBZ, a transcription factor playing a critical role in HTLV-1 associated diseases [145,146]. Because the pX region of HTLV-1 also contains the sense-overlap Tax protein/Rex protein, it constitutes a hotspot of gene origination, or gene "nursery" [147]. Its complex pattern of origin and evolution is accurately presented in [31].

The other anti-sense coding sequence, termed ASP and overlapping the gene Env, was predicted in HIV-1 by Cassan et al. [148]. Using computer simulations, they showed that conservation of ASP in HIV-1 (specifically in the group M) could not be due to chance but to selection pressure conserving the start codon and avoiding stop codons. Affram et al. demonstrated the presence of the ASP protein on the surfaces of both infected cells and viral particles, yielding evidence that this accessory protein is a new structural component of HIV-1 [149].

13. Concluding Remarks and Future Directions

Over four decades after the discovery of overlapping genes [11,12], we have an accurate knowledge of their origin and evolution. This review highlights that *de novo* protein creation by overprinting is a significant factor in viral evolution, in particular in the evolution of pathogenicity. At the same time, it is a valuable start point for future studies.

For example, factors affecting the birth of overlapping genes can be further investigated by a sequence-composition analysis of "pre-overlapping coding regions", that is the genome regions homologous to a gene overlap but lacking it. This analysis could assess if the composition bias is a contributing factor (i.e., a cause) to the existence of overlapping genes or a consequence of selection acting on overlapping genes after they are born.

The accuracy of multivariate statistics (LDA and PLS-DA) in determining whether a candidate overlapping ORF is coding or non-coding can be improved by comparing the sample set of overlapping genes to a control set of spurious overlapping genes, rather than of non-overlapping genes (a spurious overlapping gene is a protein-coding region that overlaps purely by chance an ORF not interrupted by stop codon).

Having found that a small set of mammalian overlapping genes follows a composition bias similar to viral one [35], a few prediction methods could be used to detect overlapping genes in eukaryotic genome sequences. They probably contain numerous undetected overlapping genes, as suggested by increasing experimental evidence [136]. Because stop codons (TGA, TAG, and TAA) are GC-poor, overlapping genes are expected to occur less frequently by chance in eukaryotic GC-rich sequences [150]. Theoretical studies focused on constraints (and their combinatorics) acting on the amino acid composition of paired overlapping proteins may form the basis for a quick and simple method to detect overlapping regions within proteins [151–153].

The computational methods reported in Section 10 are also a valuable tool to detect new potential overlapping genes in the NCBI Viral Genome Database (e.g., in large DNA viruses), to include in database proven overlaps overlooked during genome annotation, or to exclude hypothetical overlaps that may be artefacts of genome annotation.

The wide collection of proven overlapping genes and their homologs [35,83] can be used by others as reference datasets for further studies. They could expand our knowledge about their relative age, thus increasing the number of known cases of oldest and youngest *de novo* overlapping genes. They could test the occurrence of symmetric/asymmetric evolution in different regions of the same overlapping gene, as done for example in the overlap Tat protein/Rev protein of HIV-1 [52]. The relationship between gene overlap and evolutionary rate, investigated in RNA viruses [154], could be extended to DNA viruses.

A web server, called Coevolution in OVerlapped sequences by Tree analysis (COVTree), has been developed recently by Teppa et al. [155]. COVTree analyzes the effect of mutations in one protein over the other and detects coevolution signals in "mirrored" positions. It could be applied to the large dataset of homologous overlapping genes assembled in [83].

As viral protein synthesis is completely dependent upon the translational machinery of the eukaryotic host cell, studying overlapping genes has greatly improved our knowledge of gene expression. Indeed, non-canonical translational strategies such as leaky scanning, ribosomal frameshifting and alternative initiation are essential for expression of overlapping genes [41–45,156]. Therefore, detection of overlapping genes in eukaryotes may further improve our knowledge of gene expression by translational recoding [157].

Finally, the finding that a few de novo proteins have previously unknown 3D structural folds [158,159] and mechanisms of action [160] suggests that overlapping genes provide powerful model systems to test ideas about protein folding and evolution.

Funding: This research received no external funding.

Institutional Review Board Statement: Not applicable.

Informed Consent Statement: Not applicable.

Data Availability Statement: The data presented in this study are available on request from the corresponding author.

Acknowledgments: The author is grateful to Alberto Vianelli (University of Insubria) for helpful suggestions. This work has benefited from the equipment and framework of the COMP-HUB initiative, funded by the 'Departments of Excellence' program of the Italian Ministry for Education, University and Research (MIUR, 2018–2022).

Conflicts of Interest: The author declares no conflict of interest.

References

1. Taylor, J.S.; Raes, J. Duplication and divergence: The evolution of new genes and old ideas. *Annu. Rev. Genet.* **2004**, *38*, 615–643. [CrossRef]
2. Long, M.; Betran, E.; Thornton, K.; Wang, W. The origin of new genes: Glimpses from the young and old. *Nat. Rev. Genet.* **2003**, *4*, 865–875. [CrossRef] [PubMed]
3. Patthy, L. Genome evolution and the evolution of exon-shuffling—A review. *Gene* **1999**, *238*, 103–114. [CrossRef]
4. Treangen, T.J.; Rocha, E.P.C. Horizontal transfer, not duplication, drives the expansion of protein families in prokaryotes. *PLoS Genet.* **2011**, *7*, e1001284. [CrossRef] [PubMed]
5. Li, C.Y.; Zhang, Y.; Wang, Z.; Cao, C.; Zhang, P.W.; Lu, S.J.; Li, X.M.; Yu, Q.; Zheng, Y.; Du, Q.; et al. A human-specific de novo protein-coding gene associated with human brain functions. *PLoS Comput. Biol.* **2010**, *6*, e1000734. [CrossRef] [PubMed]
6. Sorek, R. The birth of new exons: Mechanisms and evolutionary consequences. *RNA* **2007**, *13*, 1603–1608. [CrossRef]
7. Grassé, P.P. *Evolution of Living Organisms*; Academic Press: New York, NY, USA, 1977; p. 297.
8. Normark, S.; Bergström, S.; Edlund, T.; Grundström, T.; Jaurin, B.; Lindberg, F.P.; Olsson, O. Overlapping genes. *Annu. Rev. Genet.* **1983**, *17*, 499–525. [CrossRef] [PubMed]
9. Keese, P.K.; Gibbs, A. Origin of genes: "big bang" or continuous creation? *Proc. Natl. Acad. Sci. USA* **1992**, *89*, 9489–9493. [CrossRef]
10. Gibbs, A.; Keese, P.K. *Molecular Basis of Virus Evolution*; Cambridge University Press: Cambridge, UK, 1995; pp. 76–90.
11. Barrell, B.G.; Air, G.M.; Hutchison, C.A. Overlapping genes in bacteriophage phiX174. *Nature* **1976**, *264*, 34–41. [CrossRef]
12. Sanger, F.; Air, G.M.; Barrell, B.G.; Brown, N.L.; Coulson, A.R.; Fiddes, C.A.; Hutchinson, C.A.; Slocombe, P.M.; Smith, M. Nucleotide sequence of bacteriophage phi X174. *Nature* **1977**, *265*, 687–695. [CrossRef]
13. Miyata, T.; Yasunaga, T. Evolution of overlapping genes. *Nature* **1978**, *272*, 532–535. [CrossRef]
14. Fiddes, J.C.; Godson, G.N. Evolution of the three overlapping gene systems in G4 and phi X174. *J. Mol. Biol.* **1979**, *133*, 19–43. [CrossRef]
15. Buckley, K.J.; Hayashi, M. Lytic activity localized to membrane spanning region of ΦX174 E protein. *Mol. Gen. Genet.* **1986**, *204*, 120–125. [CrossRef]
16. Bernhardt, T.G.; Roof, W.D.; Young, R. Genetic evidence that the bacteriophage phi X174 lysis protein inhibits cell wall synthesis. *Proc. Natl. Acad. Sci. USA* **2000**, *97*, 4297–4302. [CrossRef]
17. Pavesi, A.; De Iaco, B.; Granero, M.I.; Porati, A. On the informational content of overlapping genes in prokaryotic and eukaryotic viruses. *J. Mol. Evol.* **1997**, *44*, 625–631. [CrossRef]
18. Pavesi, A. Origin and evolution of overlapping genes in the family Microviridae. *J. Gen. Virol.* **2006**, *87*, 1013–1017. [CrossRef] [PubMed]
19. Sander, C.; Schultz, G.E. Degeneracy of the information contained in amino acid sequences: Evidence from overlaid genes. *J. Mol. Evol.* **1979**, *13*, 245–252. [CrossRef] [PubMed]
20. Bozarth, C.S.; Weiland, J.J.; Dreher, T.W. Expression of ORF-69 of turnip yellow mosaic virus is necessary for viral spread in plants. *Virology* **1992**, *187*, 124–130. [CrossRef]
21. Rancurel, C.; Khosravi, M.; Dunker, K.A.; Romero, P.R.; Karlin, D. Overlapping genes produce proteins with unusual sequence properties and offer insight into de novo protein creation. *J. Virol.* **2009**, *83*, 10719–10736. [CrossRef]
22. Krishnamurthy, K.; Heppler, M.; Mitra, R.; Blancaflor, E.; Payton, M.; Nelson, R.S.; Verchot-Lubicz, J. The potato virus X TGBp3 protein associates with the ER network for virus cell-to-cell movement. *Virology* **2003**, *309*, 135–151. [CrossRef]
23. Yaegashi, H.; Tamura, A.; Isogai, M.; Yoshikawa, N. Inhibition of long-distance movement of RNA silencing signals in Nicotiana benthamiana by Apple chlorotic leaf spot virus 50 kDa movement protein. *Virology* **2008**, *382*, 199–206. [CrossRef]
24. Zhou, T.; Fan, Z.F.; Li, H.F.; Wong, S.M. Hibiscus chlorotic ringspot virus p27 and its isoforms affect symptom expression and potentiate virus movement in kenaf (*Hibiscus cannabinus* L.). *Mol. Plant. Microbe Interact.* **2006**, *19*, 948–957. [CrossRef]
25. Samuilova, O.; Santala, J.; Valkonen, J.P.T. Tyrosine phosphorylation of the triple gene block protein 3 regulates cell-to-cell movement and protein interactions of Potato mop-top virus. *J. Virol.* **2013**, *87*, 4313–4321. [CrossRef]

26. Taliansky, M.; Roberts, I.M.; Kalinina, N.; Ryabov, E.V.; Raj, S.K.; Robinson, D.J.; Oparka, K.J. An umbraviral protein, involved in long-distance RNA movement, binds viral RNA and forms unique, protective ribonucleoprotein complexes. *J. Virol.* **2003**, *77*, 3031–3040. [CrossRef]
27. Skjesol, A.; Aamo, T.; Hegseth, M.N.; Robertsen, B.; Jørgensen, J.B. The interplay between infectious pancreatic necrosis virus (IPNV) and the IFN system: IFN signaling is inhibited by IPNV infection. *Virus Res.* **2009**, *143*, 53–60. [CrossRef]
28. Vargason, J.M.; Szittya, G.; Burgyan, J.; Hall, T.M. Size selective recognition of siRNA by an RNA silencing suppressor. *Cell* **2003**, *115*, 799–811. [CrossRef]
29. Brown, C.J.; Takayama, S.; Campen, A.M.; Vise, P.; Marshall, T.W.; Oldfield, C.J.; Williams, C.J.; Dunker, A.K. Evolutionary rate heterogeneity in proteins with long disordered regions. *J. Mol. Evol.* **2002**, *55*, 104–110. [CrossRef]
30. Willis, S.; Masel, J. Gene birth contributes to structural disorder encoded by overlapping genes. *Genetics* **2018**, *210*, 303–313. [CrossRef]
31. Pavesi, A.; Magiorkinis, G.; Karlin, D.G. Viral proteins originated de novo by overprinting can be identified by codon usage: Application to the "gene nursery" of Deltaretroviruses. *PLoS Comput. Biol.* **2013**, *9*, e10031632. [CrossRef] [PubMed]
32. Hidaka, M.; Inoue, J.; Yoshida, M.; Seiki, M. Post-transcriptional regulator (rex) of HTLV-1 initiates expression of viral structural proteins but suppresses expression of regulatory proteins. *EMBO J.* **1988**, *7*, 519–523. [CrossRef]
33. Iwamoto, T.; Mise, K.; Takeda, A.; Okinaka, Y.; Mori, K.I.; Arimoto, M.; Okuno, T.; Nakai, T.J. Characterization of Striped jack nervous necrosis virus subgenomic RNA3 and biological activities of its encoded protein B2. *J. Gen. Virol.* **2005**, *86*, 2807–2816. [CrossRef]
34. Sabath, N.; Wagner, A.; Karlin, D. Evolution of viral proteins originated de novo by overprinting. *Mol. Biol. Evol.* **2012**, *29*, 3767–3780. [CrossRef]
35. Pavesi, A.; Vianelli, A.; Chirico, N.; Bao, Y.; Blinkova, O.; Belshaw, R.; Firth, A.; Karlin, D. Overlapping genes and the proteins they encode differ significantly in their sequence composition from non-overlapping genes. *PLoS ONE* **2018**, *13*, e0202513. [CrossRef]
36. Ringnér, M. What is principal component analysis? *Nat. Biotechnol.* **2008**, *26*, 303–304. [CrossRef]
37. Sun, F.; Pan, W.; Gao, H.; Qi, X.; Qin, L.; Wang, Y.; Gao, Y.; Wang, X. Identification of the interaction and interaction domains of chicken anemia virus VP2 and VP3 proteins. *Virology* **2017**, *513*, 188–194. [CrossRef]
38. Mazur, I.; Anhlan, D.; Mitzner, D.; Wixler, L.; Schubert, U.; Ludwig, S. The proapoptotic influenza A virus protein PB1-F2 regulates viral polymerase activity by interaction with the PB1 protein. *Cell Microbiol.* **2008**, *10*, 1140–1152. [CrossRef]
39. Davy, C.; McIntosh, P.; Jackson, D.J.; Sorathia, R.; Miell, M.; Wang, Q.; Khan, J.; Soneji, Y.; Doorbar, J. A novel interaction between the human papillomavirus type 16 E2 and E1-E4 proteins leads to stabilization of E2. *Virology* **2009**, *394*, 266–275. [CrossRef]
40. Wieringa, R.; de Vries, A.A.; Rottier, P.J. Formation of disulfide-linked complexes between the three minor envelope glycoproteins (GP2b, GP3, and GP4) of equine arteritis virus. *J. Virol.* **2003**, *77*, 6216–6226. [CrossRef]
41. Kobayashi, T.; Watanabe, M.; Kamitani, W.; Tomonaga, K.; Ikuta, K. Translation initiation of a bicistronic mRNA of Borna disease virus: A 16-kDa phosphoprotein is initiated at an internal start codon. *Virology* **2000**, *277*, 296–305. [CrossRef]
42. Loughran, G.; Firth, A.E.; Atkins, J.F. Ribosomal frameshifting into an overlapping gene in the 2B-encoding region of the cardiovirus genome. *Proc. Natl. Acad. Sci. USA* **2011**, *108*, E1111-9. [CrossRef]
43. Ren, Q.; Wang, S.Q.; Firth, A.E.; Chan, M.M.Y.; Gouw, J.W.; Guarna, M.M.; Foster, L.J.; Atkins, J.F.; Jan, E. Alternative reading frame selection mediated by a tRNA-like domain of an internal ribosome entry site. *Proc. Natl. Acad. Sci. USA* **2012**, *109*, E630-9. [CrossRef] [PubMed]
44. Ding, S.W.; Anderson, B.J.; Haase, H.R.; Svmons, R.H. New overlapping gene encoded by the cucumber mosaic virus genome. *Virology* **1994**, *198*, 593–601. [CrossRef]
45. Olspert, A.; Carr, J.P.; Firth, A.E. Mutational analysis of the Potyviridae transcriptional slippage site utilized for expression of the P3N-PIPO and P1N-PISPO proteins. *Nucleic Acids Res.* **2016**, *44*, 7618–7629. [CrossRef] [PubMed]
46. Mizokami, M.; Orito, E.; Ohba, K.; Ikeo, K.; Lau, J.Y.; Gojobori, T. Constrained evolution with respect to gene overlap of hepatitis B virus. *J. Mol. Evol.* **1997**, *44*, S83–S90. [CrossRef]
47. Hughes, A.L.; Westover, K.; da Silva, J.; O'Connor, D.H.; Watkins, D.I. Simultaneously positive and purifying selection on overlapping reading frames of the tat and vpr genes of simian immunodeficiency virus. *J. Virol.* **2001**, *75*, 7966–7972. [CrossRef]
48. Allison, J.R.; Lechner, M.; Hoeppner, M.P.; Poole, A.M. Positive selection or free to vary? Assessing the functional significance of sequence change using molecular dynamics. *PLoS ONE* **2016**, *11*, e0147619. [CrossRef]
49. Scholthof, H.B. The Tombusvirus-encoded P19: From irrelevance to elegance. *Nat. Rev. Microbiol.* **2006**, *4*, 405–411. [CrossRef]
50. Scholthof, H.B.; Scholthof, K.B.; Kikkert, M.; Jackson, A.O. Tomato bushy stunt virus spread is regulated by two nested genes that function in cell-to-cell movement and host-dependent systemic invasion. *Virology* **1995**, *213*, 425–438. [CrossRef] [PubMed]
51. Pavesi, A. Asymmetric evolution in viral overlapping genes is a source of selective protein adaptation. *Virology* **2019**, *532*, 39–47. [CrossRef]
52. Fernandes, J.D.; Faust, T.B.; Strauli, N.B.; Smith, C.; Crosby, D.C.; Nakamura, R.L.; Hernandez, R.D.; Frankel, A.D. Functional segregation of overlapping genes. *Cell* **2016**, *167*, 1762–1773. [CrossRef]
53. Fujii, Y.; Kiyotani, K.; Yoshida, T.; Sakaguchi, T. Conserved and non-conserved regions in the Sendai virus genome: Evolution of a gene possessing overlapping reading frames. *Virus Genes* **2001**, *22*, 47–52. [CrossRef] [PubMed]
54. Guyader, S.; Ducray, D.G. Sequence analysis of Potato leafroll virus isolates reveals genetic stability, major evolutionary events and differential selection pressure between overlapping reading frame products. *J. Gen. Virol.* **2002**, *83*, 1799–1807. [CrossRef]

55. Stamenković, G.G.; Ćirković, V.S.; Šiljić, M.M.; Blagojević, J.V.; Knežević, A.M.; Joksić, I.D.; Stanojević, M.P. Substitution rate and natural selection in parvovirus B19. *Sci. Rep.* **2016**, *6*, 35759. [CrossRef] [PubMed]
56. Szklarczyk, R.; Heringa, J.; Pond, S.K.; Nekrutenko, A. Rapid asymmetric evolution of a dual-coding tumor suppressor INK4a/ARF locus contradicts its function. *Proc. Natl. Acad. Sci. USA* **2007**, *104*, 12807–12812. [CrossRef]
57. Danen-Van Oorschot, A.A.; Fischer, D.F.; Grimbergen, J.M.; Klein, B.; Zhuang, S.; Falkenburg, J.H.; Backendorf, C.; Quax, P.H.; Van der Eb, A.J.; Noteborn, M.H. Apoptin induces apoptosis in human transformed and malignant cells but not in normal cells. *Proc. Natl. Acad. Sci. USA* **1997**, *94*, 5843–5847. [CrossRef]
58. Malla, W.A.; Arora, R.; Khan, R.I.N.; Mahajan, S.; Tiwari, A.K. Apoptin as a tumor-specific therapeutic agent: Current perspective on mechanism of action and delivery systems. *Front. Cell Dev. Biol.* **2020**, *8*, 524. [CrossRef]
59. Szelechowski, M.; Bétourné, A.; Monnet, Y.; Ferré, C.A.; Thouard, A.; Foret, C.; Peyrin, J.M.; Hunot, S.; Gonzalez-Dunia, D. A viral peptide that targets mitochondria protects against neuronal degeneration in models of Parkinson's disease. *Nat. Commun.* **2014**, *5*, 5181. [CrossRef]
60. Ferré, C.A.; Davezac, N.; Thouard, A.; Peyrin, J.M.; Belenguer, P.; Miquel, M.C.; Gonzalez-Dunia, D.; Szelechowski, M. Manipulation of the N-terminal sequence of the Borna disease virus X protein improves its mitochondrial targeting and neuroprotective potential. *FASEB J.* **2016**, *30*, 1523–1533. [CrossRef] [PubMed]
61. Sorgeloos, F.; Jha, B.K.; Silverman, R.H.; Michiels, T. Evasion of antiviral innate immunity by Theiler's virus L* protein through direct inhibition of RNase L. *PLoS Pathog.* **2013**, *9*, e1003474. [CrossRef]
62. Park, S.B.; Seronello, S.; Mayer, W.; Ojcius, D.M. Hepatitis C virus frameshift/alternate reading frame protein suppresses interferon responses mediated by pattern recognition receptor retinoic-acid-inducible gene-I. *PLoS ONE* **2016**, *11*, e0158419. [CrossRef]
63. Jääskeläinen, K.M.; Kaukinen, P.; Minskaya, E.S.; Plyusnina, A.; Vapalahti, O.; Elliott, R.M.; Weber, F.; Vaheri, A.; Plyusnin, A. Tula and Puumala hantavirus NSs ORFs are functional and the products inhibit activation of the interferon-beta promoter. *J. Med. Virol.* **2007**, *79*, 1527–1536. [CrossRef]
64. Lauksund, S.; Greiner-Tollersrud, L.; Chang, C.J.; Robertsen, B. Infectious pancreatic necrosis virus proteins VP2, VP3, VP4 and VP5 antagonize IFNa1 promoter activation while VP1 induces IFNa1. *Virus Res.* **2015**, *196*, 113–121. [CrossRef]
65. Wensman, J.J.; Munir, M.; Thaduri, S.; Hörnaeus, K.; Rizwan, M.; Blomström, A.L.; Briese, T.; Lipkin, W.I.; Berg, M. The X proteins of bornaviruses interfere with type I interferon signaling. *J. Gen. Virol.* **2013**, *94*, 263–269. [CrossRef]
66. García-Rosado, E.; Markussen, T.; Kileng, O.; Baekkevold, E.S.; Robertsen, B.; Mjaaland, S.; Rimstad, E. Molecular and functional characterization of two infectious salmon anaemia virus (ISAV) proteins with type I interferon antagonizing activity. *Virus Res.* **2008**, *133*, 228–238. [CrossRef] [PubMed]
67. Silhavy, D.; Molnár, A.; Lucigli, A.; Szittya, G.; Hornyik, C.; Tavazza, M.; Burgyán, J. A viral protein suppresses RNA silencing and binds silencing-generated, 21- to 25-nucleotide double-stranded RNAs. *EMBO J.* **2002**, *21*, 3070–3080. [CrossRef]
68. Chen, J.; Li, W.X.; Xie, D.; Peng, J.R.; Ding, S.W. Viral virulence protein suppresses RNA silencing-mediated defense but upregulates the role of microrna in host gene expression. *Plant. Cell* **2004**, *16*, 1302–1313. [CrossRef] [PubMed]
69. Chellappan, P.; Vanitharani, R.; Fauquet, C.M. MicroRNA-binding viral protein interferes with Arabidopsis development. *Proc. Natl. Acad. Sci. USA* **2005**, *102*, 10381–10386. [CrossRef]
70. McFadden, N.; Bailey, D.; Carrara, G.; Benson, A.; Chaudhry, Y.; Shortland, A.; Heeney, J.; Yarovinsky, F.; Simmonds, P.; Macdonald, A.; et al. Norovirus regulation of the innate immune response and apoptosis occurs via the product of the alternative open reading frame 4. *PLoS Pathog.* **2011**, *7*, e1002413. [CrossRef]
71. Varga, Z.T.; Ramos, I.; Hai, R.; Schmolke, M.; García-Sastre, A.; Fernandez-Sesma, A.; Palese, P. The influenza virus protein PB1-F2 inhibits the induction of type I interferon at the level of the MAVS adaptor protein. *PLoS Pathog.* **2011**, *7*, e1002067. [CrossRef] [PubMed]
72. Chen, W.; Calvo, P.A.; Malide, D.; Gibbs, J.; Schubert, U.; Bacik, I.; Basta, S.; O'Neill, R.; Schickli, J.; Palese, P.; et al. A novel influenza A virus mitochondrial protein that induces cell death. *Nat. Med.* **2001**, *7*, 1306–1312. [CrossRef]
73. Noteborn, M.H.; Todd, D.; Verschueren, C.A.; de Gauw, H.W.; Curran, W.L.; Veldkamp, S.; Douglas, A.J.; McNulty, M.S.; van der EB, A.J.; Koch, G. A single chicken anemia virus protein induces apoptosis. *J. Virol.* **1994**, *68*, 346–351. [CrossRef]
74. Khaperskyy, D.A.; Schmaling, S.; Larkins-Ford, J.; McCormick, C.; Gaglia, M.M. Selective degradation of host RNA polymerase II transcripts by influenza A virus PA-X host shutoff protein. *PLoS Pathog.* **2016**, *12*, e1005427. [CrossRef]
75. Trifonov, E.N. Searching for Codes in the Sequences. In *Biomolecular Data. A Resource in Transition*; Oxford University Press: Oxford, UK, 1989; p. 199.
76. Smith, T.F. Semantic and Syntactic Patterns in the Genetic Language. In *Biomolecular Data. A Resource in Transition*; Oxford University Press: Oxford, UK, 1989; p. 211.
77. Shannon, C.E.; Weaver, W. *The Mathematical Theory of Communication*; University of Illinois Press: Urbana, IL, USA, 1949.
78. Granero-Porati, M.I.; Porati, A.; Zani, L. Informational parameters of an exact DNA base sequence. *J. Theor. Biol.* **1980**, *86*, 401–403. [CrossRef]
79. Smith, T.F.; Waterman, M.S. Overlapping genes and information theory. *J. Theor. Biol.* **1981**, *91*, 379–380. [CrossRef]
80. Luo, L.F.; Tsai, L.; Zhou, Y.M. Informational parameters of nucleic acid and molecular evolution. *J. Theor. Biol.* **1988**, *130*, 351–361. [CrossRef]

81. Tan, D.Y.; Bejo, M.H.; Aini, I.; Omar, A.R.; Goh, Y.M. Base usage and dinucleotide frequency of infectious bursal disease virus. *Virus Genes* **2004**, *28*, 41–53. [CrossRef]
82. Hughes, A.L.; Hughes, M.A.K. Patterns of nucleotide difference in overlapping and non-overlapping reading frames of papillomavirus genomes. *Virus Res.* **2005**, *113*, 81–88. [CrossRef] [PubMed]
83. Pavesi, A. New insights into the evolutionary features of viral overlapping genes by discriminant analysis. *Virology* **2020**, *546*, 51–66. [CrossRef] [PubMed]
84. Brister, J.R.; Ako-Adjei, D.; Bao, Y.; Blinkova, O. NCBI viral genomes resource. *Nucleic Acids Res.* **2015**, *43*, D571–D577. [CrossRef] [PubMed]
85. Fisher, R.A. The use of multiple measurements in taxonomic problems. *Ann. Eugen.* **1936**, *8*, 376–386. [CrossRef]
86. Lachenbruch, P.A.; Goldstein, M. Discriminant analysis. *Biometrics* **1979**, *35*, 69–85. [CrossRef]
87. Brereton, R.G.; Lloyd, G.R. Partial least squares discriminant analysis: Taking the magic away. *Chemometrics* **2014**, *28*, 213–225. [CrossRef]
88. Lee, L.C.; Liong, C.Y.; Jemain, A.A. Partial least squares-discriminant analysis (PLS-DA) for classification of high-dimensional (HD) data: A review of contemporary practice strategy and knowledge gaps. *Analyst* **2014**, *143*, 3526–3539. [CrossRef]
89. Schlub, T.E.; Holmes, E.C. Properties and abundance of overlapping genes in viruses. *Virus Evol.* **2020**, *6*, veaa009. [CrossRef]
90. Lamb, R.A.; Orvath, C.M. Diversity of coding strategies in influenza viruses. *Trends Genet.* **1991**, *7*, 261–266. [CrossRef]
91. Krakauer, D.C. Stability and evolution of overlapping genes. *Evolution* **2000**, *54*, 731–739. [CrossRef]
92. Peleg, O.; Kirzhner, V.; Trifonov, E.; Bolshoy, A. Overlapping messages and survivability. *J. Mol. Evol.* **2004**, *59*, 520–527. [CrossRef]
93. Chirico, N.; Vianelli, A.; Belshaw, R. Why genes overlap in viruses. *Proc. Biol. Sci.* **2010**, *277*, 3809–3817. [CrossRef] [PubMed]
94. Belshaw, R.; Pybus, O.G.; Rambaut, A. The evolution of genome compression and genomic novelty in RNA viruses. *Genome Res.* **2007**, *17*, 1496–1504. [CrossRef]
95. Holmes, E.C. *The Evolution and Emergence of RNA Viruses*; Oxford University Press: Oxford, UK, 2009.
96. Brandes, N.; Linial, M. Gene overlapping and size constraints in the viral world. *Biol. Direct.* **2016**, *11*, 26. [CrossRef]
97. Botstein, D. A theory of modular evolution for bacteriophages. *Ann. N.Y. Acad. Sci.* **1980**, *354*, 484–491. [CrossRef]
98. Gibbs, A. Molecular evolution of viruses; 'trees', 'clocks' and 'modules'. *J. Cell Sci.* **1987**, *1987* (Suppl. 7), 319–337. [CrossRef]
99. Lucchini, S.; Desiere, F.; Brüssow, H. Comparative genomics of Streptococcus thermophilus phage species supports a modular evolution theory. *J. Virol.* **1999**, *73*, 8647–8656. [CrossRef] [PubMed]
100. Pavesi, A. Different patterns of codon usage in the overlapping polymerase and surface genes of hepatitis B virus suggest a de novo origin by modular evolution. *J. Gen. Virol.* **2015**, *96*, 3577–3586. [CrossRef] [PubMed]
101. Lauber, C.; Seitz, S.; Mattei, S.; Suh, A.; Beck, J.; Herstein, J.; Börold, J.; Salzburger, W.; Kaderali, L.; Briggs, J.A.G.; et al. Deciphering the origin and evolution of hepatitis B viruses by means of a family of non-enveloped fish viruses. *Cell Host Microbe* **2017**, *22*, 387–399. [CrossRef]
102. Zaaijer, H.L.; van Hemert, F.J.; Koppelman, M.H.; Lukashov, V.V. Independent evolution of overlapping polymerase and surface protein genes of hepatitis B virus. *J. Gen. Virol.* **2007**, *88*, 2137–2143. [CrossRef] [PubMed]
103. Zhang, D.; Chen, J.; Deng, L.; Mao, Q.; Zheng, J.; Wu, J.; Zeng, C.; Li, Y. Evolutionary selection associated with the multi-function of overlapping genes in the hepatitis B virus. *Infect. Genet. Evol.* **2010**, *10*, 84–88. [CrossRef]
104. Campo, D.S.; Dimitrova, Z.; Lara, J.; Purdy, M.; Thai, H.; Ramachandran, S.; Ganova-Raeva, L.; Zhai, X.; Forbi, J.C.; Teo, C.G.; et al. Coordinate evolution of the hepatitis B virus polymerase. *Silico Biol.* **2011**, *11*, 175–182. [CrossRef]
105. Torres, C.; Fernández, M.D.; Flichman, D.M.; Campos, R.H.; Mbayed, V.A. Influence of overlapping genes on the evolution of human hepatitis B virus. *Virology* **2013**, *441*, 40–48. [CrossRef] [PubMed]
106. Moskovitz, D.N.; Osiowy, C.; Giles, E.; Tomlinson, G.; Heathcote, E.J. Response to long-term lamivudine treatment (up to 5 years) in patients with severe chronic hepatitis B, role of genotype and drug resistance. *J. Viral. Hepat.* **2005**, *12*, 398–404. [CrossRef]
107. Cooreman, M.P.; Leroux-Roels, G.; Paulij, W.P. Vaccine- and hepatitis B immune globulin-induced escape mutations of hepatitis B virus surface antigen. *J. Biomed. Sci.* **2001**, *8*, 237–247. [CrossRef]
108. Torresi, J. The virological and clinical significance of mutations in the overlapping envelope and polymerase genes of hepatitis B virus. *J. Clin. Virol.* **2002**, *25*, 97–106. [CrossRef]
109. Suh, A.; Weber, C.C.; Kehlmaier, C.; Braun, E.L.; Green, R.E.; Fritz, U.; Ray, D.A.; Ellegren, H. Early mesozoic coexistence of amniotes and hepadnaviridae. *PLoS Genet.* **2014**, *10*, e1004559. [CrossRef]
110. Gojobori, T.; Li, W.H.; Graur, D. Patterns of nucleotide substitution in pseudogenes and functional genes. *J. Mol. Evol.* **1982**, *18*, 360–369. [CrossRef]
111. Nei, M.; Gojobori, T. Simple methods for estimating the numbers of synonymous and nonsynonymous nucleotide substitutions. *Mol. Biol. Evol.* **1986**, *3*, 418–426. [CrossRef] [PubMed]
112. Hein, J.; Støvlbæk, J. A maximum-likelihood approach to analyzing non-overlapping and overlapping reading frames. *J. Mol. Evol.* **1995**, *40*, 181–189. [CrossRef] [PubMed]
113. de Groot, S.; Mailund, T.; Lunter, G.; Hein, J. Investigating selection on viruses: A statistical alignment approach. *BMC Bioinform.* **2008**, *9*, 304. [CrossRef] [PubMed]
114. McCauley, S.; de Groot, S.; Mailund, T.; Hein, J. Annotation of selection strengths in viral genomes. *Bioinformatics* **2007**, *23*, 2978–2986. [CrossRef] [PubMed]

115. Pedersen, A.M.; Jensen, J.L. A dependent-rates model and an MCMC-based methodology for the maximum-likelihood analysis of sequences with overlapping reading frames. *Mol. Biol. Evol.* **2001**, *18*, 763–776. [CrossRef] [PubMed]
116. Sabath, N.; Landan, G.; Graur, D. A method for the simultaneous estimation of selection intensities in overlapping genes. *PLoS ONE* **2008**, *3*, e3996. [CrossRef] [PubMed]
117. Obenauer, J.C.; Denson, J.; Mehta, P.K.; Su, X.; Mukatira, S.; Finkelstein, D.B.; Xu, X.; Wang, J.; Ma, J.; Fan, Y.; et al. Large-scale sequence analysis of avian influenza isolates. *Science* **2006**, *311*, 1576–1580. [CrossRef]
118. Pavesi, A. Pattern of nucleotide substitutions in the overlapping nonstructural genes of influenza A virus and implication for the genetic diversity of the H5N1 subtype. *Gene* **2007**, *402*, 28–34. [CrossRef]
119. Wei, X.; Zhang, J. A simple method for estimating the strength of natural selection on overlapping genes. *Genome Biol. Evol.* **2014**, *7*, 381–390. [CrossRef]
120. Firth, A.E.; Brown, C.M. Detecting overlapping coding sequences with pairwise alignments. *Bioinformatics* **2005**, *21*, 282–292. [CrossRef] [PubMed]
121. Firth, A.E.; Brown, C.M. Detecting overlapping coding sequences in virus genomes. *BMC Bioinform.* **2006**, *7*, 75. [CrossRef] [PubMed]
122. Firth, A.E. Mapping overlapping functional elements embedded within the protein-coding regions of RNA viruses. *Nucleic Acids Res.* **2014**, *42*, 12425–12439. [CrossRef] [PubMed]
123. Sealfon, R.S.; Lin, M.F.; Jungreis, I.; Wolf, M.Y.; Kellis, M.; Sabeti, P.C. FRESCo: Finding regions of excess synonymous constraint in diverse viruses. *Genome Biol.* **2015**, *16*, 38. [CrossRef] [PubMed]
124. Nelson, C.W.; Ardern, Z.; Wei, X. OLGenie: Estimating Natural Selection to predict functional overlapping genes. *Mol. Biol. Evol.* **2020**, *37*, 2440–2449. [CrossRef]
125. Schlub, T.E.; Buchmann, J.P.; Holmes, E.C. A simple method to detect candidate overlapping genes using single genome sequences. *Mol. Biol. Evol.* **2018**, *35*, 2572–2581. [CrossRef] [PubMed]
126. Zhou, P.; Yang, X.L.; Wang, X.G.; Hu, B.; Zhang, L.; Zhang, W.; Si, H.R.; Zhu, Y.; Li, B.; Huang, C.L.; et al. A pneumonia outbreak associated with a new coronavirus of probable bat origin. *Nature* **2020**, *579*, 265–269. [CrossRef]
127. Michel, C.J.; Mayer, C.; Poch, O.; Thompson, J.D. Characterization of accessory genes in coronavirus genomes. *Virology J.* **2020**, *17*, 131. [CrossRef]
128. Delaye, L.; Deluna, A.; Lazcano, A.; Becerra, A. The origin of a novel gene through overprinting in Escherichia coli. *BMC Evol. Biol.* **2008**, *8*, 31. [CrossRef] [PubMed]
129. Fellner, L.; Simon, S.; Scherling, C.; Witting, M.; Schober, S.; Polte, C.; Schmitt-Kopplin, P.; Keim, D.A.; Scherer, S.; Neuhaus, K. Evidence for the recent origin of a bacterial protein-coding, overlapping orphan gene by evolutionary overprinting. *BMC Evol. Biol.* **2015**, *15*, 283. [CrossRef]
130. Chung, W.Y.; Wadhawan, S.; Szklarczyk, R.; Pond, S.K.; Nekrutenko, A. A first look at ARFome: Dual-coding genes in mammalian genomes. *PLoS Comput. Biol.* **2007**, *3*, e91. [CrossRef]
131. Ribrioux, S.; Brungger, A.; Baumgarten, B.; Seuwen, K.; John, M.R. Bioinformatics prediction of overlapping frameshifted translation products in mammalian transcripts. *BMC Genom.* **2008**, *9*, 122. [CrossRef] [PubMed]
132. Michel, A.M.; Choudhury, K.R.; Firth, A.E.; Ingolia, N.T.; Atkins, J.F.; Baranov, P.V. Observation of dually decoded regions of the human genome using ribosome profiling data. *Genome Res.* **2012**, *22*, 2219–2229. [CrossRef]
133. Vanderperre, B.; Lucier, J.F.; Roucou, X. HAltORF: A database of predicted out-of-frame alternative open reading frames in human. *Database* **2012**, *2012*, bas025. [CrossRef] [PubMed]
134. Bergeron, D.; Lapointe, C.; Bissonnette, C.; Tremblay, G.; Motard, J.; Roucou, X. An out-of-frame overlapping reading frame in the ataxin-1 coding sequence encodes a novel ataxin-1 interacting protein. *J. Biol. Chem.* **2013**, *288*, 21824–21835. [CrossRef]
135. Vanderperre, B.; Lucier, J.F.; Bissonnette, C.; Motard, J.; Tremblay, G.; Vanderperre, S.; Wisztorski, M.; Salzet, M.; Boisvert, F.M.; Roucou, X. Direct detection of alternative open reading frames translation products in human significantly expands the proteome. *PLoS ONE* **2013**, *8*, e70698. [CrossRef] [PubMed]
136. Mouilleron, H.; Delcourt, V.; Roucou, X. Death of a dogma: Eukaryotic mRNAs can code for more than one protein. *Nucleic Acids Res.* **2016**, *44*, 14–23. [CrossRef]
137. Brunet, M.A.; Levesque, S.A.; Hunting, D.J.; Cohen, A.A. Recognition of the polycistronic nature of human genes is critical to understanding the genotype-phenotype relationship. *Genome Res.* **2018**, *28*, 609–624. [CrossRef] [PubMed]
138. Wang, R.F.; Parkhurst, M.R.; Kawakami, Y.; Robbins, P.F.; Rosenberg, S.A. Utilization of an alternative open reading frame of a normal gene in generating a novel human cancer antigen. *J. Exp. Med.* **1996**, *183*, 1131–1140. [CrossRef] [PubMed]
139. Wang, R.F.; Johnson, S.L.; Zeng, G.; Topalian, S.L.; Schwartzentruber, D.J.; Rosenberg, S.A. A breast and melanoma-shared tumor antigen: T cell responses to antigenic peptides translated from different open reading frames. *J. Immunol.* **1998**, *161*, 3598–3606.
140. Rosenberg, S.A.; Tong-On, P.; Li, Y.; Riley, J.P.; El-Gamil, L.; Parkhurst, M.R.; Robbins, P.F. Identification of BING-4 cancer antigen from an alternative open reading frame of a gene in the extended MHC class II region using lymphocytes from a patient with a durable complete regression following immunotherapy. *J. Immunol.* **2002**, *168*, 2402–2407. [CrossRef] [PubMed]
141. Mandic, M.; Almunia, C.; Vicel, S.; Gillet, D.; Janjic, B.; Coval, K.; Maillere, B.; Kirkwood, J.M.; Zarour, H.M. The alternative open reading frame of LAGE-1 gives rise to multiple promiscuous HLA-DR-restricted epitopes recognized by T-helper 1-type tumor-reactive CD4+ T cells. *Cancer Res.* **2003**, *63*, 6506–6515.

142. Slager, E.H.; Borghi, M.; van der Minne, C.E.; Aarnoudse, C.A.; Havenga, M.J.E.; Schrier, P.I.; Osanto, S.; Griffioen, M. CD4+ Th2 cell recognition of HLA-DR-restricted epitopes derived from CAMEL: A tumor antigen translated in an alternative open reading frame. *J. Immunol.* **2003**, *170*, 1490–1497. [CrossRef]
143. Smith, C.C.; Selitsky, S.R.; Chai, S.; Armistead, P.M.; Vincent, B.G.; Serody, J.S. Alternative tumour-specific antigens. *Nat. Rev. Canc.* **2019**, *8*, 465–478. [CrossRef]
144. Moles, R.; Sarkis, S.; Galli, V.; Omsland, M.; Purcell, D.F.J.; Yurick, D.; Khoury, G.; Pise-Masison, C.A.; Franchini, G. p30 protein: A critical regulator of HTLV-1 viral latency and host immunity. *Retrovirology* **2019**, *16*, 42. [CrossRef] [PubMed]
145. Gaudray, G.; Gachon, F.; Basbous, J.; Biard-Piechaczyk, M.; Devaux, C.; Mesnard, J.M. The complementary strand of the human T-cell leukemia virus type 1 RNA genome encodes a bZIP transcription factor that down-regulates viral transcription. *J. Virol.* **2002**, *76*, 12813–12822. [CrossRef]
146. Baratella, M.; Forlani, G.; Accolla, R.S. HTLV-1 HBZ viral protein: A key player in HTLV-1 mediated diseases. *Front. Microbiol.* **2017**, *8*, 2615. [CrossRef] [PubMed]
147. Nahon, J.L. Birth of 'human-specific' genes during primate evolution. *Genetica* **2003**, *118*, 193–208. [CrossRef]
148. Cassan, E.; Arigon-Chifolleau, A.M.; Mesnard, J.M.; Gross, A.; Gascuel, O. Concomitant emergence of the antisense protein gene of HIV-1 and of the pandemic. *Proc. Natl. Acad. Sci. USA* **2016**, *113*, 11537–11542. [CrossRef] [PubMed]
149. Affram, Y.; Zapata, J.C.; Gholizadeh, Z.; Tolbert, W.D.; Zhou, W.; Iglesias-Ussel, M.D.; Pazgier, M.; Ray, K.; Latinovic, O.S.; Romerio, F. The HIV-1 antisense protein ASP is a transmembrane protein of the cell surface and an integral protein of the viral envelope. *J. Virol.* **2019**, *93*, e00574-19. [CrossRef]
150. Oliver, J.L.; Marin, A.A. A relationship between GC content and coding-sequence length. *J. Mol. Evol.* **1996**, *43*, 216–223. [CrossRef]
151. Smith, T.F.; Waterman, M.S. Protein constraints induced by multiframe encoding. *Math. Biosci.* **1980**, *49*, 17–26. [CrossRef]
152. Pavesi, A. Detection of signature sequences in overlapping genes and prediction of a novel overlapping gene in hepatitis G virus. *J. Mol. Evol.* **2000**, *50*, 284–295. [CrossRef]
153. Lèbre, S.; Gascuel, O. The combinatorics of overlapping genes. *J. Theor. Biol.* **2017**, *415*, 90–101. [CrossRef]
154. Simon-Loriere, E.; Holmes, E.C.; Pagán, I. The effect of gene overlapping on the rate of RNA virus evolution. *Mol. Biol. Evol.* **2013**, *30*, 1916–1928. [CrossRef] [PubMed]
155. Teppa, E.; Zea, D.J.; Oteri, F.; Carbone, A. COVTree: Coevolution in Overlapped sequences by Tree analysis server. *Nucleic Acids Res.* **2020**, *48*, W558–W565. [CrossRef]
156. Firth, A.E.; Brierley, I. Non-canonical translation in RNA viruses. *J. Gen. Virol.* **2012**, *93*, 1385–1409. [CrossRef] [PubMed]
157. Dinman, J.D. Translational recoding signals: Expanding the synthetic biology toolbox. *J. Biol. Chem.* **2019**, *294*, 7537–7545. [CrossRef]
158. Meier, C.; Aricescu, A.R.; Assenberg, R.; Aplin, R.T.; Gilbert, R.J.; Grimes, J.M.; Stuart, D.I. The crystal structure of ORF-9b, a lipid binding protein from the SARS coronavirus. *Structure* **2006**, *14*, 1157–1165. [CrossRef] [PubMed]
159. Baulcombe, D.C.; Molnar, A. Crystal structure of p19—A universal suppressor of RNA silencing. *Trends Biochem. Sci.* **2004**, *29*, 279–281. [CrossRef] [PubMed]
160. Lingel, A.; Simon, B.; Izaurralde, E.; Sattler, M. The structure of the flock house virus B2 protein, a viral suppressor of RNA interference, shows a novel mode of double-stranded RNA recognition. *EMBO Rep.* **2005**, *6*, 1149–1155. [CrossRef] [PubMed]

genes

MDPI

Article

New Data on Comparative Cytogenetics of the Mouse-Like Hamsters (*Calomyscus* Thomas, 1905) from Iran and Turkmenistan

Svetlana A. Romanenko [1,*], Vladimir G. Malikov [2], Ahmad Mahmoudi [3], Feodor N. Golenishchev [2], Natalya A. Lemskaya [1], Jorge C. Pereira [4,5], Vladimir A. Trifonov [1,6], Natalia A. Serdyukova [1], Malcolm A. Ferguson-Smith [5], Mansour Aliabadian [7] and Alexander S. Graphodatsky [1]

1 Institute of molecular and Cellular Biology (IMCB), Siberian Branch of Russian Academy of Sciences (SB RAS), 630090 Novosibirsk, Russia; lemnat@mcb.nsc.ru (N.A.L.); vlad@mcb.nsc.ru (V.A.T.); serd@mcb.nsc.ru (N.A.S.); graf@mcb.nsc.ru (A.S.G.)
2 Zoological Institute (ZIN), Russian Academy of Sciences (RAS), 199034 Saint-Petersburg, Russia; malikovzin@mail.ru (V.G.M.); f_gol@mail.ru (F.N.G.)
3 Department of Biology, Faculty of Science, Urmia University, Urmia 5756151818, Iran; ahmad.mahmoodi1985@gmail.com
4 Animal and Veterinary Research Centre (CECAV), University of Trás-os-Montes and Alto Douro (UTAD), 5000-801 Vila Real, Portugal; jorgecpereira599@gmail.com
5 Cambridge Resource Centre for Comparative Genomics, Department of Veterinary medicine, University of Cambridge, Cambridge CB3 OES, UK; maf12@cam.ac.uk
6 Department of Natural Science, Novosibirsk State University, 630090 Novosibirsk, Russia
7 Department of Biology, Faculty of Sciences, Ferdowsi University of mashhad, mashhad 9177948974, Iran; aliabadi@um.ac.ir
* Correspondence: rosa@mcb.nsc.ru; Tel.: +7-383-363-90-63

check for updates

Citation: Romanenko, S.A.; malikov, V.G.; mahmoudi, A.; Golenishchev, F.N.; Lemskaya, N.A.; Pereira, J.C.; Trifonov, V.A.; Serdyukova, N.A.; Ferguson-Smith, m.A.; Aliabadian, m.; et al. New Data on Comparative Cytogenetics of the Mouse-Like Hamsters (*Calomyscus* Thomas, 1905) from Iran and Turkmenistan. *Genes* **2021**, *12*, 964. https://doi.org/10.3390/genes12070964

Academic Editors: Luigi Viggiano and Renè massimiliano Marsano

Received: 28 April 2021
Accepted: 23 June 2021
Published: 24 June 2021

Abstract: The taxonomy of the genus *Calomyscus* remains controversial. According to the latest systematics the genus includes eight species with great karyotypic variation. Here, we studied karyotypes of 14 *Calomyscus* individuals from different regions of Iran and Turkmenistan using a new set of chromosome painting probes from a *Calomyscus* sp. male (2n = 46, XY; Shahr-e-Kord-Soreshjan-Cheshme maiak Province). We showed the retention of large syntenic blocks in karyotypes of individuals with identical chromosome numbers. The only rearrangement (fusion 2/21) differentiated *Calomyscus elburzensis, Calomyscus mystax mystax*, and *Calomyscus* sp. from Isfahan Province with 2n = 44 from karyotypes of *C. bailwardi, Calomyscus* sp. from Shahr-e-Kord, Chahar mahal and Bakhtiari-Aloni, and Khuzestan-Izeh Provinces with 2n = 46. The individuals from Shahdad tunnel, Kerman Province with 2n = 51–52 demonstrated non-centric fissions of chromosomes 4, 5, and 6 of the 46-chromosomal form with the formation of separate small acrocentrics. A heteromorphic pair of chromosomes in a specimen with 2n = 51 resulted from a fusion of two autosomes. C-banding and chromomycin A3-DAPI staining after G-banding showed extensive heterochromatin variation between individuals.

Keywords: banding; chromosome painting; fluorescent in situ hybridization; karyotype; molecular cytogenetics; painting probes; type locality

1. Introduction

The mouse-like hamsters (genus *Calomyscus* Thomas, 1905) are distributed mosaically and strictly associated with the rocky habitats of South-Western Syria, Iran, Afghanistan, Western Pakistan, Azerbaijan (Nakhichevan), and Turkmenistan. Initially, the genus was included in the subfamily Cricetinae of the family Cricetidae [1–5], but later, according to the results of morphological [6] and comparative molecular-genetic analyses [7], the genus was considered as a taxon of a monotypic family, Calomyscidae Vorontsov and Potapova, 1979, characterized by brush-tailed mouse-like appearance [8].

The taxonomic structure of the genus *Calomyscus* is still highly questionable. The type species of the genus, *C. bailwardi*, Thomas, 1905, was described from the vicinity of mala-Imir (modern Izeh), Khuzestan Province, the South-West of Iran. Then, Thomas [9] discerned *C. hotsoni* Thomas, 1920 and *C. baluchi* Thomas, 1920 from Pakistan based on mild morphometric and external differences. Later, *C. mystax* Kaschkarov, 1925 from Bolshoy Balkhan mountain, Turkmenistan was described [10]. Ognev and Heptner considered *C. mystax* to be synonymous with *C. hotsoni* [11]. According to their view, the genus comprises only two species: *C. bailwardi* (including two subspecies—*C. bailwardi bailwardi* and *Calomyscus bailwardi baluchi*) and *C. hotsoni* [11]. In addition, the authors did not exclude that the differences between those forms could be less than interspecific. Argyropulo considered those specific names to be synonymous with *C. bailwardi* [12]. Later, *Calomyscus* was discovered in the South-East of Transcaucasia (Nakhichevan) and considered *C. bailwardi* [13]. In the same year, *C. elburzensis* Goodwin, 1939 was described from the Central Elburz (Iran). This author did not exclude *C. elburzensis* and/or *C. mystax* from being a subspecies of *C. hotsoni* [14]. Then, for a long time the genus *Calomyscus* was considered to be monotypic [15–19], including all its forms as subspecies. In that period, *C. bailwardi mustersi* Ellerman, 1948 (from the vicinity of Kabul, Afghanistan) and *C. grandis* Schlitter et Setzer, 1977 (from the vicinity of Fasham, the southern slopes of Elburz) were described [20]. *Calomyscus tsolovi* Peshev, 1991 from the southwest of Syria [21,22] was the latest species of the genus to be described, distinguished on the basis of only external characters. It was considered that *C. hotsoni* is a subspecies of *C. bailwardi*, which included all the forms from Iran [22].

The first data on the cytogenetics of the mouse-like hamsters were obtained by matthey, who described the karyotype (diploid number 2n = 32) of a specimen from the vicinity of Julfa (Nakhichevan, Azerbajan) [23]. Then, the form with a diploid number of 2n = 30 was found in the southwest of Turkmenistan. On the basis of these data the authors described the Transcaucasian form as a new species *C. urartensis* Vorontzov et Kartavtseva, 1979 and restored the specific state of *C. mystax* which, according to them, included all the forms from Turkmenistan together with *C. elburzensis* [24]. In the same publication the hypothetical composition of the genus was given: (1) *C. urartensis*—Azerbaijan: Nakhichevan (the type locality) and the adjacent territory of Iran; (2) *C. mystax*—Turkmenistan: Bolshoy Balkhan mountain (the type locality) and mali Balkhan, the Turkmenian Kopet-Dag; Iran: mountains of the North Khorasan and the eastern part of Elburz up to the vicinity of Tehran; (3) *C. bailwardi*—Iran: Khuzestan, vicinity of Izeh (the type locality), Zagros mountains in the Fars Province and the Zard-Kuh mountains in Isfaghan Province; (4) *C. hotsoni*—Pakistan: The Central makran Chain, the region of Panjgur (the type locality); (5) *C. baluchi*, including two subspecies: (a) *C. baluchi baluchi*—Pakistan, vicinity of Kelat (the type locality); (b) *C. baluchi mustersi*—Afghanistan, vicinity of Kabul (the type locality).

Later, extensive karyotype variations were shown for the mouse-like hamsters from different regions [25]. The variation concerned not only diploid chromosome numbers, but the fundamental number of autosomal arms (FNa) and the amount and distribution of heterochromatin. As a result of comparative karyology and experimental hybridization of different chromosomal forms from the former USSR, *C. urartensis* and *C. mystax* were recognized as different species. The nominative subspecies *C. mystax mystax* (diploid chromosome number 2n = 44, FNa = 46) was established and two new forms from Turkmenistan were described—*C. m. zykovi* meyer et malikov, 2000 (diploid number 2n = 30), which was for the first time karyotyped by Vorontzov et al. [24], and *C. firiuzaensis* meyer et malikov, 2000 (diploid number 2n = 44, FNa = 58), which was defined by Graphodatsky et al. [25] as karyotype 3. The "2n = 44" karyotypes of *C. m. mystax* and *C. firiuzaensis* differed from each other in the patterns of C- and G-banding. The interspecific hybrid F1 males were sterile, while the hybrids of both sexes between *C. m. mystax* and *C. m. zykovi* were fertile [26]. Nevertheless, only eight species of the genus—*C. bailwardi, C. hotsoni, C. baluchi, C. mystax, C. elburzensis, C. grandis, C. urartensis,* and *C. tsolovi*—were included in the checklist [8].

The first multivariate craniometric analysis of *Calomyscus* demonstrated distinct clusters [27,28] that did not always correspond to the karyotypic forms. For instance, *C. m. mystax* and *C. m. zykovi*, in spite of their karyotype differentiation, were lumped in the same craniometric sub-cluster. A large form from the vicinity of Tehran (presumably *C. grandis*), having the same karyotype as *C. m. mystax* [25] occurred to be apart from all the forms analyzed (unfortunately, previous karyotype comparisons were made using conventional banding techniques only). However, the craniometrics [29] and phylogenetic trees based on *CytB* [30] of the forms were in concordance. At the same time, the karyotyped *C. firiuzaensis* occurred in the same craniometric sub-cluster as the non-karyotyped *C. elburzensis* from the type locality and populations of northeastern Iran (Khorossan Province) and northwestern Afghanistan (Gerat Province), which strongly suggests them to be conspecific. However, recent molecular evidence distinguishes seven species (*C. bailwardi, C. hotsoni, C. baluchi, C. mystax, C. elburzensis, C. grandis,* and *C. urartensis*) and three additional lineages from the Zagros mountains [31].

Here, we present a detailed karyotype description of 14 specimens of *Calomyscus* spp. from five different localities in Iran, including the type localities of *C. bailwardi* and *C. elburzensis*, and *C. m. mystax* from Turkmenistan.

2. Materials and methods

2.1. Compliance with Ethical Standards

All applicable international, national, and/or institutional guidelines for the care and use of animals were followed. All experiments were approved by the Ethical committee at the ZIN RAS, Russia (permission No. 2-7/02-04-2021 of 2 April 2021) and the Ethics Committee on Animal and Human Research at the IMCB SB RAS, Russia (protocol No. 01/19 of 21 January 2019).

2.2. Species Sampled

The live individuals captured in Iran in 1998, 2014, and 2016 field seasons became the ancestors of two laboratory colonies—in the Department of Theriology of the Zoological Institute, St. Petersburg, and in moscow Zoo. Here, we analyzed three individuals from the laboratory colony of moscow Zoo (CSP1m, CSP2m, CSP3m) and two captive-born specimens form the Zoological Institute's colony (CMYS2m, CELB3m) (Table 1). Nine specimens examined were collected from free-living populations at five localities in Iran (Table 1 and Figure 1). It must be emphasized that in fact, we karyotyped natural individuals of all forms described so far, and the founder animals from the laboratory colonies were also analyzed karyologically. We consistently demonstrated stability of the karyotypes within the colonies, thus they represent a natural complexity.

Table 1. Origins of the mouse-like hamsters examined in this report.

Calomyscus Species, Diploid Chromosome Number (2n)	Geographic Position in Figure 1	Sex	Abbreviation	Locality	Coordinates/Comments
C. bailwardi (2n = 46)	1	♀	CBAI1f	Izeh, Khuzestan Province, Iran *	N 31°49′20.8″ E 49°50′09.6″, elevation: 1063 m
C. elburzensis (2n = 44)	2	♀	CELB1f	Korkhout mt., North Khorasan Province, Bojnourd, Iran *	N 37°26′35.2″ E 56°31′30.0″, elevation: 1903 m
C. elburzensis (2n = 44)	2	♂	CELB3m	Korkhout mt.,North Khorasan Province, Bojnourd, Iran *	N 37°26′35.2″ E 56°31′30.0″, elevation: 1903 m (laboratory colony of ZIN RAN)
C. mystax mystax (2n = 44)	3	♂	CMYS2m	maly Balkan Gershi (Arlandag) mt., Bolshoi Balkhan, Turkmenistan *	N 39°40′20.28″ E 54°32′35.61″, elevation: 1870 m (laboratory colony of ZIN RAN)
Calomyscus sp. 1 (2n = 52)	4	♂	CSP1m **	Shahdad tunnel, Kerman Province, Iran	N 30°10′36.41″ E 57°24′44.24″, elevation: 2664 m (laboratory colony of moscow Zoo)
Calomyscus sp. 1 (2n = 52)	4	♂	CSP2m **	Shahdad tunnel, Kerman Province, Iran	N 30°10′36.41″ E 57°24′44.24″, elevation: 2664 m (laboratory colony of moscow Zoo)
Calomyscus sp. 1 (2n = 51)	4	♂	CSP3m **	Shahdad tunnel, Kerman Province, Iran	N 30°10′36.41″ E 57°24′44.24″, elevation: 2664 m (laboratory colony of moscow Zoo)
Calomyscus sp. 2 (2n = 46)	5	♂	CSP17m	Cheshme maiak, Shahr-e-Kord—Soreshjan, Chahar mahal and Bakhtiaria Province, Iran	N 32°18′37.9″ E 50°37′34.3″, elevation 2161 m
Calomyscus sp. 2 (2n = 46)	5	♀	CSP18f	Aloni, Chahar mahal and Bakhtiaria Province, Iran	N 31°33′21.3″ E 51°11′25.3″, elevation 1833 m
Calomyscus sp. 2 (2n = 46)	5	♂	CSP20m	Cheshme maiak, Shahr-e-Kord—Soreshjan, Chahar mahal and Bakhtiaria Province, Iran	N 32°18′37.9″ E 50°37′34.3″, elevation 2161 m
C. bailwardi (2n = 46)	5	♂	CSP21m	Izeh, Khozestan Province, Iran	N 31°49′20.8″ E 49°50′09.6″, elevation: 1063 m
Calomyscus sp. 2 (2n = 46)	5	♀	CSP22f	Cheshme maiak, Shahr-e-Kord—Soreshjan, Bakhtiaria Province, Iran	N 32°18′37.9″ E 50°37′34.3″, elevation 2161 m
Calomyscus sp. 2 (2n = 46)	5	♀	CSP23f	Aloni, Chahar mahal and Bakhtiaria Province, Iran	N 31°33′21.3″ E 51°11′25.3″, elevation 1833 m
Calomyscus sp. 3 (2n = 44)	6	♂	CSP26m	S-E ridge Zard-Kuh mts., Isfahan Province, Iran	N 32°27′4.93″ E 50°2′28.36″, elevation 2885 m

* type locality; ** CSP1m, CSP2m, and CSP3m belong to the same form of unclear taxonomic rank.

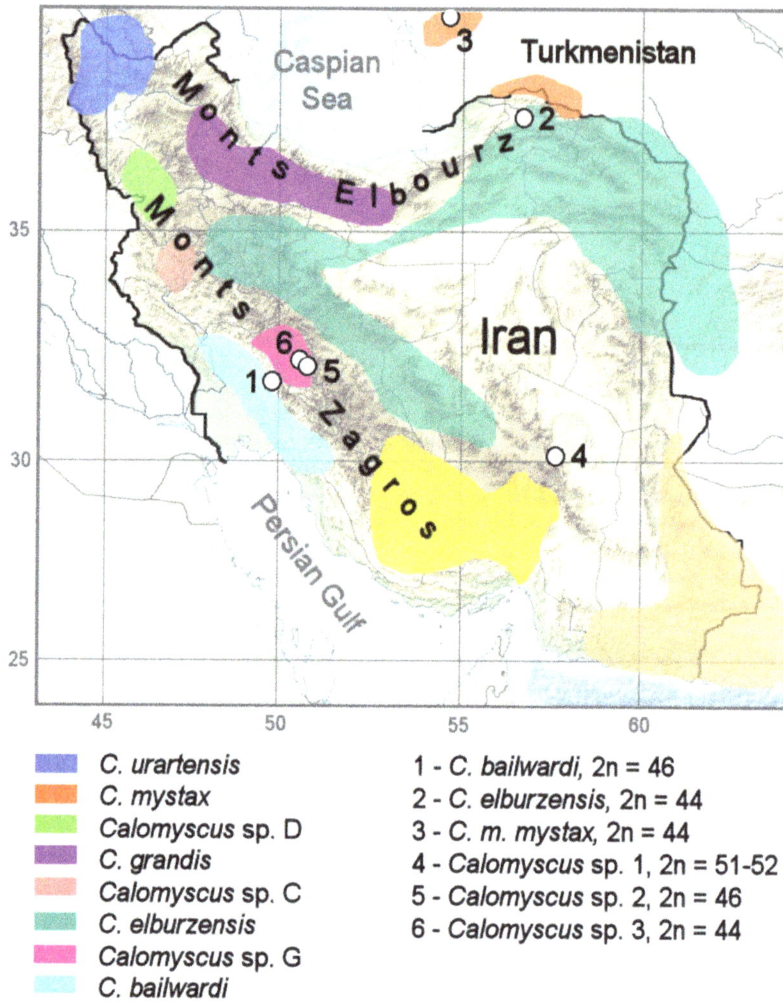

Figure 1. Geographic position of the localities studied: 1—Izeh, Khuzestan Province, Iran, 2—Korkhout mt., North Khorasan Province, Bojnourd, Iran, 3—Maly Balkan Gershi (Arlandag) mt., Bolshoi Balkhan, Turkmenistan, 4—Shahdad tunnel, Kerman Province, Iran, 5—Aloni and Cheshme maiak, Shahr-e-Kord—Soreshjan, Chahar mahal and Bakhtiaria Province, Iran, 6—southeastern ridge Zard-Kuh mts., Isfahan Province, Iran. For detailed localities description see Table 1. *Calomyscus* distributional ranges in Iran and Turkmenistan are given in accordance with the mitochondrial DNA analyses [31].

2.3. Chromosome Preparation and Chromosome Staining

Cell lines and chromosome suspensions were obtained in the Laboratory of Animal Cytogenetics, IMCB SB RAS, Russia. The fibroblast cell lines were derived from lung, breastbone, and tail biopsies using enzymatic treatment (with trypsin, collagenase, and hyaluronidase) of tissues as described previously [32,33]. All the cell lines were stored in the IMCB SB RAS cell bank ("The general collection of cell cultures", No. 0310-2016-0002).

Metaphase chromosome spreads were prepared as described previously [34]. We used colcemid to arrest the cell cycle in metaphase and ethidium bromide to directly inhibits chromosome condensation. C-banding using barium hydroxide octahydrate was made as described earlier [35]. G-banding was performed on chromosomes of all species prior to FISH by the standard trypsin/Giemsa procedure [36]. Chromomycin A3-DAPI-after G-banding (CDAG) was carried out following a previously published technique [37]. In short, after G-banding slides were heat denatured in the presence of formamide with consecutive fluorochrome staining.

2.4. Fluorescent in Situ Hybridization (FISH)

Calomyscus Cot-6 DNA extraction and FISH with the Cot DNA was performed following previously published protocols [38]. The set of chromosome-specific *Calomyscus* sp. (CSP17m) male probes was generated in the Cambridge Resource Centre for Comparative Genomics by DOP-PCR amplification of flow sorted chromosomes. Probes were labeled with biotin and digoxigenin by DOP-PCR amplification as described previously [39–42] (Figure 2).

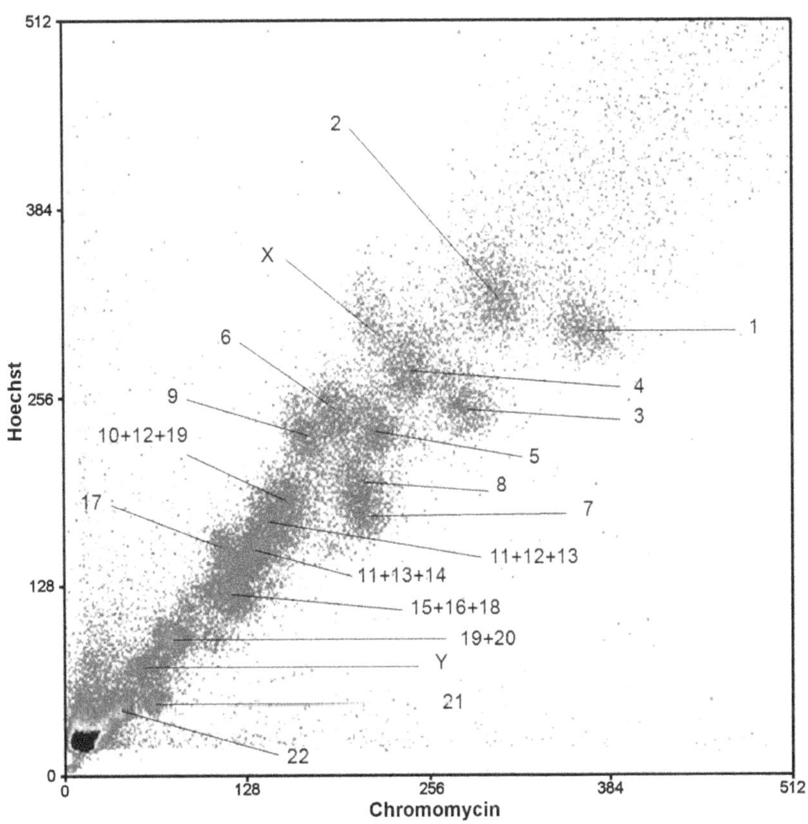

Figure 2. Bivariate flow karyotype of *Calomyscus* sp. (CSP17m) cell line with chromosomal assignments.

2.5. Microdissection, Probe Amplification, and Labeling

Glass needle-based microdissection of small autosomes of *Calomyscus* sp. 3m (CSP3m) was performed on G-banded chromosomes as described [43]. One copy of each chromosome was collected. Chromosomal DNA was amplified and labeled using WGA kits (Sigma-Aldrich, Saint Louis, mO, USA).

2.6. Image Acquisition and Processing

Images were captured using the VideoTest-FISH software (VideoTesT) with a CCD camera (JenOptic) mounted on an Olympus BX53 microscope. Hybridization signals were assigned to specific chromosome regions identified by means of G-banding patterns photographed by the CCD camera. All images were processed in Corel Paint Shop Photo Pro X3 (Corel Corporation).

3. Results

3.1. Karyotype Descriptions

The *C. bailwardi* (CBAI1f) and *Calomyscus* sp. 21m (CSP21m) females from Khuzestan-Izeh, the specimens CSP17m, CSP20m, and CSP22f from Shahr-e-Kord, and the females *Calomyscus* sp. (CSP18f) and CSP23f from Chahar mahal and Bakhtiari-Aloni have 2n = 46 (Figure 3). C-banding detected small heterochromatic blocks in pericentromeric regions of all chromosomes and a small interstitial heterochromatic block in the distal region of acrocentric X chromosomes (Figure 3). The block on X chromosome is AT-rich (see CDAG description below). A heteromorphic pair of chromosomes 4 composed of an acrocentric and a submetacentric was detected in karyotype of CBAI1f, CSP18f, CSP20m, CSP21m, and CSP22f, so FNa = 45. The individuals CSP17m and CSP23f have FNa = 44. The Y chromosome of CSP17m (metacentric) has a heterochromatic short arm (p-arm).

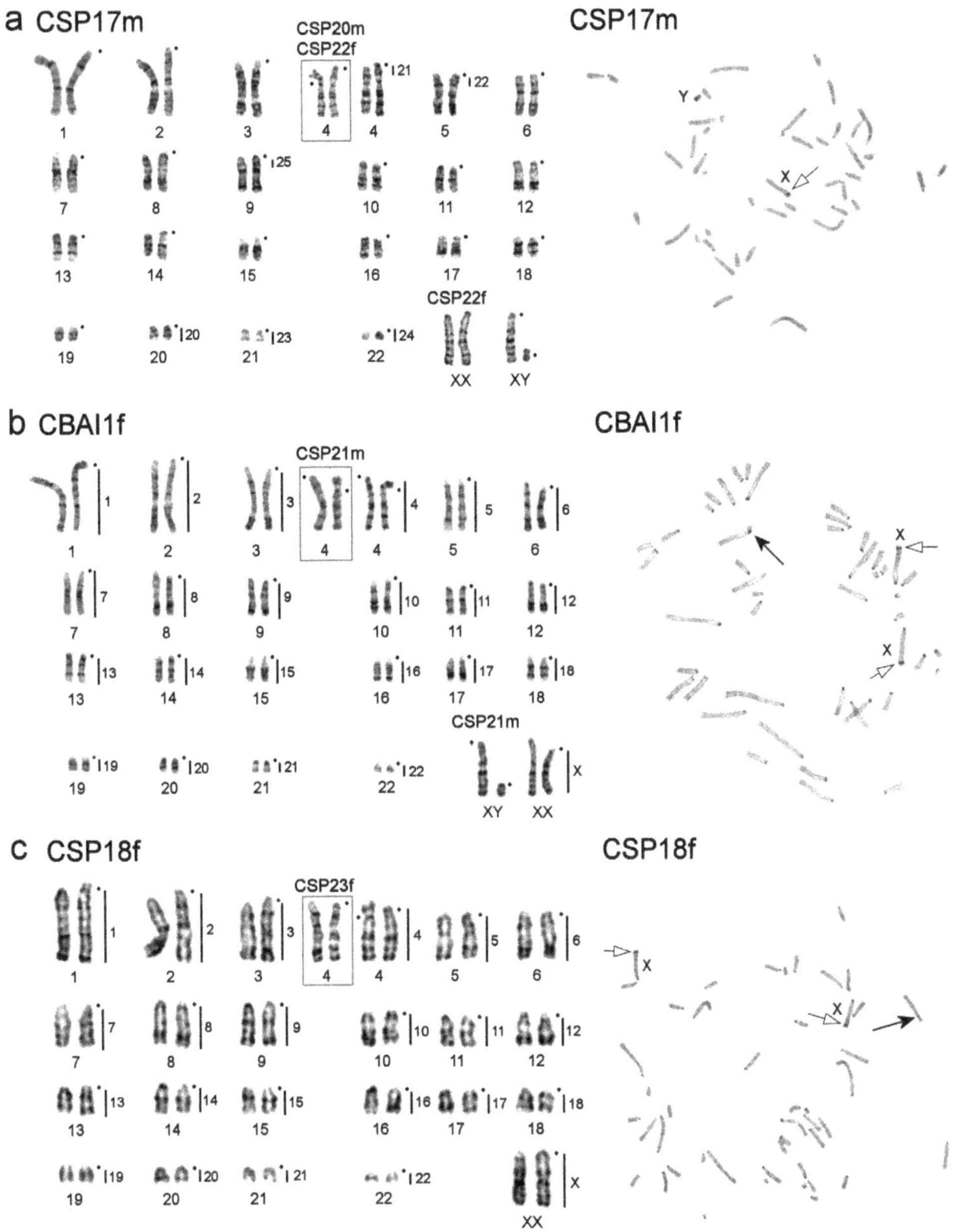

Figure 3. G-banded karyotypes (on the left) and C-banded metaphases (on the right) of the studied specimens with 2n = 46 and FNa = 44–45: (**a**) *Calomyscus* sp. 17m, *Calomyscus* sp. 20m (CSP20m), and *Calomyscus* sp. 22f (CSP22f), vertical black lines and figures on the right of G-banded chromosomes show localization of CSP3m probes; (**b**) *C. bailwardi* (the karyotype was presented previously in [44]) and *Calomyscus* sp. 21m (CSP21m); (**c**) *Calomyscus* sp. 18f and *Calomyscus* sp. 23f (CSP23f). In part b and c vertical black lines and figures on the right of G-banded chromosomes show localization of CSP17m probes. Black dots mark the position of centromeres. In C-banding photos the black arrows mark the submetacentric homolog of chromosome 4 and sex chromosomes (X and Y) are shown. The white arrows with black stroke mark the interstitial heterochromatic block in the distal region of X chromosomes.

The karyotype of the male *Calomyscus* sp. 2m (CSP2m) was described and published previously [45]. It had 2n = 52 and FNa = 56. A similar karyotype structure is revealed for CSP1m with 2n = 52, FNa = 56 (Figure 4). The only difference between these two karyotypes is the heteromorphism of chromosome 1 due to heterochromatin variation. *Calomyscus* sp. 3m (CSP3m) has 2n = 51, FNa = 57, and carries a heteromorphic pair of chromosomes 1 consisting of a large submetacentric chromosome and two acrocentrics homologous to its short (p-) and long (q-) arms, respectively (Figure 4). C-banding reveals a large block of heterochromatin in the pericentromeric part of the biggest submetacentric chromosome (Figure 4).

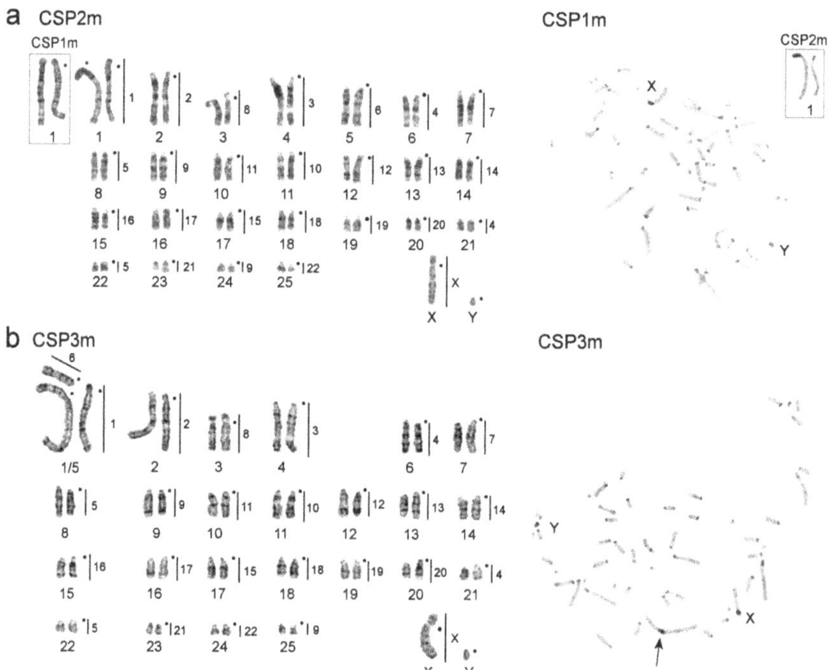

Figure 4. G-banded karyotypes (on the left) and C-banded metaphases (on the right) of the studied specimens with 2n = 51–52 and FNa = 56–57: (**a**) *Calomyscus* sp. 1m (CSP1m) and *Calomyscus* sp. 2m (CSP2m, corrected karyotype from Romanenko et al. [45], published as "*Calomyscus* sp. 1" in [46]) with 2n = 52, FNa = 5 6; (**b**) *Calomyscus* sp. 3m (CSP3m) with 2n = 51, FNa = 57, vertical black lines and figures on the right of G-banded chromosomes show localization of *Calomyscus* sp. 17m (CSP17m) probes. Black dots mark the position of centromeres. In C-banding photos the black arrow indicates a pericentromeric heterochromatin segment in the largest chromosome and sex chromosomes (X and Y) are shown.

The karyotypes of *C. elburzensis* individuals (CELB1f and CELB3m), *C. mystax mystax*, and *Calomyscus* sp. 26m (CSP26m) have 2n = 44 (Figure 5). As CELB1f carries a heteromorphic chromosome 11 (Figure 5a), the number of autosome arms differ in the two individuals examined: FNa = 61 for CELB1f and FNa = 62 for CELB3m. C-banding reveals additional heterochromatic arms in autosomes 1–9, X chromosomes, and the submetacentric homolog of chromosome 11 of *C. elburzensis* (Figure 5). G- and C-banded karyotypes of CELB3m were presented previously [47]. The *C. m. mystax* karyotype shows acrocentric chromosomes carrying heterochromatic blocks in pericentromeric regions (Figure 5b). The karyotype of CSP26m has FNa = 67. Heterochromatic blocks were revealed in pericentromeric regions of all chromosomes forming additional p-arms in pairs 1–13 (Figure 5c).

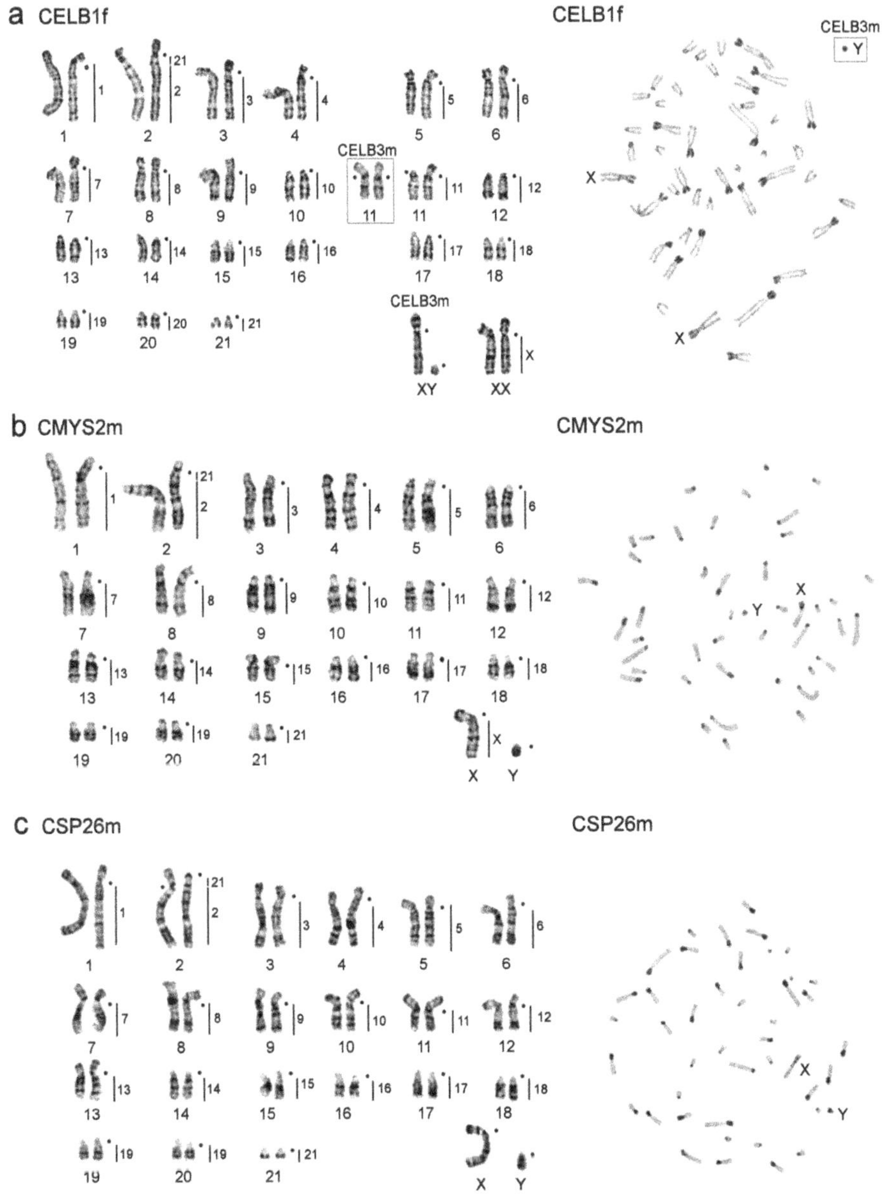

Figure 5. G-banded karyotypes (on the left) and C-banded metaphases (on the right) of the studied specimens with 2n = 44: (**a**) *C. elburzensis* 1f (CELB1f) and *C. elburzensis* 3m (CELB3m); (**b**) *C. m. mystax* 2m (CMYS2m, the karyotype was presented previously in [48]); (**c**) *Calomyscus* sp. 26m (CSP26m), vertical black lines and figures on the right of G-banded chromosomes show localization of *Calomyscus* sp. 17m probes. Black dots mark the position of centromeres. In C-banding photos sex chromosomes (X and Y) are shown.

3.2. *The Results of Chromomycin A3-DAPI-after G-banding (CDAG)*

Pericentromeric regions of CSP1m, CSP3m, and CBAI1f are AT-rich (Figure 6a–c). There is no clear AT-rich pericentromeric heterochromatin in CSP17m and CSP18f

(Figure 6d,e). The interstitial heterochromatic blocks on the X chromosomes of CSP17m, CSP18f, and CBAI1f are AT-rich (Figure 6c–e). Although the p-arms of autosomes are uniformly stained with chromomycin A3 in CSP26m and both CELB individuals, p-arms of their X chromosomes show alternation of AT- and GC-rich heterochromatin (Figure 6f–i); pericentromeric regions in these individuals are AT-rich. The pericentromeric regions of CMYS2m chromosomes are both GC- and AT-enriched with higher GC-content (Figure 6g). An AT-rich area is observed in the distal region of the p-arms of the X chromosomes. Weak staining with chromomycin A3 is characteristic of pericentromeric, interstitial, and distal regions of some chromosomes in all individuals. The Y chromosomes of all individuals are predominantly AT-rich.

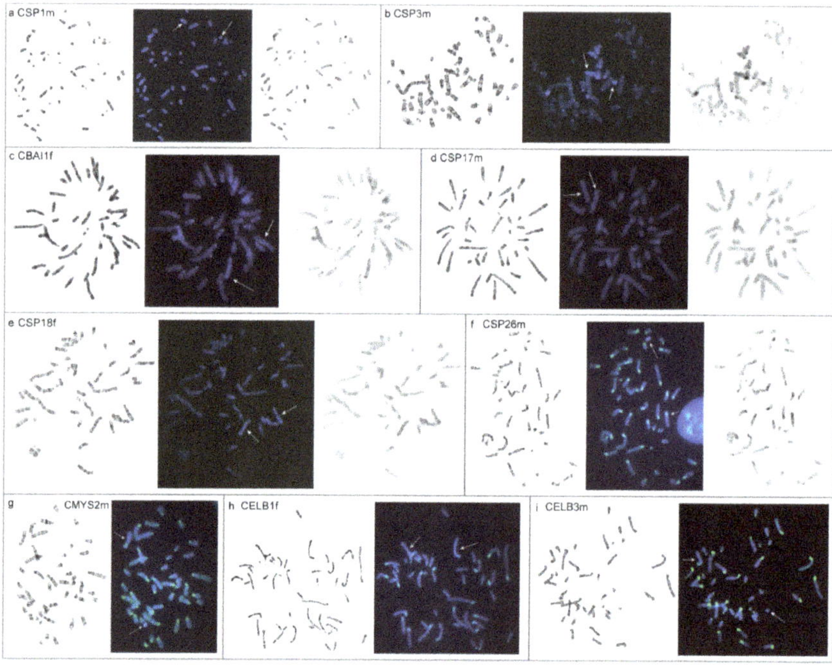

Figure 6. Results of CDAG: (**a**) *Calomyscus* sp. 1m (CSP1m); (**b**) *Calomyscus* sp. 3m (CSP3m); (**c**) *C. bailwardi* 1f (CBAI1f); (**d**) *Calomyscus* sp. 17m (CSP17m); (**e**) *Calomyscus* sp. 18f (CSP18f); (**f**) *Calomyscus* sp. 26m (CSP26m); (**g**) *C. m. mystax* 2m (CMYS2m); (**h**) *C. elburzensis* 1f (CELB1f); (**i**) *C. elburzensis* 3m (CELB3m). From left to right: G-banding, CDAG, inverted DAPI-banding (for some individuals). Blocks of AT-rich heterochromatin (blue) and blocks of GC-rich heterochromatin (green). Sex chromosomes marked by arrows.

3.3. *The Flow Karyotype of CSP17m*

The male CSP17m has a diploid number of 46 (Figure 3a). The chromosome complement of CSP17m is resolved into 21 separate peaks by flow cytometry (Figure 2). The chromosomal content of each peak was identified by hybridizing the probes derived from each peak onto G-banded chromosomes of the source specimen (Figure 7). Fifteen single chromosome-specific painting probes were obtained (1–9, 16, 17, 21, 22, X, and Y). Two probes each paint two pairs of autosomes (17 + 18, 19 + 20) and four peaks contain a mixture of three autosomes (10 + 12 + 19, 11 + 12 + 13, 11 + 13 + 14, and 15 + 16 + 18). We need to stress that the standard karyotype is not arranged in order of chromosome size as revealed by the flow karyotype. The presence of chromosomes 11, 12, and 13 in two distinct peaks can be explained by slight heteromorphism of the homologs detected by cytogenetic analysis of G-banded chromosomes. In the case of chromosome 19 the possible description is more complicated. A strong signal on chromosome 19 is revealed when we localized the painting probe (without Cot-6) derived from the peak containing chromosomes 10 + 12 (Figure 7e). The intensity of signals is much weaker when we made FISH with Cot-6 DNA, which indicates that this is due to cross hybridization of the repetitive elements of heterochromatin. FISH with the probe containing chromosomes 19 + 20 produced slight background signals on chromosomes 10 and 12. We suggest that microsatellite repeats could give such a signal distribution.

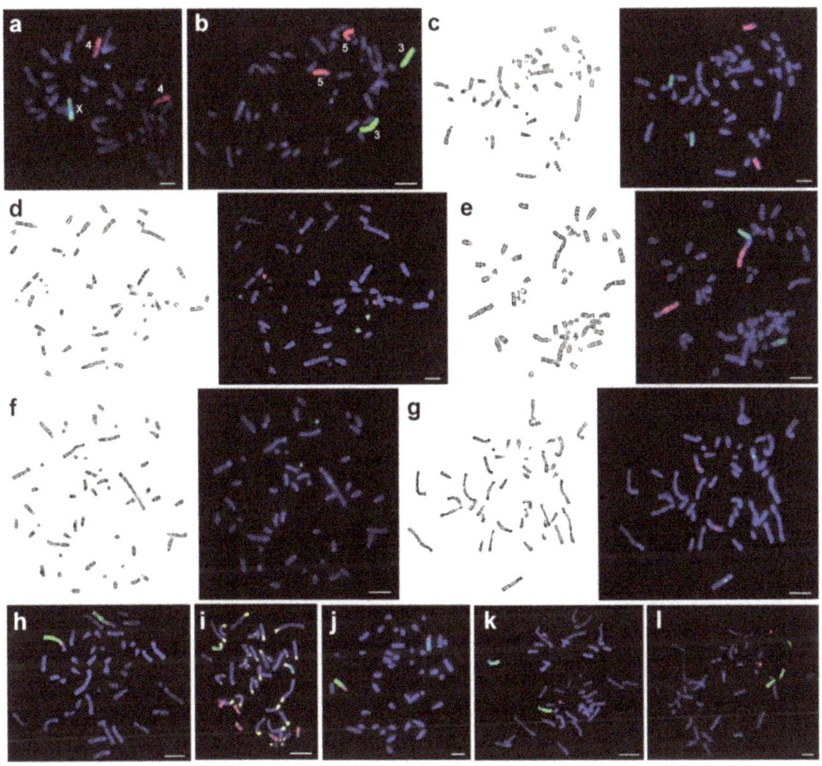

Figure 7. Examples of fluorescent in situ hybridization of chromosome-specific probes onto chromosomes of different *Calomyscus* specimens: (**a**) chromosomes X (green) and 4 (red) of CSP17m onto CSP17m; (**b**) chromosomes 3 (green) and 5 (red) of CSP17m onto CSP17m; (**c**) chromosomes 6 (green) and 4 (red) of CSP17m onto CSP1m; (**d**) microdissection-derived probes C7 (red) and C8 (green) onto CSP1m; (**e**) chromosomes 1 (red) and 6 (green) of CSP17m onto CSP3m; (**f**) microdissection-derived probes C5 (green) and C6 (red) onto CSP3m; (**g**) microdissection-derived probes C7 (red) and C8 (green) onto CELB3m; (**h**) chromosomes 2 (green) and 21 (red) onto CELB1f; (**i**) chromosomes 7 (green) and 11 + 13 + 14 (red) onto CELB1f; (**j**) chromosomes 1 (green) and 21 (red) onto CMYS2m; (**k**) chromosomes 7 (green) and 19 + 20 (red) onto CBAI1f; (**l**) chromosomes 10 + 12 + 19 (green) and 21 (red) onto CSP18f. CBAI—*C. bailwardi*, CELB—*C. elburzensis*, CMYS—*C. m. mystax*, CSP—*Calomyscus* sp. Scale bar is 10 μm.

Microdissection-derived probes of small autosomes 20–25 of CSP3m were made (Figure 4b). The chromosomal content of each probe was identified by hybridizing these probes onto G-banded chromosomes of the CSP17m source specimen (Figure 7). The localization of these probes on CSP17m chromosomes identified the corresponding regions (Figures 3a and 4b).

According to the results of the localization of painting-probes, chromosome painting using CSP17m paints show that hamster karyotypes with the same diploid chromosome numbers do not differ from each other. Regarding the 46-chromosome forms (CBAI1f, CSP17m, CSP18f, CSP20m, CSP21m, CSP22f, and CSP23f), individuals with 2n = 44 (CELB1f, CELB3m, CMYS2m, and CSP26m) showed fusion of 2/21, and individuals with 2n = 51–52 (CSP1m, CSP2m, CSP3m) demonstrated fissions of chromosomes 4, 5, and 6 with the formation of separate small acrocentrics. The heteromorphic pair of chromosomes in CSP3m was formed by the fusion of autosomes 1 and 6 of CSP17m.

4. Discussion

The mouse-like hamsters of the genus *Calomyscus* are widely spread but the recognition of species is complicated due to the morphological similarity of individuals, the possible intraspecific chromosomal variation and the occurrence of natural hybrids [25]. Therefore, the taxonomic analysis by morphometric structure and molecular cytogenetics of individuals from the localities of the named forms is particularly important.

The specimens, presumably belonging to "true" *C. bailwardi,* were initially recorded only from the Zagros mountains in west Iran [8]. At the same time, a high chromosomal variability of *Calomyscus* samples from different parts of the Zagros mountains (with diploid numbers varying from 37 to 52) [25,49] and molecular data [31], suggest that Zagros is colonized by more than one species of the genus. musser and Carleton [8] stressed that Graphodatsky et al. [25] had karyotyped individuals from the *C. bailwardi* type locality (it seems that the authors meant karyotype 7 with 2n = 50 from the region of Persepolis). moreover, the karyotype of another sample with unknown origin identified as *C. bailwardi* (2n = 44) was published previously [50].

Two specimens, *C. bailwardi,* CBAI1f, and CSP21m both from Izeh (type locality), have cytogenetically completely identical autosomal sets and X chromosomes (Figure 3b). The similarity of karyotypes and the geographical origin allow us, with a certain degree of caution, to classify CSP21m as *C. bailwardi.* The karyotype of both individuals differ from the karyotype 7 published by Graphodatsky et al. [25] by a lower diploid chromosome number (46 and 50, respectively) and chromosome morphology. The karyotypes also differ from the *C. bailwardi* karyotype presented by Radjabli et al. [50] not only by the diploid chromosome number and chromosome morphology, but also by the amount and distribution of heterochromatin. It is possible that the differences between the *C. bailwardi* karyotypes described here and presented by Radjabli et al. [50] can be explained by incorrect species identification in the previously published case. However, the high variability of chromosomal characteristics may indicate the need for additional karyological and molecular studies to determine the systematic status and composition of *C. bailwardi.*

The karyotypes of individuals with 2n = 46 from different habitats (CBAI1f and CSP21m from Khuzestan-Izeh; CSP17m, CSP20m, and CSP22f from Shahr-e-Kord; CSP18 and CSP23f from Chahar mahal and Bakhtiari-Aloni) did not actually differ from each other (Figure 3). The only exception is that the specimens from Khuzestan-Izeh displayed clearer pericentromeric blocks of heterochromatin. Interestingly, the only analyzed individual from closely located Isfahan, CSP26m, has a different karyotype, characterized by a smaller diploid number and the presence of pronounced heterochromatic p-arms (Figure 5c).

Three individuals studied here from Shahdad tunnel, Kerman Province, have similar karyotypes with 2n = 51–52 (Figure 4). The slight differences are caused by variations in heterochromatin. The karyotype structure and heterochromatic composition of these individuals is similar to karyotype 6, described earlier by Graphodatsky et al. [25] in a female specimen from Kerman Province, Iran. All that can be assumed is the presence of larger blocks of heterochromatin in the pericentromeric regions of all chromosomes in the individual described earlier by Graphodatsky et al. [25].

The karyotypes of both *C. elburzensis* (CELB1f was captured in their type locality in Iran) studied here have 2n = 44 (Figure 5a). The karyotypes show good correspondence with the one published by Radjabli et al. [50] and with the karyotype 3 by Graphodatsky et al. [25], both by C- and G-banding. Nevertheless, in the article by Graphodatsky et al. [25], chromosomes 9 and 11 pairs are designated acrocentric, consistent with FNa = 58. Although, according to their C-banding figure, it can be seen that the pairing of chromosome 9 is submetacentric, which makes FNa = 60. In our case, we find FNa = 61 for CELB1f and FNa = 62 for CELB3m due to heteromorphism of chromosome 11. The obvious disagreement in chromosome morphology between C- and G-banded karyotypes (e.g., pair 7 is acrocentric in G-banded and submetacentric in C-banded karyotypes) can be explained by possible intraspecific variability in heterochromatin amount and distribution [50]. Considering the similarity of *C. elburzensis* type karyotype to karyotype 3, *C. firiuzaensis* should be a junior synonym of *C. elburzensis* taxonomically. This conclusion is fully consistent with the results of multivariate craniometric analysis [27,28]. In addition, despite the different level of resolution, it seems that the karyotypes of CELB1f and CELB3m are similar to the karyotype of *C. elburzensis* with FNa = 62 described by Shahabi et al. [49], despite their different collection sites.

Previously it was proposed, based on cytogenetic analysis with different banding techniques, that centric and tandem fusions and heterochromatin variations play a major role in the karyotype evolution of *Calomyscus* [25]. The use of chromosome specific painting probes allows us to support the hypothesis of high rates of chromosomal transformations by translocation and variation of heterochromatin. Karyotypes of *C. elburzensis*, *C. m. mystax*, and CSP26m differ from other karyotypes studied here by the only tandem fusion 2/21 (Figure 5). This fact is especially interesting because the respective individuals were caught in places distant from one another (Figure 1). moreover, the karyotypes characterized by a large amount of heterochromatin have additional chromosome arms. The description of *C. m. mystax* karyotype indicates a correspondence to the karyotype of this species described earlier by Graphodatsky et al. [25].

All specimens studied here have stable karyotypes. Despite this, we reveal heteromorphic chromosome pairs in karyotypes of CSP2m (pair 1), CBAI1f, CSP18f, CSP20m, CSP21m, CSP22f (pair 4) and CELB1f (pair 11) (Figures 3 and 5a). However, heteromorphism caused by different chromosomal rearrangements is quite often observed, especially among relatively young and fast-evolving species [51,52]. Heteromorphism in pairs of chromosomes 1 in CSP1m and 11 in CELB1f is caused by an additional heterochromatic arm on one of the homologs. The identification of the type of chromosomal rearrangement in pair 4 of CBAI1f and CSP18f is complicated. The different morphology of homologs could be the result of either pericentromeric inversion or centromere shift and additional investigations are needed. moreover, the presence of some chromosomes in different peaks (e.g., 11, 12, and 13) of the flow sorted karyotype probably indicates heterochromatin variation in homologs or tiny chromosomal rearrangements (Figure 2).

5. Conclusions

Karyotype analysis plays an important role in the identification of morphologically similar species. Some chromosomal characteristics suggest the high plasticity of the *Calomyscus* genome and/or currently ongoing speciation process. However, neither similarity nor differentiation between the forms of *Calomyscus* in karyotype or molecular genetic markers can serve as an unambiguous indicator of their species rank. The results of morphometric and molecular genetic analysis of the diversity of *Calomyscus* [28,30,53–55], in combination with the previously published data, show that the genus demonstrates a lack of correlation between its karyotypic and morphometric structure and the reproductive incompatibility of the various forms. more integrative studies are required for an improved understanding of speciation in this genus, among which the data on reproductive isolation between genetically marked forms will be the most instructive.

Author Contributions: Conceptualization, A.S.G.; methodology, S.A.R., N.A.L., J.C.P., V.A.T., and N.A.S.; validation, S.A.R., V.G.M., and F.N.G.; formal analysis, S.A.R.; investigation, S.A.R., V.G.M., N.A.L., and F.N.G.; resources, V.G.M., A.M., m.A., and F.N.G.; data curation, S.A.R.; writing—original draft preparation, S.A.R.; writing—review and editing, V.G.M., F.N.G., N.A.L., V.A.T., and M.A.F.-S.;

visualization, S.A.R.; supervision, A.S.G.; project administration, A.S.G. and F.N.G.; funding acquisition, A.S.G. and F.N.G. All authors have read and agreed to the published version of the manuscript.

Funding: This research was funded by the Russian Science Foundation (RSF), grant number No. 19-14-00034 (A.S.G.). Significant funding was also provided by the RFBR grants № 19-04-00557a and Federal State Program ZIN RAS № AAAA-A19-119032590102-7 (F.N.G.) for field sampling.

Institutional Review Board Statement: The study was conducted according to the guidelines of the Declaration of Helsinki, and approved by the Ethical committee of the Zoological institute, Russian Academy of Sciences (ZIN RAS), Russia (permission №2-7/02-04-2021 from 2 April 2021) and the Ethics Committee on Animal and Human Research of the Institute of molecular and Cellular Biology, Siberian Branch of the Russian Academy of Sciences (IMCB, SB RAS), Russia (protocol №01/19 from 21 January 2019).

Acknowledgments: The authors gratefully acknowledge the resources provided by the "Molecular and Cellular Biology" core facility of the IMCB SB RAS (0310-2018-0011 grant). We thank Vladimir S. Lebedev from Zoological museum, moscow State University, moscow, Russia, for providing animals from the laboratory colony of moscow Zoo. We are also highly obliged to Touraj Sayyadpour, Kordiyeh Hamidi, and Safieh Akbarirad (researchers at the faculty of biology, Ferdowsi University of mashhad), morteza Radmanesh (Laboratory technician of the faculty of biology, Ferdowsi University of mashhad), Ahmad Abedipour (driver of the Ferdowsi University of mashhad) for assistance in obtaining the materials.

Conflicts of Interest: The authors declare no conflict of interest.

References

1. Simpson, G.G. The principles of classification and a classification of mammals. *Bull. Am. museum Nat. Hist.* **1945**, *1*, 350. [CrossRef]
2. Chaline, J.; mein, P.; Petter, F. Les grandes lignes d'une classification évolutive des muroidea. *Mammalia* **1977**, *41*. [CrossRef]
3. Hooper, E.T. The male Phallus in mice of the Genus Peromyscus. *Misc. Publ. mus. Zool.* **1958**, *105*, 1–24.
4. Manville, R.H. The families of mammals. *Science* **1967**, *157*, 1421–1422. [CrossRef]
5. Grassé, P.; Dekeyser, P.L. Ordre des Rongeurs. *Trait. Zool. Anat. Syst. Biol.* **1955**, *17*, 1321–1573.
6. Vorontzov, N.N.; Potapova, E.G. Taxonomy of the genus Calomyscus (Cricetidae). Status of Calomyscus in the system of Cricetinae. *Russ. J. Zool.* **1979**, *58*, 1391–1397.
7. Michaux, J.; Reyes, A.; Catzeflis, F. Evolutionary History of the most Speciose mammals: molecular Phylogeny of muroid Rodents. *Mol. Biol. Evol.* **2001**, *18*, 2017–2031. [CrossRef]
8. Musser, G.G.; Carleton, m.D. Superfamily muroidea. In *Mammal Species of the World: A Taxonomie and Geographic Reference*; John Hopkins University Press: Baltimore, mD, USA, 2005; pp. 894–1532.
9. Thomas, O. Two new species of Calomyscus, (part B) scientific results from the mammal survey no. *J. Bombay Nat. Hist. Soc.* **1920**, *26*, 938–940.
10. Kaschkarov, D. materials to the knowledge of the rodents of the Turkestan. *Trans. Sci. Soc. Turkestan middle Asiat. Univ.* **1925**, *2*, 43–56.
11. Ognev, S.I.; Heptner, V.G. The mammals of the Central Kopet-Dag and the adjacent plain. *Proc. Zool. mus. mosc. Univ.* **1929**, *3*, 46–172.
12. Argyropulo, A.I. The genera and species of Palearctic hamsters (Cricetinae). *Proc. Zool. Inst. Acad. Sci. USSR* **1933**, *1*, 239–248.
13. Kalabuhov, N.I. The discovery of the mouse-like hamster in Transcaucasia. *Nature* **1939**, *12*, 83–84.
14. Goodwin, G.G. *Five New Rodents from the Eastern Elburz mountains and a New Race of Hare from Teheran. American museum Novitates*; The American museum of Natural History: New York, NY, USA, 1939.
15. Ellerman, J.R. Key to the Rodents of South-West Asia in the British museum Collection. *Proc. Zool. Soc. London* **1948**, 765–816. [CrossRef]
16. Ellerman, J.R.; morrison-Scott, T.C.S. *Checklist of Palaearctic and Indian mammals 1758 to 1946*; British museum (Natural History): London, UK, 1951; 810p.
17. Gromov, I.M.; Gureev, A.A.; Novikov, G.A.; Sokolov, I.I.; Strelkov, P.P.; Chaptskii, K.K. Order Rodentia. In *Mammals of the USSR Fauna*; AN SSSR: moscow/Leningrad, Russia, 1963; pp. 244–638.
18. Sokolov, V. Order Lagomorpha, Rodentia. In *Systematics of mammals*; Vysshaya Shkola Press: moscow, Russia, 1977.
19. Corbet, G.B.; Hill, J.E. *A World List of mammalian Species*; British museum (Natural History) and Cornell University Press: London, UK; Ithaca: New York, NY, USA, 1980.
20. Schlitter, D.A.; Setzer, H.W. New rodents (Mammalia: Cricetidae, muridae) from Iran and Pakistan. *Proc. Biol. Soc. Washingt.* **1973**, *86*, 163–174.
21. Peshev, D. The mouse like hamster (Calomyscus bailwardi Thomas, 1905), a new mammal, 1905), a new mammal for the Syrian fauna and the Arab peninsula. *Mammalia* **1989**, *53*, 109–112.

22. Peshev, D. On the systematic position of the mouse-like hamster Calomyscus bailwardi Thomas, 1905 (Cricetidae, Rodentia) from the Near East and middle Asia. *Mammalia* **1991**, *55*. [CrossRef]
23. Matthey, R. Cytologie comparee des cricetinae palearctiques et d'Americanes. *Rev. Suisse Zool.* **1961**, *68*, 41–61.
24. Vorontsov, N.N.; Kartavtseva, I.V.; Potapova, E.G. The systematic of mouse-like hamsters from Calomyscus (Cricetidae). The karyological differentiation of sibling species from Transcaucasia and Turkmenistan and review of species of Calomyscus genus. *Russ. J. Zool.* **1979**, *58*, 1213–1224.
25. Graphodatsky, A.S.; Sablina, O.V.; meyer, m.N.; malikov, V.G.; Isakova, E.A.; Trifonov, V.A.; Polyakov, A.V.; Lushnikova, T.P.; Vorobieva, N.V.; Serdyukova, N.A.; et al. Comparative cytogenetics of hamsters of the genus Calomyscus. *Cytogenet. Genom. Res.* **2000**, *88*, 296–304. [CrossRef] [PubMed]
26. Meyer, m.; malikov, V.G. New species and subspecies of mouse-like hamsters of the genus Calomyscus (Rodentia, Cricetidae) from southern Turkmenistan. *Russ. J. Zool.* **2000**, *79*, 219–223.
27. Lebedev, V.S.; Pavlinov, I. Ya.; meyer, m.N.; malikov, V.G. Craniometric analysis of mouse-like hamsters of the genus Calomyscus (Critecidae). *Zool. Zhurnal* **1998**, *77*, 721–731.
28. Shahabi, S.; Aliabadian, m.; Darvish, J.; Kilpatrick, C.W. molecular phylogeny of brush-tailed mice of the genus Calomyscus (Rodentia: Calomyscidae) inferred from mitochondrial DNA sequences (Cox1 gene). *Mammalia* **2013**, *77*, 425–431. [CrossRef]
29. Lebedev, V.S.; Bannikova, A.A.; Neumann, K.; Ushakova, m.V.; Ivanova, N.V.; Surov, A.V. molecular phylogenetics and taxonomy of dwarf hamsters Cricetulus milne-Edwards, 1867 (Cricetidae, Rodentia): Description of a new genus and reinstatement of another. *Zootaxa* **2018**, *4387*, 331–339. [CrossRef]
30. Norris, R.W.; Woods, C.A.; Kilpatrick, C.W. morphological and molecular definition of Calomyscus hotsoni (Rodentia: muroidea: Calomyscidae). *J. Mammal.* **2008**, *89*, 306–315. [CrossRef]
31. Rezazadeh, E.; Aliabadian, m.; Darvish, J.; Ahmadzadeh, F. Diversification and evolutionary history of brush-tailed mice, Calomyscidae (Rodentia), in southwestern Asia. *Org. Divers. Evol.* **2020**, *20*, 155–170. [CrossRef]
32. Stanyon, R.; Galleni, L. A rapid fibroblast culture technique for high resolution karyotypes. *Bollet. Zool.* **1991**, *58*, 81–83. [CrossRef]
33. Romanenko, S.A.; Biltueva, L.S.; Serdyukova, N.A.; Kulemzina, A.I.; Beklemisheva, V.R.; Gladkikh, O.L.; Lemskaya, N.A.; Interesova, E.A.; Korentovich, m.A.; Vorobieva, N.V.; et al. Segmental paleotetraploidy revealed in sterlet (Acipenser ruthenus) genome by chromosome painting. *Mol. Cytogenet.* **2015**, *8*, 90. [CrossRef] [PubMed]
34. Yang, F.; Graphodatsky, A.S. Animal probes and Zoo-FISH. In *Fluorescence In Situ Hybridization (FISH)—Application Guide;* Springer: Berlin/Heldelberg, Germany, 2017; pp. 395–416.
35. Gladkikh, O.L.; Romanenko, S.A.; Lemskaya, N.A.; Serdyukova, N.A.; O'Brien, P.C.M.; Kovalskaya, J.M.; Smorkatcheva, A.V.; Golenishchev, F.N.; Perelman, P.L.; Trifonov, V.A.; et al. Rapid karyotype evolution in Lasiopodomys involved at least two autosome-Sex chromosome translocations. *PLoS ONE* **2016**, *11*, e0167653. [CrossRef]
36. Seabright, m. A rapid banding technique for human chromosomes. *Lancet* **1971**, *2*, 971–972. [CrossRef]
37. Lemskaya, N.A.; Kulemzina, A.I.; Beklemisheva, V.R.; Biltueva, L.S.; Proskuryakova, A.A.; Hallenbeck, J.M.; Perelman, P.L.; Graphodatsky, A.S. A combined banding method that allows the reliable identification of chromosomes as well as differentiation of AT- and GC-rich heterochromatin. *Chromosom. Res.* **2018**, *26*, 307–315. [CrossRef]
38. Trifonov, V.A.; Vorobieva, N.V.; Serdyukova, N.A.; Rens, W. FISH with and Without COT1 DNA. In *Fluorescence In Situ Hybridization (FISH)—Application Guide*; Springer: Berlin/Heldelberg, Germany, 2017; pp. 123–135.
39. Telenius, Håk.; Ponder, B.A.J.; Tunnacliffe, A.; Pelmear, A.H.; Carter, N.P.; Ferguson-Smith, m.A.; Behmel, A.; Nordenskjöld, m.; Pfragner, R. Cytogenetic analysis by chromosome painting using dop-PCR amplified flow-sorted chromosomes. *Genes Chromosom. Cancer* **1992**, *4*, 257–263. [CrossRef]
40. Yang, F.; Carter, N.P.; Shi, L.; Ferguson-Smith, m.A. A comparative study of karyotypes of muntjacs by chromosome painting. *Chromosoma* **1995**, *103*, 642–652. [CrossRef]
41. Sitnikova, N.A.; Romanenko, S.A.; O'Brien, P.C.M.; Perelman, P.L.; Fu, B.; Rubtsova, N.V.; Serdukova, N.A.; Golenishchev, F.N.; Trifonov, V.A.; Ferguson-Smith, m.A.; et al. Chromosomal evolution of Arvicolinae (Cricetidae, Rodentia). I. The genome homology of tundra vole, field vole, mouse and golden hamster revealed by comparative chromosome painting. *Chromosom. Res.* **2007**, *15*, 447–456. [CrossRef] [PubMed]
42. Romanenko, S.A.; Lemskaya, N.A.; Trifonov, V.A.; Serdyukova, N.A.; O'Brien, P.C.M.; Bulatova, N.S.; Golenishchev, F.N.; Ferguson-Smith, m.A.; Yang, F.; Graphodatsky, A.S. Genome-wide comparative chromosome maps of *Arvicola amphibius, Dicrostonyx torquatus*, and *Myodes rutilus*. *Chromosom. Res.* **2016**, *24*, 145–159. [CrossRef] [PubMed]
43. Yang, F.; Trifonov, V.; Ng, B.L.; Kosyakova, N.; Carter, N.P. Generation of paint probes from flow-sorted and microdissected chromosomes. In *Fluorescence In Situ Hybridization (FISH)*; Springer: Berlin/Heldelberg, Germany, 2017; pp. 63–79.
44. Romanenko, S. Calomyscus bailwardi. In *Atlas of mammalian Chromosomes*; Graphodatsky, A.S., Perelman, P.L., O'Brien, S.J., Eds.; Wiley: Hoboken, NJ, USA, 2020; p. 410, ISBN 9781119418030.
45. Romanenko, S.A.; Volobouev, V.T.; Perelman, P.L.; Lebedev, V.S.; Serdukova, N.A.; Trifonov, V.A.; Biltueva, L.S.; Nie, W.; O'Brien, P.C.M.; Bulatova, N.S.; et al. Karyotype evolution and phylogenetic relationships of hamsters (Cricetidae, muroidea, Rodentia) inferred from chromosomal painting and banding comparison. *Chromosom. Res.* **2007**, *15*, 283–297. [CrossRef] [PubMed]
46. Romanenko, S. Calomyscus sp. In *Atlas of mammalian Chromosomes;* Graphodatsky, A.S., Perelman, P.L., O'Brien, S.J., Eds.; Wiley: Hoboken, NJ, USA, 2020; p. 408, ISBN 9781119418030.

47. Romanenko, S. Calomyscus elburzensis. In *Atlas of mammalian Chromosomes*; Graphodatsky, A.S., Perelman, P.L., O'Brien, S.J., Eds.; Wiley: Hoboken, NJ, USA, 2020; p. 411, ISBN 9781119418030.
48. Romanenko, S. Calomyscus mystax. In *Atlas of mammalian Chromosomes*; Graphodatsky, A.S., Perelman, P.L., O'Brien, S.J., Eds.; Wiley: Hoboken, NJ, USA, 2020; p. 409, ISBN 9781119418030.
49. Shahabi, S.; Zarei, B.; Sahebjam, B. Karyologic Study of Three Species of Calomyscus (Rodentia: Calomyscidae) from Iran. *Iran. J. Anim. Biosyst. IJAB* **2010**, *6*, 55–60.
50. Radjabli, S.I.; Sablina, O.V.; Graphodatsky, A.S. Calomyscus bailwardi. In *Atlas of mammalian Chromosomes*; Wiley Online Library: Hoboken, NJ, USA, 2006; p. 202.
51. Lemskaya, N.A.; Kartavtseva, I.V.; Rubtsova, N.V.; Golenishchev, F.N.; Sheremetyeva, I.N.; Graphodatsky, A.S. Chromosome Polymorphism in microtus (Alexandromys) mujanensis (Arvicolinae, Rodentia). *Cytogenet. Genome Res.* **2015**, *146*, 238–242. [CrossRef] [PubMed]
52. Meyer, m.N.; Golenishchev, F.N.; Radjably, S.I.; Sablina, O.V. Voles (subgenus microtus Schrank) of Russia and adjacent territories. *Proc. Zool. Inst. RAS* **1996**, *232*, 1–320.
53. Shahabi, S.; Darvish, J.; Aliabadian, m.; mirshamsi, O.; mohammadi, Z. Cranial and dental analysis of mouse-like hamsters of the genus Calomyscus (Rodentia: Calomyscidae) from plateau of Iran. *Hystrix It. J. mamm.* **2012**, *22*, 311–323.
54. Akbarirad, S.; Darvish, J.; Aliabadian, m. Phylogeography of Calomyscus elburzensis (Calomyscidae, Rodentia) around the Central Iranian Desert with Description of a New Subspecies in Center of Iranian Plateau. *J. Sci. Islam. Repub. Iran* **2016**, *27*, 5–21.
55. Rawson, B. Phylogenomic and Species Delimitation in the Brush-Tailed mouse Genus Calomyscus Using DDRADSEQ data. In *An Undergraduate Research Thesis*; The Ohio State University: Columbus, OH, USA, 2019.

genes

MDPI

Article

Mosaic Evolution of Molecular Pathways for Sex Pheromone Communication in a Butterfly

Caroline M. Nieberding [1,*], Patrícia Beldade [2], Véronique Baumlé [1], Gilles San Martin [1], Alok Arun [1], Georges Lognay [1], Nicolas Montagné [3], Lucie Bastin-Héline [3], Emmanuelle Jacquin-Joly [3], Céline Noirot [4], Christophe Klopp [4] and Bertanne Visser [5]

[1] Evolutionary Ecology and Genetics Group, Earth and Life Institute, UC Louvain, 1348 Louvain-la-Neuve, Belgium; veroniquebaumle@hotmail.com (V.B.); gilles.sanmartin@gmail.com (G.S.M.); alok_arun@br.inter.edu (A.A.); georges.lognay@uliege.be (G.L.)
[2] Center for Ecology, Evolution and Environmental Changes (cE3c) & Global Change and Sustainability Institute (CHANGE), Faculty of Sciences, University of Lisbon (FCUL), 1749-016 Lisboa, Portugal; pbeldade@fc.ul.pt
[3] INRAE, CNRS, IRD, UPEC, Sorbonne Université, Institute of Ecology and Environmental Sciences of Paris, Université de Paris, 78000 Versailles, France; nicolas.montagne@sorbonne-universite.fr (N.M.); lucie.bastin@laposte.net (L.B.-H.); emmanuelle.joly@inrae.fr (E.J.-J.)
[4] Plateforme Bio-Informatique GenoToul, MIAT, INRAE, UR875 Mathématiques et Informatique Appliquées Toulouse, 31326 Castanet-Tolosan, France; celine.noirot@inrae.fr (C.N.); christophe.klopp@inrae.fr (C.K.)
[5] Evolution and Ecophysiology Group, Department of Functional and Evolutionary Entomology, Gembloux Agro-Bio Tech, University of Liège, 5030 Gembloux, Belgium; bertanne.visser@uliege.be
* Correspondence: caroline.nieberding@uclouvain.be

Citation: Nieberding, C.M.; Beldade, P.; Baumlé, V.; San Martin, G.; Arun, A.; Lognay, G.; Montagné, N.; Bastin-Héline, L.; Jacquin-Joly, E.; Noirot, C.; et al. Mosaic Evolution of Molecular Pathways for Sex Pheromone Communication in a Butterfly. *Genes* **2022**, *13*, 1372. https://doi.org/10.3390/genes13081372

Academic Editors: Luigi Viggiano and Renè Massimiliano Marsano

Received: 21 June 2022
Accepted: 18 July 2022
Published: 31 July 2022

Abstract: Unraveling the origin of molecular pathways underlying the evolution of adaptive traits is essential for understanding how new lineages emerge, including the relative contribution of conserved ancestral traits and newly evolved derived traits. Here, we investigated the evolutionary divergence of sex pheromone communication from moths (mostly nocturnal) to butterflies (mostly diurnal) that occurred ~119 million years ago. In moths, it is the females that typically emit pheromones to attract male mates, but in butterflies males emit pheromones that are used by females for mate choice. The molecular bases of sex pheromone communication are well understood in moths, but they have remained relatively unexplored in butterflies. We used a combination of transcriptomics, real time qPCR, and phylogenetics to identify genes involved in the different steps (i.e., production, regulation, and reception) of sex pheromone communication of the butterfly *Bicyclus anynana*. Our results show that the biosynthesis and reception of sex pheromones relies both on moth-specific gene families (reductases) and on more ancestral insect gene families (desaturases, olfactory receptors, odorant binding proteins). Interestingly, *B. anynana* appears to use what was believed to be the moth-specific neuropeptide Pheromone Biosynthesis Activating Neuropeptide (PBAN) for regulating sex pheromone production. Altogether, our results suggest that a mosaic pattern best explains how sex pheromone communication evolved in butterflies, with some molecular components derived from moths, and others conserved from more ancient insect ancestors. This is the first large-scale investigation of the genetic pathways underlying sex pheromone communication in a butterfly.

Keywords: olfactory communication; desaturase; reductase; odorant receptor; odorant binding protein; chemosensory protein; PBAN; phylogeny

1. Introduction

The evolution of new life forms occurs through the transition from an ancestral to a descendant clade, where the new lineage generally shows a mosaic phenotype of conserved and newly evolved traits. Mosaic evolution is indeed a recurring pattern in paleontology [1–3]. For example, *Tiktaalik roseae*, believed to represent the transition from fishes to amphibians (the "fishapod"; ~375 Mya), shares some traits with more primitive sarcopterygian fishes (e.g., body scales, fin rays, lower jaw, and palate) and other traits more typical of tetrapods (e.g., a shortened skull roof, modified ear region, mobile neck, functional wrist joint, and other features) [4]. Investigating the genetic bases of ancestral and derived phenotypic traits is essential for obtaining a mechanistic explanation of how mosaic evolution takes place. Studies investigating the mechanistic basis of mosaic evolution have increased in the last decade, including recent genomic evolution analyses identifying patterns of gene loss, retention, and *de novo* evolution [5–7]. Other patterns in the genetic bases of trait evolution have suggested a role for hybridization between species [8–10] or co-option of molecular pathways that acquired new functions [11,12]. Derived phenotypic traits can thus be generated through different molecular mechanisms that need to be identified case-by-case.

Here, we focused on the genetic bases of divergence in sex pheromone communication during the evolutionary transition from moths to butterflies, which occurred ~119 Mya [13]. Sex pheromone communication is used by most insects, including butterfly species, for finding, identifying, and assessing the quality of potential mating partners [14–18]. Sex pheromone communication is under strong selection, because it determines mating success and consequently an individual's contribution to the next generation [19–23]. Molecular pathways for sex pheromone biosynthesis, its regulation, and pheromone reception have been identified in several moth species [24–30]. Compared to other insects, moths appear to have evolved moth-specific genes or gene lineages involved in sex pheromone communication [25,27,30–35]. For biosynthesis, most moths use a limited number of enzymes for desaturation, chain-shortening, reduction, acetylation or oxidation of *de novo* synthesized saturated fatty acids [25,30,31,36], which generate tremendous chemical diversity of pheromone components and high species-specificity of pheromone blends [24,25]. Some of these enzyme families comprise Lepidoptera-specific subclades, such as the Δ9- and Δ11-desaturases [25,30,31]. Similarly, novel odorant receptor and odorant binding protein subclades have evolved in moths that bind specifically to sex pheromone chemicals [27,32–35]. We aimed to investigate whether butterflies use moth-specific molecular pathways, more ancestral insect pathways, and/or have evolved butterfly-specific molecular pathways for pheromone communication.

Butterfly sex pheromones have some ecological specificities unlike those of moths [17]. Female moths, for example, use sex pheromones to signal their location to mating partners in the dark or at dusk [22]. In contrast, butterfly sex pheromones are generally produced by males and the importance of sex pheromone communication in butterflies was less clear due to their diurnal lifestyle (see [17] for studies on male sex pheromones in moths). Studies on some butterfly species have, however, revealed that sex pheromone communication is important for determining mating success [17,37] and sex pheromones play an important role in speciation events [38]. Butterfly sex pheromones can indeed convey refined information as to the identity or quality of potential mating partners and can be critical for mate choice and species recognition [39,40]. We focused on the butterfly *B. anynana*, whose sex pheromone composition was previously identified and functionally validated [17,41]. Moreover, experimental manipulation, including the addition of synthetic sex pheromone perfumes [17,41], and artificial induction of "anosmia" [37,42] confirmed the importance of *B. anynana* sex pheromone for mating success. More than a hundred chemical components have been identified on *B. anynana* adult male and female bodies [43], but the composition of the male sex pheromone ("MSP" hereafter) consists of three main volatile components: (Z)-9-tetradecenol (MSP1), hexadecanal (MSP2) and 6,10,14-trimethylpentadecan-2-ol (MSP3) [17]. The identification and functional validation of these three MSPs and their role in *B. anynana* chemical communication set this species apart from other lepidopterans, where the role of specific chemical components often remains elusive in relation to fitness (but see [17,44] where specific components were identified). MSP1 and MSP2 are derived from fatty acids and are typically found in the pheromone blends of many moth species [25], which led us to hypothesize that the same genes as in moths are involved in sex pheromone communication of *B. anynana*.

To investigate if butterflies use what were believed to be moth-specific gene families, we used RNA-seq to identify genes that could be involved in *B. anynana* pheromone production, regulation, and reception (Figure 1). We compared transcript abundance of sex pheromone-related adult tissues (male pheromone producing structures, heads and antennae) with control tissues (female wings and heads), and validated our findings with RT-qPCR. We identified specific candidate genes involved in the different olfactory communication functions and used phylogenetic analyses to identify the molecular origin of those genes. Our results reveal that sex pheromone communication in *B. anynana* evolved through a mosaic of ancestral insect genes, and more derived lepidoptera-specific genes.

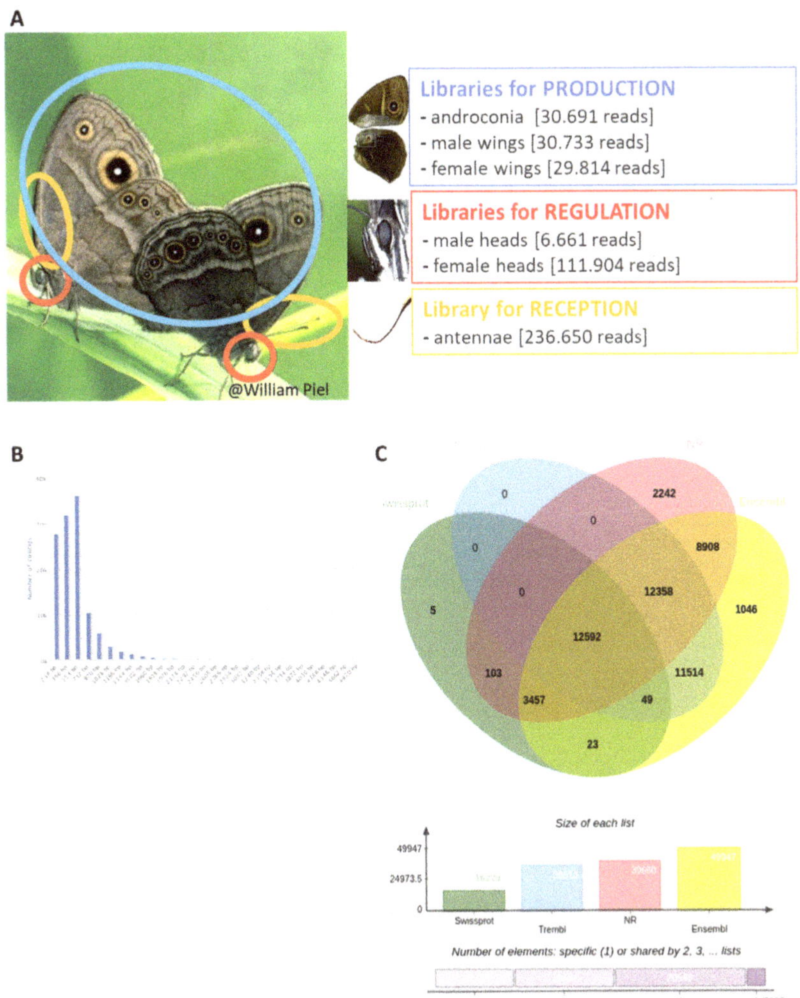

Figure 1. (**A**) Experimental design for the transcriptomics experiment showing the 3 steps involved in male sex pheromone communication and the corresponding tissues sampled to produce the RNA libraries (also including developmental libraries, not shown here). The number of sequenced reads per library is provided. (**B**) Information about the number of contigs in the transcriptome. (**C**) Venn diagram of annotated contig with regard to databases: Swissprot, Trembl, NR and 10 species of Ensembl Lepbase.

2. Materials and Methods

2.1. Insects

B. anynana (Butler, 1879) (Lepidoptera: Nymphalidae) originated from an outbred wild type population that was collected in Malawi, Africa, in 1988 using 80 gravid females. Since 1988, several hundred individuals have been used each generation to maintain high levels of heterozygosity [45] in a climate-controlled room at a temperature of 27 °C, a relative humidity of 70%, and a photoperiod of L:D 12:12. Larvae were kept under these conditions on maize plants, *Zea mays*, and adults were fed mashed banana, *Musa acuminata*, for all experiments, except when stated otherwise.

2.2. Transcriptome

2.2.1. Tissue Collection

For the transcriptome dataset, several hundreds of virgin males and females were separated at the pupal stage in different cages, and tissues collected in March 2010. Pupal tissues were collected from male and female pupae 1 to 7 days after pupation (1 or 2 pupae per day after pupation and per sex), after which the wing imaginal discs were dissected as described in [46]. Tissues for adult libraries (wings, heads and antennae) were collected from adult virgin males and females aged 1, 3, 5, 8, 10 and 14 days after emergence (Supplementary Figure S1): ~50 adults were used per age category and per library for wing libraries, ~10 adults were used per age category and per library for head tissues, and ~5 adult females and 5 males were used per age category for the antennae library. Brain tissue was obtained by cutting the head and cutting off the eyes, the proboscis and the antennae; antennal tissue was collected for a similar number of adult males and females (Figure 1). Dissected tissues were conserved immediately at −20°C in RNAlater (Sigma-Aldrich, Hoeilaart, Belgium).

2.2.2. RNA Extraction

RNA of all dissections was extracted in April 2010, within a month after collection of tissues, in an RNA-free environment, on ice, and using the RNeasy Mini kit and the RNAase free DNAase kit (Qiagen, Venlo, The Netherlands). After RNA extraction, 1 μL of each RNA extract was used for testing RNA quality and quantity using a Bioanalyzer System (Agilent, Machelen, Belgium) at the LUMC hospital in Leiden (The Netherlands, courtesy of Dr Jeroen Pijpe), and the RiboGreen RNA quantification kit, respectively. The remaining RNA extract was stored at −80 °C for cDNA synthesis. For cDNA synthesis, we first pooled all RNA extracts dedicated to the same library in one tube per library, in such a way that: (*i*) the same amount of RNA was present for each sex (male and female), (*ii*) each life stage was represented by similar RNA amounts (days 1 to 7 after pupation for pupal tissue libraries; days 1 to 14 after emergence for adult tissue libraries; Supplementary Table S1). Total RNA yield was 27 to 40 μg per library as requested for sequencing.

2.2.3. mRNA Isolation, cDNA Synthesis and Sequencing

mRNA capture, cDNA synthesis, and tagging for Titanium 454-sequencing was performed by Biogenomics, a KU Leuven Research & Development Division of the Laboratory of Animal Diversity and Systematics (Leuven, Belgium). Between 370 and 1340 ng (0.3 and 1.6%) mRNA yield was obtained for each library, providing enough mRNA (minimum 200 ng per library) for cDNA construction and tagging. Yet, cDNA synthesis failed when started from mRNA, which is why a SMART cDNA synthesis was performed from total RNA. A custom normalization step (based on the EVROGEN Trimmer kit) was optimized in collaboration with the Roche R&D department and applied to the cDNA libraries, as no validated normalization protocol was available from Roche in 2010 for Titanium cDNA sequencing. Each normalized library was quality checked for fragment length and integrity before sequencing. Each library was subjected to GS FLX Titanium Emulsion PCR and Sequencing, and each library was sequenced 5 times. After sequencing, data were processed through certified Roche software (GS Transcriptome Assembler/Mapper) and custom scripts for advanced analysis. Basic data analysis included read quality trimming and assembly into contigs, including potential alternative splicing products. The sequences were trimmed by removing low quality sequences, ambiguous nucleotides, adapter sequences, and sequences with lengths less than 20 nucleotides. The 454-sequencing generated 824,439 reads, with an average length of 293 base pairs and a total of 242,005,027 nucleotides (Supplementary Figure S2).

2.2.4. Transcriptome Assembly, Quantification, and Annotation

Adaptors were removed with smartkitCleaner and adaptorCleaner. Raw sequences (reads) were cleaned with the software Pyrocleaner ([47], using the following criteria: *(i)* complexity/length ratio less than 40 (using a sliding window approach based on 100 bp sequence length, and a step of 5 bp); *(ii)* duplicate read removal (see bias associated with pyrosequencing, due to the random generation of duplicate reads); *(iii)* removal of too long/too short reads (maximum and minimum read length = mean read length ± 2 SD); *(iv)* removal of reads with too many undetermined bases (more than 4%). Contaminations were discarded by searching hits against *Escherichia coli*, phage and yeasts.

The reads were assembled de novo in 43,149 contigs of 488 base pairs on average with a total of 21,087,824 nucleotides (Supplementary Figure S2). The average GC content was 36.44%. The assembly was performed with tgic l.

(https://academic.oup.com/bioinformatics/article/19/5/651/239299) version 2.1 using standard parameters. The reads where realigned to the contigs and singletons with bwa aln version 0.7.2 using standard parameters and transformed in bam format, sorted and indexed with samtools version 0.1.19 with default parameters. The bam files were then processed with samtools idxstats to extract expression measures in the form of numbers of reads aligned on each contig for every condition. These measures were than merged to produce the quantification file using unix cut and paste commands. Diamond was used to search for sequence homology between contig and the following generalist databases: UniProtKB/Swiss-Prot, UniProtKB/TrEMBLRelease of April, NR release of end of March.

The following species from the ensemble database were queried: *B. anynana* (nBA.0.1), *Calycopis cecrops* (v1.1), *Danaus plexippus* (v3), *Heliconius melpomene melpomene* (Hmel2.5), *Junonia coenia* (JC v1.0), *Lerema accius* (v1.1), *Melitaea cinxia*, *Papilio machaon* (Pap_ma_1.0), *Phoebis sennae* (v1.1), and *Pieris napi* (v1.1).

2.2.5. Candidate Gene Identification Using Transcriptome Sequencing

Numerous publications document gene expression studies focusing on chemical communication in Lepidoptera, but only three of these studies focused on butterflies [48–50], and butterfly sex pheromone communication has rarely been studied in this context [48]. Here, we produced six RNA libraries from different adult tissues that were specifically chosen to cover the different steps of male pheromone communication (Figure 1): pheromone biosynthesis (which occurs in dedicated structures on male wings, called androconia) [17], its neuro-regulation (in the brain), and pheromone reception (in antennae). Approximately 500 male and female *B. anynana* adults were dissected and relevant tissues assigned to different libraries (Figure 1A). For pheromone synthesis, we compared transcripts in male androconia (Library "androconia") with those in remaining adult male wing parts (Library "male wings") and adult female wings (Library "female wings") as controls. For regulation of pheromone communication, we compared transcript abundance between adult male heads (where the regulation of pheromone synthesis takes place; Library "adult male heads") and adult female heads (Library "adult female heads", control). For pheromone reception, we compared transcripts between adult male and female antennae (the tissue where pheromone reception takes place) [17]; Library "antennae") and adult heads (Libraries "male heads" and "female heads") as controls. Two other libraries were also analyzed, corresponding to pupal wings in males (Library "pupal male wings"), and females (Library "pupal female wings"), but these data will not be discussed here. We focus solely on adults, the stage during which pheromone communication takes place. A total of

737,206 reads were obtained from the different tissues sampled in *B. anynana* and were assembled into 43,149 contigs, with 76,818 remaining non-assembled singletons (Figure 1B,C, Supplementary Table S2). Transcripts were annotated based on reference genomes for several butterfly species (including *B. anynana*; [51]), as well as other relevant insect species. Using the digital differential display (DDD) tool (of NCBI's UniGene database; $p < 0.05$), a total of 422 contigs were found to be differentially expressed when tissue-specific libraries were compared (Supplementary Table S2). Expression differences were validated by real time quantitative PCR analyses on 10 selected candidate chemical communication genes, showing that relative differences in expression levels in our transcriptome dataset matched those quantified by RT-qPCR (Supplementary Figure S3).

2.2.6. Identification of Specific Gene Families

We also mined the transcriptome for specific families of genes supposedly involved in sex pheromone communication based on the available evidence in moths: desaturases, reductases, odorant receptors (OR), odorant binding proteins (OBP), and chemosensory proteins (CSP). To do so:

(i) we downloaded the DNA sequence of every *B. anynana* contig named as a desaturase, reductase, OR, OBP or CSP in our transcriptome;

(ii) we checked the homology of the sequence of each candidate contig with gene members of the same family in other Lepidoptera by performing a blastx in Genbank;

(iii) every *B. anynana* contig that showed significant homology in step *ii* was blasted in the transcriptome, allowing us to find more *B. anynana* ESTs of the same gene family, even if some had not been annotated as such. All these contigs and ESTs were then "candidate members of each respective gene family". If no significant homology was found using blastx in step *ii*, the sequence was removed from the list of candidate members of the gene family;

(iv) every *B. anynana* contig and EST candidate was then translated into an amino acid sequence using Expasy (https://web.expasy.org/translate/). When necessary, cdd analyses of domains were done. Using this procedure, 27 OR, 44 OBP and 70 CSP candidate members were found in the *B. anynana* transcriptome (Supplementary Tables S3–S5 for OR, OBP and CSP, respectively; for reductases and desaturases see Results). For example, 40 contigs were initially annotated as "odorant-binding protein" in our transcriptome, based on the characteristic hallmarks of the OBP protein families, including six highly conserved cysteines, i.e., "C" (in Lepidoptera C1-X25-30-C2-X3-C3-X36-42-C4-X8-14-C5-X8-C6, with "X" being any amino acid) [52]. As sequence conservation between OBPs is low, i.e., between 25 and 50% identity for amino acid sequences, manually mining the transcriptome allowed us to find another seven OBP candidate members (Supplementary Tables S3–S5 for OR, OBP and CSP, respectively).

(v) Candidate members were then manually aligned in Bioedit to group them into distinct expressed gene units, or unigenes: 17 Bany_OR unigenes (Supplementary Table S3), 9 Bany_OBP unigenes, including in some cases more "gene subunits" when contigs were similar enough to suggest that they represented different allelic variants of the same gene, such as Bany_OBP3, Bany_OBP4, Bany_OBP6 (Supplementary Table S4) and 8 Bany_CSP unigenes with some more gene subunits as well (Supplementary Table S5).

(vi) The expression level of each candidate unigene across libraries was then obtained by pooling the number of copies in the *B. anynana* transcriptome of each EST and contig forming the unigene.

2.3. *Phylogenies*

For the OR phylogeny, amino acid sequences found in the *B. anynana* transcriptome were aligned with OR sequences previously identified in the genomes of *Bombyx mori* and *H. melpomene* [53] and in antennal transcriptomes of *Cydia pomonella* [54] and *Spodoptera littoralis* [55]. Alignment was performed with MAFFT v7 (https://mafft.cbrc.jp/alignment/server/), and the maximum-likelihood phylogeny was built using PhyML 3.0 [56]. Branch support was assessed using a likelihood-ratio test [57]. Published datasets of Lepidoptera protein sequences from previous phylogenetic studies were used for testing the phylogenetic position of *B. anynana* ORs [58] (Supplementary File S1), OBPs [59] (Supplementary File S2), desaturases and reductases [36] (Supplementary Files S3 and S4).

2.4. Real Time Quantitative PCR

For biological replicates, mRNA was extracted either from a single individual or formed by pooling 3 to 5 individuals of various ages in experiments for the "reception" and the "production" communication steps, respectively. Each treatment is represented by 3 to 7 biological replicates. The protocol used for quantitative real time PCR is described in [60]. Briefly, total RNA was extracted using the RNeasy Mini kit following manufacturer's instructions. Residual DNA was removed after treating extracted RNA using a DNase enzyme. A nanodrop ND-1000 spectrophotometer was then used to assess the integrity of the RNA before conversion into cDNA. qRT-PCR was carried out using the SYBR green dye in a 96-well thermocycler with parameters described in [60]. Primer sequences for all genes are available in Supplementary Table S6. Relative transcript abundance was calculated using the $2^{-\Delta\Delta Ct}$ method. Statistical significance of differences in expression levels expressed in Rq values after log-transformation was tested using nested ANOVA with technical replicates nested with biological replicates; the model was log (Rq) ~ treatment/biological replicate/technical replicate + Error (tissue/biological replicate/technical replicate). R version 3.6.1 [61] was used for statistical analyses.

2.5. Behavioral Experiments

2.5.1. Mating Experiments for Quantifying Odorant Receptor Expression Levels

Naïve virgin females were reared in isolated conditions (devoid of the male secondary sexual traits putatively involved in sexual communication, i.e., olfaction, vision and audition) directly after egg collection. The virgin sensitized females were reared in a MSP-containing environment near cages containing males (and thus exposed to the sex pheromones of males). The sensitized mated females were reared in a MSP-containing environment and mated at an age of 3 days. All females were sacrificed at day 5 and the antennal tissues were used for RNA extraction and RT-qPCR analysis (described in Section 2.4).

2.5.2. Daily Variation in Courtship Activity

We tested whether courtship activity in *B. anynana* males varies throughout the day. A large number of individuals were reared and age after emergence was recorded. The day before the experiment, 5 males and 4 females between 2 and 12 days old were randomly chosen and grouped in a cage (40 cm × 30 cm, cylindrical). The cages were placed in a room with a temperature of ~27 °C with natural light, and a 14:10 day-night regime. The butterflies were fed with banana slices and had access to water during the course of the experiment. We used 5 cages per trial and produced 3 trials with different individuals. A generalized mixed model with binomial error distribution was used to characterize the courting activity of males during the day. The presence/absence of courtship behavior for each male during 15 min of observations per hour was used as the dependent variable. As we expected courtship activity to peak at some time point during the day, we used "time of the day" (in the number of hours after natural sunrise) and its second order polynomial as a fixed explanatory variable. The age of males (in days) was also included as a fixed cofactor to control for the effect of age. The identity of each individual, cage and trial were used as random effects with individual nested within cage and cage nested within trial. We tested the model parameters with type III likelihood ratio tests, in which a model without the explanatory variable of interest is compared to the full model, both models being estimated by Maximum Likelihood.

2.5.3. Daily Variation in Male Sex Pheromone Production

A number of butterfly couples were set up using adult virgin stock males and females. Three families were started from 3 couples that produced over 200 offspring. The 3 families were each partly reared into 2 different climate rooms that differed in the onset of artificial daylight (one at 9:00 a.m., the other at 12:00 p.m.). This allowed us to control for the potential effect climate cell-specific conditions on variation in MSP production. Forty to 80 males that emerged the same day were selected per family. MSP production of 8-day old males was sampled, an age at which each MSP component is produced in measurable quantities [17]. Four to 7 males of each family were killed and conserved at −80° for subsequent pheromone analysis at 7 sampling points during the day: 1, 4, 8, 11, 13, 18 and 23 h after the onset of daylight. MSP production was measured as described below in the Section 2.6. We used mixed models with normal error distribution to characterize the variation of MSP production during the day. The titre of each MSP component and the ratios between pairs of MSP components were used as dependent variables. MSP titres were square root transformed and MSP ratios were log-transformed to improve the normality and homoscedasticity of the residuals. As we suspected MSP production to peak at some time point during the day, we used a second order polynomial equation with time and time^2 as a fixed explanatory variables and family as a random effect. We tested model parameters with type III likelihood ratio tests, in which a model without the explanatory variable of interest is compared to the full model, both models being estimated by Maximum Likelihood. We estimated the percentage of variation explained by the models and each of their components with pseudo R^2 based on ratios of sums of squared residuals. We followed [62] for the variance decomposition procedure.

2.6. *Quantification of Male Sex Pheromone Production*

MSP concentrations were determined as previously described [17,63]. In short, one forewing and one hindwing of each male were soaked in 600 ul of hexane during 5 min. One ng/µL of the internal standard (palmitic acid) was then added. Extracts were analyzed on a Hewlett-Packard 6890 series II gas chromatograph (GC) equipped with a flame-ionization detector (FID) and interfaced with a HP-6890 series integrator, using nitrogen as carrier gas. The injector temperature was set at 240 °C and the detector temperature at 250 °C. A HP-1 column was used and temperature increased from the initial temperature of 50 °C by 15 °C/min up to a final temperature of 295 °C, which was maintained for 6 min.

3. Results and Discussion

3.1. *B. anynana Sex Pheromone Biosynthesis*

B. anynana was the first butterfly for which molecular pathways underlying sex pheromone biosynthesis were investigated and compared to those of moths [36]. In this study, one gene related to pheromone communication was highly expressed in *B. anynana* male androconial wing tissues compared to male and female control wing samples (i.e., 'type 1' contigs in Supplementary Table S2): an aldose reductase-like gene. This gene was also highly expressed in the male androconial wing tissue alone. Moreover, a Δ9-desaturase gene was also found to be highly expressed in this library. In contrast to earlier findings in *B. anynana* [36], no fatty-acyl reductase (FAR), nor Δ11-desaturase were found to be highly expressed in male androconial wing tissue (Supplementary Table S2).

Desaturases (that add a double bond to fatty acid substrates) were previously found to be involved in *B. anynana* MSP biosynthesis [36]. Therefore, we extended our search for desaturase genes for each of the libraries separately. We focused specifically on Δ9 and Δ11-desaturases, because these enzymes play an important role in moth pheromone biosynthesis [30,36]. Previous work with *B. anynana* suggested that a Δ11-desaturase is involved in the production of MSP1 [36]. Both Δ9- and Δ11-desaturase were present in the transcriptome, mainly in antennae. A phylogenetic tree containing our Δ9- and Δ11-desaturase contigs revealed a similar position within the larger desaturase phylogenetic tree, compared to earlier work [36] (Supplementary Figure S4).

To get more insight into the role played by the Δ9 and Δ11 desaturase gene, we used RT-qPCR (as in [60]) to compare transcript abundance between different adult wing tissues, the main tissue producing MSP1 (using RNA extracted from new samples). Δ9-desaturase transcript abundance was approximately four-fold higher than that of Δ11-desaturase (Supplementary Figure S3). When comparing the spatial distribution of MSP1 on *B. anynana* body parts with our RT-qPCR data for the two Δ-desaturase genes (Supplementary Figure S5A), the expression profile of the Δ9-desaturase gene, but not the Δ11-desaturase gene, matched MSP1 distribution (Supplementary Figure S5B,C, respectively). Indeed, the Δ9-desaturase gene showed overall significant variation in transcript abundance across tissues that correlated with the distribution pattern of MSP1 (Supplementary Figure S5). Specifically, Δ9-desaturase was found to be significantly expressed in male wing parts containing the androconia that produce MSP1, compared to remaining male wing tissues and female wings. Moreover, Δ9-desaturase gene expression was also found to be significantly expressed in male head tissue containing MSP1. No such match between gene expression and MSP1 abundance was found for the Δ11-desaturase gene, which showed no significant variation in transcript abundance across tissues known to contain MSP1 (Supplementary Figure S5). Altogether, these findings suggest that a Δ9 desaturase plays a role in *B. anynana* pheromone biosynthesis.

We then searched for genes from a second gene family known to be involved in sex pheromone production in *B. anynana*: fatty acyl reductases (*far*) that convert fatty-acyl pheromone precursors to alcohol [36]. While more than 20 FARs have been experimentally characterized from 23 moth and butterfly species, all FARs implicated in moth and butterfly sex pheromone biosynthesis are restricted to a single clade, suggesting that one FAR group was exclusively recruited for pheromone biosynthesis [64,65]. In our transcriptome, two reductase contigs were annotated and identified in male and female antennae: enoyl-CoA reductase and fatty-acyl reductase 1, *far 1*. As *far1* and another fatty-acyl reductase, *far2*, were previously found to be involved in MSP2 and MSP1 biosynthesis, respectively [36], we manually mined our transcriptome for *far1* and *far2* contigs by n-blasting *far1* and *far2* specific gene sequences. Contigs matching *far1* were largely expressed in androconia (171 copies), compared to wing controls (0 copies; Supplementary Table S7). While contigs matching *far2* showed an overall low expression level in wing tissues (Supplementary Table S8), a previous qRT-PCR study revealed that *far2* gene expression matched MSP1 biosynthesis [60], highlighting the potential importance of *far2* for *B. anynana* pheromone production.

The low expression level of *far2* is surprising given the amount of MSP1 present on male wings (2 μg/individual on average); hence we suggest that alternative candidates for MSP1 biosynthesis could be aldo-keto reductases, two of which are among the most expressed genes in androconial male wing tissues (Supplementary Table S2). Indeed, fatty-acyl reductases are usually associated with the reduction of aldehyde into alcohols producing various sex pheromone components in moths, but aldo-keto reductases are regularly found highly expressed in sex pheromone transcriptomes of moth species [66–69]. Guo et al., (2014) [70] and Yamamoto et al. (2016) [71] have proposed that aldo-keto reductases are involved in sex pheromone biosynthesis of the moths *Helicoverpa armigera* and *B. mori* by reducing 9-hexadecenal, 11-hexadecenal and 10E,12Z-hexadecadienal into alcohol. Our expression data suggest that an aldo-keto reductase, with or without *far2*, may be involved in MSP1 biosynthesis.

3.2. *B. anynana* Sex Pheromone Reception

The genomes of the butterflies *Danaus plexippus* and *H. melpomene* (i.e., species for which phylogenies of odorant receptor genes were available) have revealed a large number of genes belonging to families involved in olfactory reception in moths, including odorant receptors and odorant binding proteins [48,53,72]. Specifically, the odorant receptor and odorant binding protein gene families contain lineages specialized in the detection of sex pheromones in moths, the so-called pheromone receptors (PRs) and pheromone-binding proteins (PBPs) [35,59,73,74]. ORs are transmembrane receptors that bind volatile chemicals and are responsible for signal transduction in insect olfactory sensory neurons. They exhibit various response tuning breadths, and moth ORs involved in pheromone detection are often (but not always) highly specific to one or a few pheromone components [74]. Therefore, we expected to identify ORs binding to each of the three known chemical components of the *B. anynana* male sex pheromone: MSP1, 2, and 3 [17]. We identified the obligatory co-receptor "Orco" and 16 ORs in the transcriptome, some of which were highly expressed in antennae compared to other adult tissues (Supplementary Table S3).

Phylogenetic analysis revealed that ORs expressed in *B. anynana* were distributed among various lepidopteran OR lineages [53], but none were located in the classically defined sex pheromone receptor clade [35,75] (Figure 2). This suggests that *B. anynana* sex pheromone reception may have evolved from lepidopteran OR lineages other than the sex pheromone lineage.

Recent studies have revealed that moth PRs do not constitute a monophyletic clade and, instead, evolved several times during OR evolution [35,76]. Functional PRs that have been found outside of the PR clade in some moth species were identified based on their sex-biased expression. We, therefore, searched for potential *B. anynana* PRs by quantifying the mRNA expression levels between sexes using RT-qPCR, expecting that PR in *B. anynana* should show higher expression in male compared to female antennae. We further expected that gene expression levels would correlate with temporally varying physiological and biological needs. In moth species, PRs are critical for detecting the female sex pheromone and the male's behavioral and physiological responses to female sex pheromones were shown to be affected by moth age and mating status [77,78]. Therefore, we tried to identify *B. anynana* candidate PRs by comparing RNA expression levels in females with different mating status (using RT-qPCR). We expected that virgin females that had developed either in isolation (naive "virgin") or in the presence of male scent ("virgin sensitized") would exhibit higher expression levels for OR genes responsible for detecting the male sex pheromone, compared to mated females ("mated") [79,80]. This difference would be due to virgin females taking information about the composition of the male sex pheromone for choosing mates regarding their inbreeding level or their age, and because recently mated females are much less receptive to courtship attempts in *B. anynana* [41,42]. The candidate genes Ban_OR1, Ban_OR2 and Ban_Orco were selected for RT-qPCR experiments because these genes displayed the highest expression among the 16 identified candidate ORs and were significantly expressed in antennae compared to control libraries (Supplementary Table S3). Orco expression was significantly decreased in mated compared to virgin (naïve or sensitized) females, but Bany_OR1 or Bany_OR2 were not (Figure 3), suggesting that regulation of the expression of Orco could be a mediator of sex pheromone receptivity. Orco, and not specific parts of the odorant receptor dimer, such as OR1, OR2 or other ORs that we did not test here, could be regulated by sex pheromone communication, similar to what was previously found in cockroaches [77,81].

In addition to the work described above, we aimed to functionally investigate if some specific OR candidate genes were responsible for the detection of male pheromone components using heterologous expression in *Drosophila melanogaster* olfactory sensory neurons coupled to electrophysiological recordings. These experiments did not lead to functional validation, but the procedures followed and results obtained are described in Supplementary File S5.

A second gene family specific to insects, the Odorant Binding Protein or OBP family, is involved in olfaction by solubilizing semiochemicals once they have entered the aqueous lymph within olfactory sensilla [27]. OBPs were proposed to play an important role in response sensitivity. In Lepidoptera, a dedicated lineage of OBPs (the so-called "pheromone-binding proteins" or PBPs) has evolved high affinity towards pheromone components [59]. We identified 46 contigs assembled into 13 OBP unigenes expressed in our *B. anynana* transcriptome (Supplementary Table S4), a number lower than what has been described in various transcriptomes from moth species (49 predicted OBPs in *S. littoralis* and *Manduca sexta;* [55]) and in the genomes of two butterfly species (32 in

D. plexippus, 51 in *H. melpomene*; [82–84]). *B. anynana* expressed OBPs were found in most subclades of the phylogenetic tree of lepidopteran OBPs, including general odorant binding protein 1 and 2 lineages, as well as classic, minus-C, plus-C and duplex OBP lineages (with categories based on the level of sequence homology and conserved amino acid signatures; Supplementary Figure S6). In Lepidoptera, the OBP gene family also includes a lineage of the PBPs, thought to transport pheromone molecules [59]. In moths, such as *M. sexta* and *B. mori*, trichoid sensilla are associated with pheromone perception and express specifically PBP-A. No *B. anynana* expressed OBP clustered in the pheromone-binding protein-A or -B lineages (Supplementary Figure S6). This is similar to findings in other butterfly species: the PBP-A lineage is lacking in the genome of *D. plexippus* and the PBP-A and PBP-B lineages are also absent from the genomes of *H. melpomene* and *M. cinxia* [59].

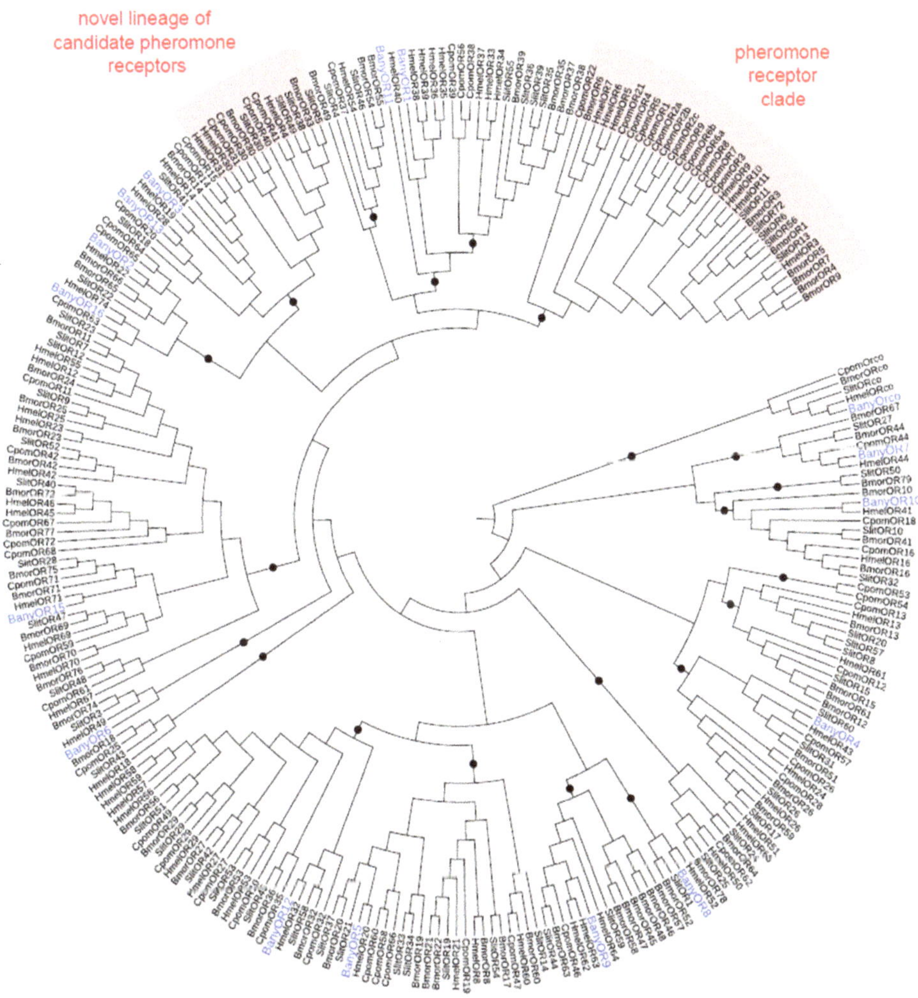

Figure 2. Maximum-likelihood phylogeny of Lepidopteran odorant receptors (OR), including the 16 ORs found in the *B. anynana* transcriptome (BanyOR, in blue). Bmor, *Bombyx mori*; Cpom, *Cydia pomonella*; Hmel, *Heliconius melpomene*; Slit, *Spodoptera littoralis*. Black circles indicate branchings highly supported by the approximate likelihood-ratio test (aLRT > 0.95).

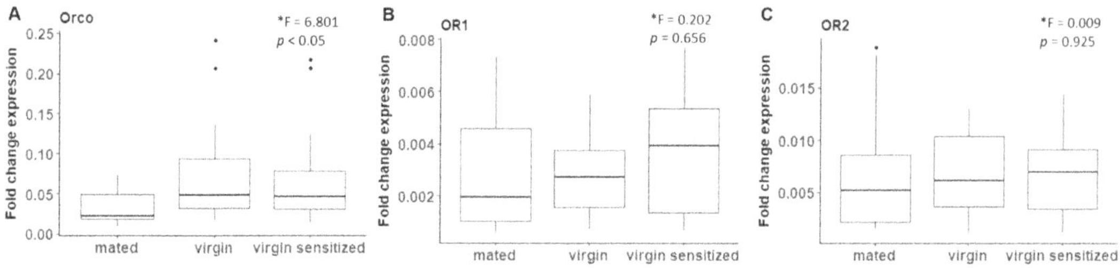

Figure 3. RT-qPCR mRNA expression level of olfactory receptors (ORs) in antennae of female *B. anynana* with different mating status. Orco (**A**), but not OR1 (**B**) or OR2 (**C**), mRNA level differed significantly in virgin naïve (middle) and virgin sensitized (right) compared to mated (left) females. Each treatment is the mean of 3 to 7 biological replicates. A nested ANOVA was used to test for differences between groups. F and p values are included for each graph. * log transformed data.

In contrast, we did find two candidate PBPs (Supplementary Table S4) expressed in *B. anynana* antennae that belong to the PBP-C and -D lineages present in all butterfly genomes investigated to date [59]. These candidate PBPs indeed correspond to the two sole candidate PBP genes identified in the *B. anynana* genome, and are both most similar to two PBPs found in the antennae of *H. melpomene* [53] (Supplementary Figure S6). In most moths, PBP-C and PBP-D OBPs are expressed in basiconic sensilla and are associated with foraging [59]. Although we cannot exclude that we missed BanOBPs in our transcriptome, the lack of a PBP-A subgene family in *B. anynana,* as in four other butterflies studied (*H. melpomene, D. plexippus, M. cinxia, P. polytes*), suggests that butterflies lost this gene lineage (at least in Nymphalidae to which the sampled species belong), and possibly use other PBP lineages to functionally aid the OR-pheromone connection. The transcriptome was also mined for Chemosensory Proteins (CSPs), a third gene family potentially implicated in olfaction in insects [85,86] (Supplementary Table S5).

3.3. *B. anynana Sex Pheromone Regulation*

Eleven contigs were found to be highly expressed in male compared to female brains, but their role in the regulation of sex pheromone processing remains open (Supplementary Table S2). Given its role as a key regulator of female sex pheromone biosynthesis in many moth species [87], we focused our attention on Pheromone Biosynthesis Activating Neuropeptide (PBAN). We hypothesized that PBAN could be involved in male sex pheromone regulation in *B. anynana,* and looked for it in our transcriptome database. We identified one unigene annotated as PBAN (BA_PBAN.1.1), which was expressed in adult heads. We used this sequence to obtain the complete cDNA sequence of PBAN in *B. anynana* (RACE), Ban_PBAN (Figure 4A). The phylogenetic reconstruction of PBAN across Lepidoptera shows monophyly of butterfly PBANs, with *B. anynana* full length PBAN encoding the typical five peptides (diapause hormone, α, β, PBAN, and γ), containing the signature FXPRL conserved amino acid sequence. We next investigated the PBAN cDNA tissue distribution using semi-quantitative and quantitative PCR. PBAN was found to be expressed in adult heads, but not in other tissues, and expression was higher in males than in females (Figure 4B). PBAN in male moths is suspected to be involved in male pheromone biosynthesis: the PBAN receptor of the moth *H. armigera* was found expressed in male hairpencils, and PBAN stimulation of the hairpencils was found to be responsible for the production and release of male pheromonal components [88]. Next, using RT-qPCR we found that PBAN expression level in male brains correlated with the amount of male sex pheromone found on male wings during the adult male's lifetime, with maximum content around 15 days of age (Figure 4C) [41].

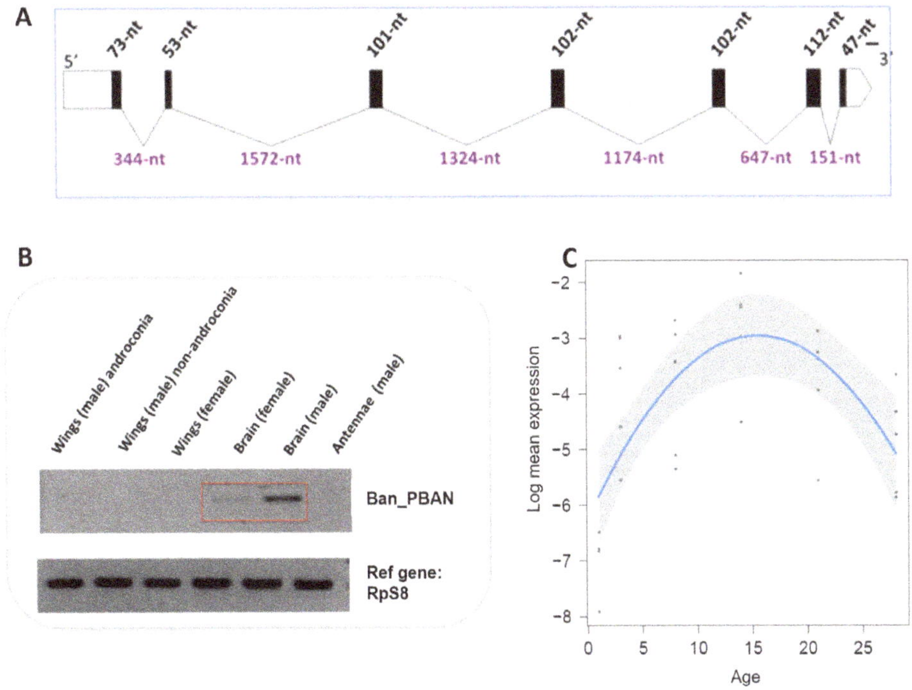

Figure 4. Pheromone biosynthesis activating neuropeptide (PBAN) expression in *B. anynana*. (**A**) Structure of Bany_PBAN full length gene sequence. The 7 exons are represented by black boxes and the 6 introns by lines. (**B**) PBAN expression level quantified by reverse transcriptase qPCR in adult tissues (brains, wings, antennae) of males and females ranging from 3 to 14 days of age. Higher levels of PBAN are observed in male brains compared to the other adult tissues. (**C**) PBAN expression level quantified by real time qPCR in adult male brains from 1 to 28 days of age.

In moths, production of volatile sex pheromones usually shows a circadian pattern that is regulated by PBAN and correlates with the female "calling" behavior (extrusion of the sex pheromone gland) during specific hours of the scotophase [87,89]. A circadian rhythm of male sex pheromone production was also found in the moth *Aphomia sabella* [90]. We tested whether *B. anynana* displayed daily variation in courtship activity, MSP production, and PBAN expression in 8-day old adult males. We found that courtship activity peaked 7 to 12 h after sunrise, and courtship activity was significantly higher in the afternoon compared to the rest of the day (Figure 5A; Supplementary Table S9). Similarly, MSP production significantly varied during the course of the day, and peaked around maximum courtship activity, with MSP1/MSP2 and MSP2/MSP3 ratios displaying significant reversed changes during the day (Figure 5D; Supplementary Table S10). MSP amounts also displayed a slight, but non-significant, variation with time of the day (Figure 5C; Supplementary Table S10). MSP titers were estimated to be minimal around 11 h after sunrise for MSP1 and MSP3, while the MSP2 titer was estimated to be at a maximum 12.4 h after sunrise. We further found that PBAN expression significantly varied throughout the day (Figure 5B; Supplementary Table S9), with the highest expression 11 to 14 h after sunrise. Daily variation in PBAN expression thus correlates both to male courtship activity and to male sex pheromone quantities found on male wings: all three traits peak during the afternoon and PBAN expression is maximal just before the peak in MSP2/MSP1 and MSP2/MSP3 ratios and MSP2 amount. This suggests that the daily regulation of male sex pheromone may be associated to circadian variation in PBAN expression, a neuropeptide that is specific to sex pheromone regulation in moths [91].

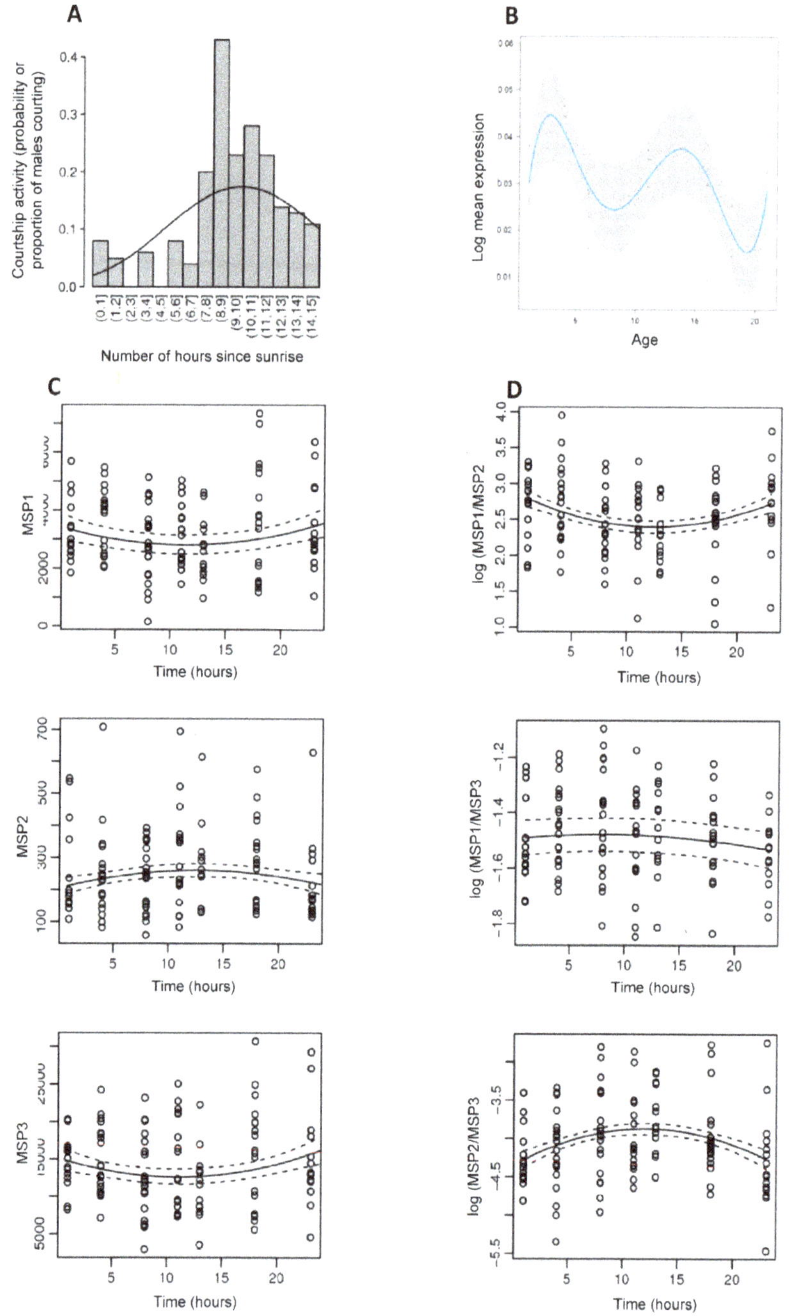

Figure 5. Daily variation in *B. anynana* male courtship activity (**A**), PBAN expression by RT-qPCR (**B**), MSP production (**C**) and log MSP ratio production (**D**). Statistics are provided in Supplementary Tables S9 and S10.

In addition to the work described above, we aimed to functionally demonstrate the role of PBAN expression in regulating male sex pheromone biosynthesis in *B. anynana*. These experiments did not lead to functional validation of the role of PBAN, but all procedures followed and results obtained are described in Supplementary File S6.

4. Conclusions and Perspectives

Mosaic evolution appears to have taken place at the molecular level based on our investigation of the pathways involved in the production, reception, and regulation of the sex pheromone in *B. anynana*. Our data suggest that the biosynthesis of the three chemical components forming the male sex pheromone (MSP1, 2, and 3) could be partly due to moth-specific genes (*far1* and *far2* for the MSP2 and MSP1 components, respectively) and partly due to genes present in insects other than moths (Δ9-desaturase, aldo-keto reductase for the MSP1 component). This is also likely the case for the MSP3 component whose synthesis is not expected to rely on moth-specific gene families, as this pheromone component is not derived from fatty acids. None of the expressed ORs or OBPs in *B. anynana* belonged to Lepidoptera-specific gene lineages responsible for sex pheromone reception in moths, suggesting that sex pheromone reception in this butterfly may have evolved independently from their moth ancestors. In contrast, we found that sex pheromone biosynthesis could be regulated by the neuropeptide PBAN in both moths and butterflies, an evolutionarily shared derived trait for Lepidoptera. Recently, the genomes of 250 species of skippers (Hesperiidae) [92] and 845 North American butterfly species [93] have been sequenced. A systematic comparative analysis of major gene families involved in moth sex pheromone communication in these ~1100 butterfly genomes would provide important information on the level of conservation of molecular pathways when butterflies diverged from moths about 119 million years ago.

Supplementary Materials: The following supporting information can be downloaded at: https://www.mdpi.com/article/10.3390/genes13081372/s1, Figure S1: Tissues used for RNA adult wing libraries.; Figure S2: Contig and read length distribution; Figure S3: RT-qPCR validation of transcriptome data comparing the relative expression level of a subset of candidate genes between treatment and control libraries; Figure S4: Phylogenetic tree of desaturases.; Figure S5: Boxplots showing the spatial distribution of MSP1 (Z9-14:OH) and fold change expression of candidate genes Δ9- and Δ11-desaturase across the *B. anynana* body.; Figure S6: Phylogenetic tree of OBP gene family from a representative set of other moths and the butterflies.; Table S1: Yield of total RNA in μg per library and per life stage; Table S2: List of contigs significantly higher expressed in treatment compared to control library(ies) for olfactory communication in *B. anynana* butterflies; Table S3: List of Odorant Receptor unigenes expressed in *B. anynana*; Table S4: List of odorant binding protein (OBP) contigs and supposed unigenes expressed in the *B. anynana* transcriptome; Table S5: List of chemosensory protein (CSP) contigs and supposed unigenes expressed in the *B. anynana* transcriptome; Table S6: Primers used for the validation of our transcriptomic study using qRT-PCR; Table S7: Overexpression of FAR-1 (fatty acyl reductase 1) in androconial male wing tissues compared to control wing tissues of males and females; Table S8: Overexpression of FAR-2 (fatty acyl reductase 2) in androconial male wing tissues compared to control wing tissues of males and females; Table S9: Daily variation in courtship activity; Table S10: Daily variation in MSP production; File S1: Phylogenetic reconstruction of odorant receptors; File S2: Phylogenetic reconstruction of odorant binding proteins; File S3: Phylogenetic reconstructions for the delta desaturase gene family; File S4: Phylogenetic reconstructions for the reductase gene family; File S5: Functional expression of *B. anynana* odorant receptor candidate genes in transgenic flies; File S6: Manipulation of male sex pheromone production with Ban_PBAN synthetic peptide.

Author Contributions: C.M.N. conceived and designed the research, collected data, analyzed data, and prepared the manuscript; P.B. designed the transcriptomic analysis and discussed the manuscript; A.A., V.B., G.S.M., N.M., E.J.-J. and L.B.-H. collected data; C.N., G.S.M., C.K. and G.L. analyzed data; B.V. collected data, analyzed data, and prepared the manuscript. All authors edited the manuscript. All authors have read and agreed to the published version of the manuscript.

Funding: CMN's team was supported by the Fonds National de la Recherche Scientifique (FNRS), FRFC grants 2.4600.10 and 2.4560.11 and CR grant 29109376, as well as UCLouvain ARC grant no 10/15-031 and FSR grant no372 605031. BV was supported by FNRS CR grant 24905063. E.J.-J., N.M. and L.B.-H. were funded by the French National Research Agency (ANR-16-CE02-0003-01 and ANR-16-CE21-0002-01 grants).

Data Availability Statement: Data are available at http://ngspipelines.toulouse.inra.fr:9011/.

Acknowledgments: We would like to thank Ken-Ichi Moto and Tetsu Ando for discussions about sex pheromone biosynthesis in Lepidoptera; Jeroen Pijpe for mentoring us about, and providing access to, a Bioanalyzer Systems (Agilent) at the LUMC hospital in Leiden (The Netherlands); Marleen van Eijk for assistance with preparing the butterfly tissues.

Conflicts of Interest: The authors declare no competing interest.

Abbreviations

MSP, male sex pheromone; OR, odorant receptor; OBP, odorant binding protein; PBAN, pheromone biosynthesis activating neuropeptide.

References

1. Cracraft, J. Mandible of Archaeopteryx Provides an Example of Mosaic Evolution. *Nature* **1970**, *226*, 1268. [CrossRef] [PubMed]
2. Gómez-Robles, A.; Hopkins, W.D.; Sherwood, C.C. Modular Structure Facilitates Mosaic Evolution of the Brain in Chimpanzees and Humans. *Nat. Commun.* **2014**, *5*, 4469. [CrossRef] [PubMed]
3. Xu, X.; Currie, P.; Pittman, M.; Xing, L.; Meng, Q.; Lü, J.; Hu, D.; Yu, C. Mosaic Evolution in an Asymmetrically Feathered Troodontid Dinosaur with Transitional Features. *Nat. Commun.* **2017**, *8*, 14972. [CrossRef] [PubMed]
4. Daeschler, E.B.; Shubin, N.H.; Jenkins, F.A. A Devonian Tetrapod-like Fish and the Evolution of the Tetrapod Body Plan. *Nature* **2006**, *440*, 757–763. [CrossRef]
5. van Gestel, J.; Ackermann, M.; Wagner, A. Microbial Life Cycles Link Global Modularity in Regulation to Mosaic Evolution. *Nat. Ecol. Evol.* **2019**, *3*, 1184–1196. [CrossRef] [PubMed]
6. Fernández, R.; Gabaldón, T. Gene Gain and Loss across the Metazoan Tree of Life. *Nat. Ecol. Evol.* **2020**, *4*, 524–533. [CrossRef]
7. Guijarro-Clarke, C.; Holland, P.W.H.; Paps, J. Widespread Patterns of Gene Loss in the Evolution of the Animal Kingdom. *Nat. Ecol. Evol.* **2020**, *4*, 519–523. [CrossRef]
8. Berner, D.; Salzburger, W. The Genomics of Organismal Diversification Illuminated by Adaptive Radiations. *Trends Genet.* **2015**, *31*, 491–499. [CrossRef]
9. Stryjewski, K.F.; Sorenson, M.D. Mosaic Genome Evolution in a Recent and Rapid Avian Radiation. *Nat. Ecol. Evol.* **2017**, *1*, 1912–1922. [CrossRef] [PubMed]
10. Marques, D.; Meier, J.; Seehausen, O. A Combinatorial View on Speciation and Adaptive Radiation. *Trends Ecol. Evol.* **2019**, *34*, 531–544. [CrossRef] [PubMed]
11. Shubin, N.; Tabin, C.; Carroll, S. Deep Homology and the Origins of Evolutionary Novelty. *Nature* **2009**, *457*, 818–823. [CrossRef]
12. Shirai, L.T.; Saenko, S.V.; Keller, R.A.; Jeránimo, M.A.; Brakefield, P.M.; Descimon, H.; Wahlberg, N.; Beldade, P. Evolutionary History of the Recruitment of Conserved Developmental Genes in Association to the Formation and Diversification of a Novel Trait. *BMC Evol. Biol.* **2012**, *12*, 1–11. [CrossRef] [PubMed]
13. Espeland, M.; Breinholt, J.; Willmott, K.R.; Warren, A.D.; Vila, R.; Toussaint, E.F.A.; Maunsell, S.C.; Aduse-Poku, K.; Talavera, G.; Eastwood, R.; et al. A Comprehensive and Dated Phylogenomic Analysis of Butterflies. *Curr. Biol.* **2018**, *28*, 770–778.e5. [CrossRef]
14. Myers, J. Pheromones and Courtship Behavior in Butterflies. *Am. Zool.* **1972**, *12*, 545–551. [CrossRef]
15. Birch, M.C.; Poppy, G.M.; Baker, T.C. Scents and Eversible Scent Structures of Male Moths. *Annu. Rev. Entomol.* **1990**, *35*, 25–58. [CrossRef]
16. Vane-Wright, R.I.; Boppre, M. Visual and Chemical Signalling in Butterflies: Functional and Phylogenetic Perspectives. *Philos. Trans. R. Soc. London B* **1993**, *340*, 197–205.
17. Nieberding, C.M.; de Vos, H.; Schneider, M.V.; Lassance, J.M.; Estramil, N.; Andersson, J.; Bång, J.; Hedenström, E.; Löfstedt, C.; Brakefield, P.M. The Male Sex Pheromone of the Butterfly Bicyclus Anynana: Towards an Evolutionary Analysis. *PLoS ONE* **2008**, *3*, e2751. [CrossRef] [PubMed]
18. Sarto Monteys, V.I.; Quero, C.; Santa-Cruz, M.C.; Rosell, G.; Guerrero, A. Sexual Communication in Day-Flying Lepidoptera with Special Reference to Castniids or "Butterfly-Moths. " *Bull. Entomol. Res.* **2016**, *106*, 421–431. [CrossRef]
19. Smadja, C.; Butlin, R.K. On the Scent of Speciation: The Chemosensory System and Its Role in Premating Isolation. *Heredity (Edinb).* **2009**, *102*, 77–97. [CrossRef]

20. Wyatt, T.D. *Pheromones and Animal Behaviour: Chemical Signals and Signatures*, 2nd ed.; Cambridge University Press: Cambridge, UK, 2014.
21. Moore, P.J.; Reagan-Wallin, N.L.; Haynes, K.F.; Moore, A.J. Odour Conveys Status on Cockroaches. *Nature* **1997**, *389*, 25. [CrossRef]
22. Svensson, M. Sexual Selection in Moths: The Role of Chemical Communication. *Biol. Rev.* **1996**, *71*, 113–135. [CrossRef]
23. Hansson, B.S.; Stensmyr, M.C. Evolution of Insect Olfaction. *Neuron* **2011**, *72*, 698–711. [CrossRef] [PubMed]
24. Tillman, J.A.; Seybold, S.J.; Jurenka, R.A.; Blomquist, G.J. Insect Pheromones—An Overview of Biosynthesis and Endocrine Regulation. *Insect Biochem. Mol. Biol.* **1999**, *29*, 481–514. [CrossRef]
25. Jurenka, R. Insect Pheromone Biosynthesis. In *The Chemistry of Pheromones and Other Semiochemicals I. Topics in Current Chemistry*; Schulz, S., Ed.; Springer: Berlin/Heidelberg, Germany, 2004; Volume 239, pp. 97–132.
26. Blomquist, G.; Jurenka, R.; Schal, C.; Tittiger, C. 12—Pheromone Production: Biochemistry and Molecular Biology. In *Insect Endocrinology*; Gilbert, L., Ed.; Academic Press: London, UK, 2012; pp. 523–567.
27. Leal, W.S. Odorant Reception in Insects: Roles of Receptors, Binding Proteins, and Degrading Enzymes. *Annu. Rev. Entomol.* **2013**, *58*, 373–391. [CrossRef]
28. Rafaeli, A. Revelations on the Regulatory Mechanisms in Moth Sex-Pheromone Signals. In *Management of Insect Pests to Agriculture: Lessons Learned from Deciphering their Genome, Transcriptome and Proteome*; Czosnek, H., Ghanim, M., Eds.; Springer: Berlin/Heidelberg, Germany, 2016; pp. 115–129, ISBN 9783319240497.
29. Zhang, J.; Walker, W.B.; Wang, G. Pheromone Reception in Moths: From Molecules to Behaviors. In *Progress in Molecular Biology and Translational Science*; Glatz, R., Ed.; Academic Press: London, UK, 2015; Volume 130, pp. 109–128.
30. Yew, J.Y.; Chung, H. Insect Pheromones: An Overview of Function, Form, and Discovery. *Prog. Lipid Res.* **2015**, *59*, 88–105. [CrossRef]
31. Helmkampf, M.; Cash, E.; Gadau, J. Evolution of the Insect Desaturase Gene Family with an Emphasis on Social Hymenoptera. *Mol. Biol. Evol.* **2015**, *32*, 456–471. [CrossRef] [PubMed]
32. Pelosi, P.; Zhou, J.J.; Ban, L.P.; Calvello, M. Soluble Proteins in Insect Chemical Communication. *Cell. Mol. Life Sci.* **2006**, *63*, 1658–1676. [CrossRef] [PubMed]
33. Vieira, F.G.; Sánchez-Gracia, A.; Rozas, J. Comparative genomic analysis of the odorant-binding protein family in 12 Drosophila genomes: Purifying selection and birth-and-death evolution. *Genome Biol.* **2007**, *8*, R235. [CrossRef]
34. de Fouchier, A.; Montagné, N.; Mirabeau, O.; Jacquin-Joly, E. Current Views on the Function and Evolution of Olfactory Receptors in Lepidoptera. *Short Views Insect Biochem. Mol. Biol. S* **2014**, *2*, 385–408.
35. Bastin-Héline, L.; de Fouchier, A.; Cao, S.; Koutroumpa, F.; Caballero-Vidal, G.; Robakiewicz, S.; Monsempes, C.; François, M.C.; Ribeyre, T.; Maria, A.; et al. A Novel Lineage of Candidate Pheromone Receptors for Sex Communication in Moths. *eLife* **2019**, *8*, e49826. [CrossRef]
36. Liénard, M.A.; Wang, H.L.; Lassance, J.M.; Löfstedt, C. Sex Pheromone Biosynthetic Pathways Are Conserved between Moths and the Butterfly Bicyclus Anynana. *Nat. Commun.* **2014**, *5*, 3957. [CrossRef] [PubMed]
37. Costanzo, K.; Monteiro, A. The Use of Chemical and Visual Cues in Female Choice in the Butterfly Bicyclus Anynana. *Proc. R. Soc. B Biol. Sci.* **2007**, *274*, 845–851. [CrossRef] [PubMed]
38. Bacquet, P.M.B.; Brattström, O.; Wang, H.L.; Allen, C.E.; Löfstedt, C.; Brakefield, P.M.; Nieberding, C.M. Selection on Male Sex Pheromone Composition Contributes to Butterfly Reproductive Isolation. *Proc. R. Soc. B Biol. Sci.* **2015**, *282*, 20142734. [CrossRef]
39. Andersson, M. *Sexual Selection*; Princeton University Press: Princeton, NJ, USA, 1994.
40. Johansson, B.G.; Jones, T.M. The Role of Chemical Communication in Mate Choice. *Biol. Rev.* **2007**, *82*, 265–289. [CrossRef] [PubMed]
41. Nieberding, C.M.; Fischer, K.; Saastamoinen, M.; Allen, C.E.; Wallin, E.A.; Hedenström, E.; Brakefield, P.M. Cracking the Olfactory Code of a Butterfly: The Scent of Ageing. *Ecol. Lett.* **2012**, *15*, 415–424. [CrossRef] [PubMed]
42. van Bergen, E.; Brakefield, P.M.; Heuskin, S.; Zwaan, B.J.; Nieberding, C.M. The Scent of Inbreeding: A Male Sex Pheromone Betrays Inbred Males. *Proc. R. Soc. B Biol. Sci.* **2013**, *280*, 20130102. [CrossRef]
43. Heuskin, S.; Vanderplanck, M.; Bacquet, P.; Holveck, M.-J.; Kaltenpoth, M.; Engl, T.; Pels, C.; Taverne, C.; Lognay, G.; Nieberding, C.M. The Composition of Cuticular Compounds Indicates Body Parts, Sex and Age in the Model Butterfly Bicyclus Anynana (Lepidoptera). *Front. Ecol. Evol.* **2014**, *2*, 37. [CrossRef]
44. Darragh, K.; Vanjari, S.; Mann, F.; Gonzalez-Rojas, M.F.; Morrison, C.R.; Salazar, C.; Pardo-Diaz, C.; Merrill, R.M.; McMillan, W.O.; Schulz, S.; et al. Male Sex Pheromone Components in *Heliconius* Butterflies Released by the Androconia Affect Female Choice. *PeerJ* **2017**, *5*, e3953. [CrossRef]
45. Brakefield, P.M.; Filali, E.E.; Van Der Laan, R.; Brueker, C.J.; Saccheri, I.J.; Zwaan, B. Effective Population Size, Reproductive Success and Sperm Precedence in the Butterfly, Bicyclus Anynana, in Captivity. *J. Evol. Biol.* **2001**, *14*, 148–156. [CrossRef] [PubMed]
46. Brakefield, P.; Beldade, P.; Zwaan, B. Dissection of Larval and Pupal Wings from the African Butterfly *Bicyclus Anynana. Cold Spring Harb. Protoc.* **2009**, *2009*. [CrossRef]
47. Jérôme, M.; Noirot, C.; Klopp, C. Assessment of Replicate Bias in 454 Pyrosequencing and a Multi-Purpose Read-Filtering Tool. *BMC Res. Notes* **2011**, *4*, 2–5. [CrossRef]

48. Briscoe, A.D.; Macias-Muñoz, A.; Kozak, K.M.; Walters, J.R.; Yuan, F.; Jamie, G.A.; Martin, S.H.; Dasmahapatra, K.K.; Ferguson, L.C.; Mallet, J.; et al. Female Behaviour Drives Expression and Evolution of Gustatory Receptors in Butterflies. *PLoS Genet.* **2013**, *9*, e1003620. [CrossRef] [PubMed]
49. van Schooten, B.; Jiggins, C.D.; Briscoe, A.D.; Papa, R. Genome-Wide Analysis of Ionotropic Receptors Provides Insight into Their Evolution in Heliconius Butterflies. *BMC Genomics* **2016**, *17*, 254. [CrossRef] [PubMed]
50. Ernst, D.A.; Westerman, E.L. Stage- and Sex-Specific Transcriptome Analyses Reveal Distinctive Sensory Gene Expression Patterns in a Butterfly. *BMC Genomics* **2021**, *22*, 584. [CrossRef] [PubMed]
51. Nowell, R.W.; Elsworth, B.; Oostra, V.; Zwaan, B.J.; Wheat, C.W.; Saastamoinen, M.; Saccheri, I.J.; van't Hof, A.E.; Wasik, B.R.; Connahs, H.; et al. A High-Coverage Draft Genome of the Mycalesine Butterfly *Bicyclus Anynana*. *Gigascience* **2017**, *6*, gix035. [CrossRef] [PubMed]
52. Xu, Y.L.; He, P.; Zhang, L.; Fang, S.Q.; Dong, S.L.; Zhang, Y.J.; Li, F. Large-Scale Identification of Odorant-Binding Proteins and Chemosensory Proteins from Expressed Sequence Tags in Insects. *BMC Genomics* **2009**, *10*, 632. [CrossRef]
53. Dasmahapatra, K.K.; Walters, J.R.; Briscoe, A.D.; Davey, J.W.; Whibley, A.; Nadeau, N.J.; Zimin, A.V.; Salazar, C.; Ferguson, L.C.; Martin, S.H.; et al. Butterfly Genome Reveals Promiscuous Exchange of Mimicry Adaptations among Species. *Nature* **2012**, *487*, 94–98.
54. Walker, W.B.; Gonzalez, F.; Garczynski, S.F.; Witzgall, P. The Chemosensory Receptors of Codling Moth *Cydia Pomonella*-Expression in Larvae and Adults. *Sci. Rep.* **2016**, *6*, 23518. [CrossRef]
55. Walker, W.B.; Roy, A.; Anderson, P.; Schlyter, F.; Hansson, B.S.; Larsson, M.C. Transcriptome Analysis of Gene Families Involved in Chemosensory Function in *Spodoptera Littoralis* (Lepidoptera: Noctuidae). *BMC Genomics* **2019**, *20*, 428. [CrossRef]
56. Guindon, S.; Dufayard, J.F.; Lefort, V.; Anisimova, M.; Hordijk, W.; Gascuel, O. New Algorithms and Methods to Estimate Maximum-Likelihood Phylogenies: Assessing the Performance of PhyML 3.0. *Syst. Biol.* **2010**, *59*, 307–321. [CrossRef]
57. Anisimova, M.; Gascuel, O. Approximate Likelihood-Ratio Test for Branches: A Fast, Accurate, and Powerful Alternative. *Syst. Biol.* **2006**, *55*, 539–552. [CrossRef] [PubMed]
58. Poivet, E.; Gallot, A.; Montagné, N.; Glaser, N.; Legeai, F.; Jacquin-Joly, E. A Comparison of the Olfactory Gene Repertoires of Adults and Larvae in the Noctuid Moth *Spodoptera Littoralis*. *PLoS ONE* **2013**, *8*, e60263. [CrossRef] [PubMed]
59. Vogt, R.G.; Große-Wilde, E.; Zhou, J.J. The Lepidoptera Odorant Binding Protein Gene Family: Gene Gain and Loss within the GOBP/PBP Complex of Moths and Butterflies. *Insect Biochem. Mol. Biol.* **2015**, *62*, 142–153. [CrossRef] [PubMed]
60. Arun, A.; Baumlé, V.; Amelot, G.; Nieberding, C.M. Selection and Validation of Reference Genes for QRT-PCR Expression Analysis of Candidate Genes Involved in Olfactory Communication in the Butterfly Bicyclus Anynana. *PLoS ONE* **2015**, *10*, e0120401. [CrossRef] [PubMed]
61. R Development Core Team. *R: A Language and Environment for Statistical Computing*; R Foundation for Statistical Computing: Vienna, Austria, 2020.
62. Legendre, P.; Legendre, L. *Numerical Ecology*, 2nd ed.; Elsevier: Amsterdam, The Netherlands, 1998.
63. Nieberding, C.M.; San Martin, G.; Saenko, S.; Allen, C.E.; Brakefield, P.M.; Visser, B. Sexual Selection Contributes to Partial Restoration of Phenotypic Robustness in a Butterfly. *Sci. Rep.* **2018**, *8*, 14315. [CrossRef] [PubMed]
64. Tupec, M.; Buček, A.; Valterová, I.; Pichová, I. Biotechnological Potential of Insect Fatty Acid-Modifying Enzymes. *Zeitschrift für Naturforschung C* **2017**, *72*, 387–403. [CrossRef]
65. Tupec, M.; Buček, A.; Janoušek, V.; Vogel, H.; Prchalová, D.; Kindl, J.; Pavlíčková, T.; Wenzelová, P.; Jahn, U.; Valterová, I.; et al. Expansion of the Fatty Acyl Reductase Gene Family Shaped Pheromone Communication in Hymenoptera. *eLife* **2019**, *8*, e39231. [CrossRef]
66. Zhang, Y.N.; Xia, Y.H.; Zhu, J.Y.; Li, S.Y.; Dong, S.L. Putative Pathway of Sex Pheromone Biosynthesis and Degradation by Expression Patterns of Genes Identified from Female Pheromone Gland and Adult Antenna of *Sesamia Inferens* (Walker). *J. Chem. Ecol.* **2014**, *40*, 439–451. [CrossRef]
67. Gu, S.H.; Wu, K.M.; Guo, Y.Y.; Pickett, J.A.; Field, L.M.; Zhou, J.J.; Zhang, Y.J. Identification of Genes Expressed in the Sex Pheromone Gland of the Black Cutworm *Agrotis Ipsilon* with Putative Roles in Sex Pheromone Biosynthesis and Transport. *BMC Genomics* **2013**, *14*, 636. [CrossRef]
68. Moto, K.; Yoshiga, T.; Yamamoto, M.; Takahashi, S.; Okano, K.; Ando, T.; Nakata, T.; Matsumoto, S. Pheromone Gland-Specific Fatty-Acyl Reductase of the Silkmoth, *Bombyx Mori*. *Proc. Natl. Acad. Sci. USA* **2003**, *100*, 9156–9161. [CrossRef]
69. Ando, T.; Inomata, S.; Yamamoto, M. Lepidopteran Sex Pheromones. In *The Chemistry of Pheromones and Other Semiochemicals I. Topics in Current Chemistry*; Schulz, S., Ed.; Springer: Berlin/Heidelberg, Germany, 2004.
70. Guo, H.; Del Corso, A.; Huang, L.Q.; Mura, U.; Pelosi, P.; Wang, C.Z. Aldehyde Reductase Activity in the Antennae of Helicoverpa Armigera. *Insect Mol. Biol.* **2014**, *23*, 330–340. [PubMed]
71. Yamamoto, K.; Higashiura, A.; Suzuki, M.; Shiotsuki, T.; Sugahara, R.; Fujii, T.; Nakagawa, A. Structural Characterization of an Aldo-Keto Reductase (AKR2E5) from the Silkworm *Bombyx Mori*. *Biochem. Biophys. Res. Commun.* **2016**, *474*, 104–110. [CrossRef]
72. Zhan, S.; Merlin, C.; Boore, J.L.; Reppert, S.M. The Monarch Butterfly Genome Yields Insights into Long-Distance Migration. *Cell* **2011**, *147*, 1171–1185. [CrossRef] [PubMed]
73. Sakurai, T.; Nakagawa, T.; Mitsuno, H.; Mori, H.; Endo, Y.; Tanoue, S.; Yasukochi, Y.; Touhara, K.; Nishioka, T. Identification and Functional Characterization of a Sex Pheromone Receptor in the Silkmoth *Bombyx Mori*. *Proc. Natl. Acad. Sci. USA* **2004**, *101*, 16653–16658. [CrossRef]

74. Zhang, D.D.; Löfstedt, C. Moth Pheromone Receptors: Gene Sequences, Function, and Evolution. *Front. Ecol. Evol.* **2015**, *3*, 105. [CrossRef]
75. Shen, S.; Cao, S.; Zhang, Z.; Kong, X.; Liu, F.; Wang, G.; Zhang, S. Evolution of Sex Pheromone Receptors in Dendrolimus Punctatus Walker (Lepidoptera: Lasiocampidae) Is Divergent from Other Moth Species. *Insect Biochem. Mol. Biol.* **2020**, *122*, 103375. [CrossRef] [PubMed]
76. Yuvaraj, J.K.; Corcoran, J.A.; Andersson, M.N.; Newcomb, R.D.; Anderbrant, O.; Löfstedt, C. Characterization of Odorant Receptors from a Non-Ditrysian Moth, Eriocrania Semipurpurella Sheds Light on the Origin of Sex Pheromone Receptors in Lepidoptera. *Mol. Biol. Evol.* **2017**, *34*, 2733–2746. [CrossRef]
77. Soques, S.; Vásquez, G.M.; Grozinger, C.M.; Gould, F. Age and Mating Status Do Not Affect Transcript Levels of Odorant Receptor Genes in Male Antennae of *Heliothis Virescens* and *Heliothis Subflexa*. *J. Chem. Ecol.* **2010**, *36*, 1226–1233. [CrossRef]
78. Saveer, A.M.; Kromann, S.H.; Birgersson, G.; Bengtsson, M.; Lindblom, T.; Balkenius, A.; Hansson, B.S.; Witzgall, P.; Becher, P.G.; Ignell, R. Floral to Green: Mating Switches Moth Olfactory Coding and Preference. *Proc. R. Soc. B Biol. Sci.* **2012**, *279*, 2314–2322. [CrossRef]
79. Zeng, F.F.; Sun, X.; Dong, H.B.; Wang, M.Q. Analysis of a CDNA Library from the Antenna of Cnaphalocrocis Medinalis and the Expression Pattern of Olfactory Genes. *Biochem. Biophys. Res. Commun.* **2013**, *433*, 463–469. [CrossRef]
80. Immonen, E.; Ritchie, M.G. The Genomic Response to Courtship Song Stimulation in Female Drosophila Melanogaster. *Proc. R. Soc. B Biol. Sci.* **2012**, *279*, 1359–1365. [CrossRef] [PubMed]
81. Latorre-Estivalis, J.M.; Omondi, B.A.; DeSouza, O.; Oliveira, I.H.R.; Ignell, R.; Lorenzo, M.G. Molecular Basis of Peripheral Olfactory Plasticity in *Rhodnius Prolixus*, a Chagas Disease Vector. *Front. Ecol. Evol.* **2015**, *3*, 74. [CrossRef]
82. Pelosi, P.; Iovinella, I.; Felicioli, A.; Dani, F.R. Soluble Proteins of Chemical Communication: An Overview across Arthropods. *Front. Physiol.* **2014**, *5*, 320. [CrossRef] [PubMed]
83. Pelosi, P.; Zhu, J.; Knoll, W. Odorant-Binding Proteins as Sensing Elements for Odour Monitoring. *Sensors* **2018**, *18*, 3248. [CrossRef] [PubMed]
84. Venthur, H.; Zhou, J.J. Odorant Receptors and Odorant-Binding Proteins as Insect Pest Control Targets: A Comparative Analysis. *Front. Physiol.* **2018**, *9*, 1163. [CrossRef]
85. Sánchez-Gracia, A.; Vieira, F.G.; Rozas, J. Molecular Evolution of the Major Chemosensory Gene Families in Insects. *Heredity (Edinb).* **2009**, *103*, 208–216. [CrossRef]
86. Vieira, F.G.; Rozas, J. Comparative Genomics of the Odorant-Binding and Chemosensory Protein Gene Families across the Arthropoda: Origin and Evolutionary History of the Chemosensory System. *Genome Biol. Evol.* **2011**, *3*, 476–490. [CrossRef]
87. Bloch, G.; Hazan, E.; Rafaeli, A. Circadian Rhythms and Endocrine Functions in Adult Insects. *J. Insect Physiol.* **2013**, *59*, 56–69. [CrossRef]
88. Bober, R., Rafaeli, A. Gene-Silencing Reveals the Functional Significance of Pheromone Biosynthesis Activating Neuropeptide Receptor (PBAN-R) in a Male Moth. *Proc. Natl. Acad. Sci. USA* **2010**, *107*, 16858–16862. [CrossRef]
89. Rosén, W.Q. Endogenous Control of Circadian Rhythms of Pheromone Production in the Turnip Moth, Agrotis Segetum. *Arch. Insect Biochem. Physiol.* **2002**, *50*, 21–30. [CrossRef]
90. Levi-Zada, A.; David, M.; Fefer, D.; Seplyarsky, V.; Sadowsky, A.; Dobrinin, S.; Ticuchinski, T.; Harari, D.; Blumberg, D.; Dunkelblum, E. Circadian Release of Male-Specific Components of the Greater Date Moth, *Aphomia (Arenipses) Sabella*, Using Sequential SPME/GC/MS Analysis. *J. Chem. Ecol.* **2014**, *40*, 236–243. [CrossRef] [PubMed]
91. Cheng, Y.; Luo, L.; Jiang, X.; Zhang, L.; Niu, C. Expression of Pheromone Biosynthesis Activating Neuropeptide and Its Receptor (PBANR) MRNA in Adult Female *Spodoptera Exigua* (Lepidoptera: Noctuidae). *Arch. Insect Biochem. Physiol.* **2010**, *75*, 13–27. [CrossRef] [PubMed]
92. Li, W.; Cong, Q.; Shen, J.; Zhang, J.; Hallwachs, W.; Janzen, D.H.; Grishin, N.V. Genomes of Skipper Butterflies Reveal Extensive Convergence of Wing Patterns. *Proc. Natl. Acad. Sci. USA* **2019**, *116*, 6232–6237. [CrossRef]
93. Zhang, J.; Cong, Q.; Shen, J.; Opler, P.; Grishin, N. Genomics of a Complete Butterfly Continent. *bioRxiv* **2019**, *845*, 829887.

genes
MDPI

Review

The Evolutionary Relevance of Social Learning and Transmission in Non-Social Arthropods with a Focus on Oviposition-Related Behaviors

Caroline M. Nieberding [1,*], Matteo Marcantonio [1], Raluca Voda [1], Thomas Enriquez [2] and Bertanne Visser [2]

[1] Evolutionary Ecology and Genetics Group, Earth and Life Institute, UCLouvain, 1348 Louvain-la-Neuve, Belgium; matteo.marcantonio@uclouvain.be (M.M.); raluca.voda@uclouvain.be (R.V.)
[2] Evolution and Ecophysiology Group, Earth and Life Institute, UCLouvain, 1348 Louvain-la-Neuve, Belgium; thomas.enriquez@uclouvain.be (T.E.); bertanne.visser@uclouvain.be (B.V.)
* Correspondence: caroline.nieberding@uclouvain.be

Abstract: Research on social learning has centered around vertebrates, but evidence is accumulating that small-brained, non-social arthropods also learn from others. Social learning can lead to social inheritance when socially acquired behaviors are transmitted to subsequent generations. Using oviposition site selection, a critical behavior for most arthropods, as an example, we first highlight the complementarities between social and classical genetic inheritance. We then discuss the relevance of studying social learning and transmission in non-social arthropods and document known cases in the literature, including examples of social learning from con- and hetero-specifics. We further highlight under which conditions social learning can be adaptive or not. We conclude that non-social arthropods and the study of oviposition behavior offer unparalleled opportunities to unravel the importance of social learning and inheritance for animal evolution.

Keywords: behavioral plasticity; communication; culture; *Drosophila*; fitness; herbivores; oviposition site selection; natural selection

Citation: Nieberding, C.M.; Marcantonio, M.; Voda, R.; Enriquez, T.; Visser, B. The Evolutionary Relevance of Social Learning and Transmission in Non-Social Arthropods with a Focus on Oviposition-Related Behaviors. *Genes* **2021**, *12*, 1466. https://doi.org/10.3390/genes12101466

Academic Editors: Luigi Viggiano and Renè Massimiliano Marsano

Received: 31 July 2021
Accepted: 21 September 2021
Published: 22 September 2021

1. Introduction

The emergence and spread of novel behaviors through social learning, or "learning from others", has been documented in a wide variety of animals, mainly in social vertebrates [1–5]. In recent years, social learning has been demonstrated to act as the "second inheritance system", called "social inheritance", that functions in parallel with classical genetic inheritance in a number of social vertebrates in the wild. Social inheritance entails the perception of behaviors performed by others that are subsequently taken over (e.g., by imitation, imprinting or teaching) and spread throughout a population and subsequent generations [6–9] (see Figure 1 depicting the steps leading to social inheritance). Aside from examples in humans, remarkable evidence for cultural evolution includes the transmission of tool use in apes and song communication in birds and whales [8,10–14].

Social vertebrates have been at the forefront of research on social learning, but studies using small-brained and short-lived social invertebrates are increasing in number. In an exceptional experiment with *Bombus terrestris* bumblebees, Alem et al. [15] showed that some individuals can innovate by acquiring a non-natural, novel behavior for feeding: string pulling. Once demonstrator individuals (previously trained to pull a string to reach a sugar source) were observed by unexperienced individuals, these bees learned how to perform string pulling themselves. The authors further showed that string pulling behavior could spread from a single experienced individual (i.e., that perceived a social cue leading to a behavioral change) to other bees, even when the original demonstrator was no longer present (completing step 1 to 4 that demonstrate social inheritance as depicted in Figure 1; [15,16]). For invertebrates, most work has been done with social insects and recent findings support the idea that insects have the cognitive abilities necessary for transmission of socially learned behaviors [17–20].

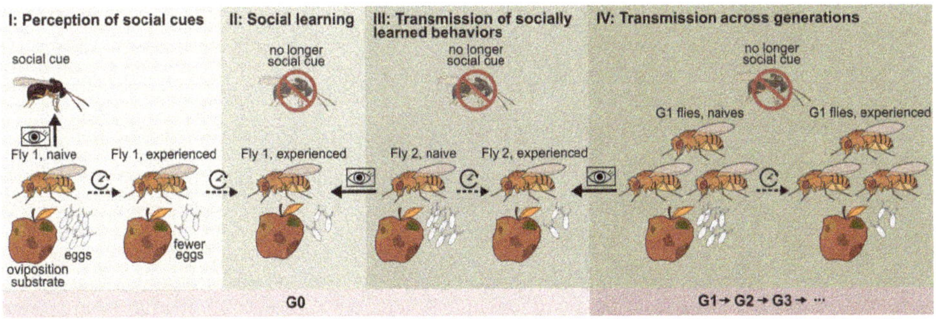

Figure 1. The steps involved in social inheritance. Step I Perception of social cues: Fly 1 perceives a social cue, e.g., the presence of a parasitic wasp that can parasitize and kill the larvae of *Drosophila melanogaster* (based on [21]). In response to the social cue, fly 1 changes its behavior, e.g., the female *D. melanogaster* reduces oviposition (fewer eggs are laid). The behavioral change proves that the cue is perceived. **Step II** Social learning: Fly 1 has learned about the social cue and is now experienced, meaning that the behavioral adjustment persists in time even when the social cue is no longer present, e.g., *D. melanogaster* females continue laying fewer eggs even when the wasp has left the patch. **Step III** Transmission: The socially learned behavior is taken over by naive fly 2 from experienced fly 1 (i.e., through visual and olfactory cues) that then changes its behavior. **Step IV** Transmission across generations: The socially learned behavior spreads throughout the population and over subsequent generations, e.g., other *Drosophila* females (including those belonging to other species) perceive the behavioral change of individuals 1 or 2 and also reduce their egg numbers (based on [22]). For social inheritance, naive flies belonging to the next generation should acquire behaviors from experienced flies exhibiting socially learned behaviors. This remains to be tested explicitly in the example of social learning of wasp threats in *Drosophila.* Of note, social inheritance can produce culture, based on additional criteria for transmission of socially learned behaviors as described in [23].

In an intricate study by Danchin et al. [23], the authors use the fly *D. melanogaster* to show that social inheritance (producing basic traditions or culture) can arise and spread throughout subsequent generations. Female *D. melanogaster* made similar mate choice decisions as the female fly they observed earlier when offered a choice between males with contrasting phenotypes (colored pink or green) themselves. Transmission of color-based mate preference also occurred when younger females observed older females, meaning that the acquired preference could spread to a potential future generation as a tradition (i.e., step 4 in Figure 1). The authors further showed that long-term memory was involved, that mate preferences can be transferred repeatedly over time, and that conformism was involved (i.e., taking over the most common behavior), leading to a stable, cultural, mate choice preference in the population. This study provides a rare example of social inheritance in non-social insects (but see [24] that consider *D. melanogaster* as moderately social; and [22,25,26] provide evidence for transmission of socially learned behaviors, step 3 in Figure 1). While the potential fitness advantages of mate-copying are clear [27,28], pink and green males do not occur in nature, meaning that there is no ecological relevance and adaptive value of the artificial cue used in this study [29].

Social inheritance may play an important role in the evolution of non-social arthropods. In this perspective, we discuss relevant examples of social learning in the context of oviposition and related behaviors to illustrate the taxonomic diversity of observations in non-social arthropods. We also highlight why studying non-social arthropods is both relevant and timely. While learning of foraging, mating, host finding and other behaviors have been discussed elsewhere [30–34], here we focus on the social transmission of oviposition site selection. Oviposition site selection is a behavioral trait of key ecological significance for the relationship between organisms and their habitat, as the decision on where to lay eggs can have massive consequences for fitness and demography ([35] and references therein). This is particularly true for herbivorous arthropods with limited mobility as juveniles, because the egg-laying site is often also the offspring's food source. Oviposition is a critical behavior with which colonization of new suitable habitats is initiated [35]. We start our perspective by illustrating the complementarities between genetic and social heredity using the hypothetical example of oviposition site selection. Next, we show that social learning related to oviposition is reported by an increasing number of studies (Table 1), adding to the ample evidence for non-social learning (i.e., learning solely from previous experience, or "autonomous" learning) for oviposition in wasps, flies, moths and butterflies (e.g., [36–44]). We then extend our discussion to cases where social learning of oviposition-related behaviors occurs not only from interactions between con-specifics, but also from hetero-specifics. Finally, we are paying particular attention to the evidence for, and quantification of, the adaptive value of social learning using existing empirical evidence for fitness effects.

2. Genetics, Epigenetics and Social Inheritance in the Context of Oviposition Site Selection

There are two non-mutually exclusive mechanisms by which socially learned behaviors can be transmitted to successive generations in a population. In his review, Whiten [7] puts forth the parallels between genetic and social inheritance, where the former encompasses genetic changes that spread throughout populations, and the latter pertains to the spread of socially learned behaviors over generations [8]. Genetic or epigenetic inheritance is based on DNA, RNA or protein materials present in the parental germ cells that are passed to the offspring when zygotes are formed. Social learning is transmitted independently from the germ line material through perception and acquisition of behaviors between individuals belonging to successive, yet overlapping generations. Genetic and social inheritance can thus function alone or interact and act simultaneously ([45–48]; see Figure 2 using oviposition site selection as an example).

There is evidence that most behaviors and behavioral variation between individuals have some genetic basis [49–53]. For example, several candidate loci were identified and associated with phenotypic variation for memorizing locations in the fly *Drosophila melanogaster* [54]. The identification of candidate loci paves the way for finding the genetic basis of complex behavioral traits, including spatial exploration ability and memory retention of spatial location (e.g., of suitable resources, including host plants for oviposition). Genetic variants with higher learning capacity and memory retention may thus become more numerous in successive generations, when there is positive selection for oviposition site selection (Figure 2). There is further evidence that learning ability itself has a genetic basis and that there is genetic variation in learning ability between individuals in various invertebrate, non-social taxa (reviewed in [54–58]). One gene whose allelic variation and expression is associated with differential learning rate and memory retention is the *foraging* gene ("*for*"), a pleiotropic gene that produces a cyclic GMP-dependent protein kinase (PKG), a protein involved in many regulatory functions, including energy homeostasis [59–62]. Although the exact function of *for* in learning (and social learning) remains to be understood [63], the existence of genetic variation for learning ability suggests that genetically "better" learners can proportionally increase in subsequent generations, for example if social learning of oviposition site selection from con-specifics is locally adaptive.

Behaviors can also be transmitted epigenetically from parents to offspring, as was found for multiple behaviors and species [64–66]. For example, mice exposed to a neutral fruity odor while receiving a mild electric shock adopt a startle behavior later in life while only experiencing the odor, a behavior that is subsequently passed on to their children and grandchildren when sensing the odor without ever experiencing the shock [67,68]. These results pointed to the fixation of epigenetic variation affecting the expression of olfactory genes [67,69–71]. There is, to the best of our knowledge, no evidence yet for epigenetic transmission of spatial localization and memory of suitable resources, as depicted in our example of Figure 2, nor for other behaviors typically related to oviposition site selection in arthropods, such as transmission of preference for novel specific host plant species across generations [68,72]. It will be important to tease apart the contribution of the genome, epigenome, and social inheritance (described below), to understand how insects track and potentially adapt to rarefying suitable habitats through oviposition site selection behavior [73].

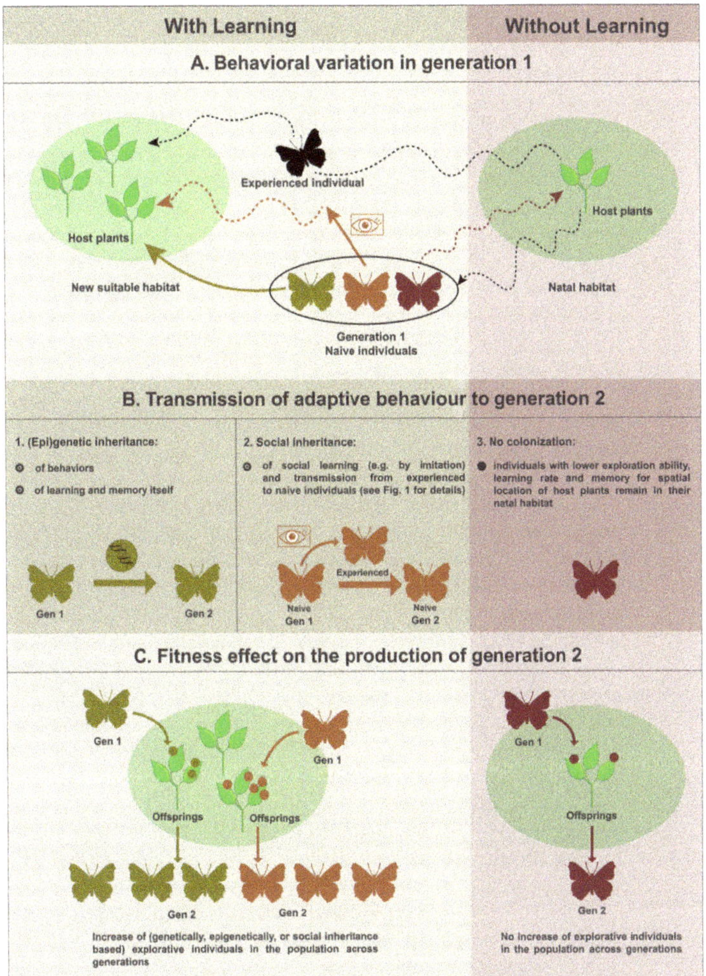

Figure 2. (Epi)genetic and social inheritance for oviposition site selection can affect the colonization of new suitable habitats with better host plant resources. **A**: Variation between individuals in oviposition site selection on host plants can be due either to (epi)genetic variation or variation in social learning skills. Social learning can lead to the colonization of new suitable habitats by naive individuals, for example by following experienced individuals towards a new habitat patch. Here, social learning is based on imitation and can occur through horizontal, oblique or (more rarely so) vertical transmission. Individuals not relying on social learning from conspecifics have a lower probability of finding new suitable habitats for oviposition. **B**: More adaptive behavioral variants for finding a new suitable habitat for oviposition can be transmitted through genetic or (epi)genetic variants (1). Transmission of social learning ability from parents to offspring can be genetically based or (epi)genetically transmitted. In addition, social learners outperform individuals not using social cues to learn about resource distribution in their environment (2). Social inheritance allows younger individuals to locate new habitats based on social information provided by older conspecifics. When there is no (epi)genetic basis for exploration, and learning and social learning does not occur, individuals have a lower probability of colonizing new habitats (3). **C**: The increasing ability of individuals within a population to learn and remember the spatial location of resources, such as host plants for oviposition, can be due to selection of (epi)genetic variants of the adaptive behavior, including learning rate and memory retention, or due to social transmission of the spatial location of resources from older to younger individuals leading to social inheritance. The accumulation of advantageous modifications of behavior in populations across generations may produce differential local adaptation between populations in socially learned traits, based on local environmental conditions and geography in much the same way as local adaptation through genetic differentiation does.

The second main inheritance mechanism, social inheritance, is based on social learning of behaviors between interacting individuals, such that learned behaviors can also be propagated without a genetic or epigenetic material basis across generations (Figures 1 and 2). Social inheritance has so far mainly been observed in social vertebrates and more recently in social insects (e.g., [15]) and non-social insects (e.g., *Drosophila;* [23]). Social learning can increase local adaptation of individuals relying on socially acquired information by increasing their chance of finding a resource, or reducing the time or energetic cost these individuals need for finding and remembering the location of a resource, such as host plants for oviposition in a new suitable habitat (i.e., oviposition site selection; Figure 2A,B). Social learners may thus have overall quicker and/or more access to suitable resources for survival and reproduction compared to conspecifics that are not using or remembering social information. This, in turn, may lead to increased reliance on social information across generations (Figure 2C), whether socially acquired traits are transmitted over longer evolutionary times and multiple generations by culture or not.

Learning the location of a suitable plant for oviposition from a skilled con-specific may represent an important evolutionary advantage compared to non-social learning of host plant location. This is because non-social learners can be in a coevolutionary arms race (i.e., Red Queen dynamics) with their host plants, given that plants are under strong selection to avoid larval feeding using elusive traits for herbivorous arthropods (e.g., a similar shape and color as non-host plants, and distinct morphologies such as "butterfly egg mimicry" or apostatic selection) [35,74,75]. Social learners can thus avoid having to "reinvent the wheel" when it comes to finding suitable host plants by following, copying or imitating others. Two key aspects of social inheritance now need to be examined and tested both in the laboratory and in the field. First, it will be important to quantify to what extent social inheritance occurs throughout the diversity of evolving life, compared to genetic inheritance (all living species have DNA or RNA and cell division), including in non-social animals. Second, quantifying the adaptive value of social learning is of central importance (as depicted in steps A and B of Figure 2), whether socially acquired traits are transmitted over longer evolutionary times, or not.

3. Relevance of Social Inheritance in Non-Social Arthropods

Socially acquired behaviors cause social inheritance only if they are transmitted over multiple generations. It is now timely to examine the extent of the transmission of socially acquired behaviors as a second inheritance system in nature (step 4 in Figure 1, Figure 2C). Small-brained, non-social invertebrates are particularly relevant to study, because they make up at least half of the species diversity on Earth [76,77]. The transmission of socially acquired behaviors across generations requires that individuals of different life stages or age groups live in contact with each other (Figure 1) [78]. For social inheritance to occur, generations must therefore be overlapping. This is indeed the case for eusocial species (i.e., with a clear reproductive division) that have overlapping generations by definition, but many non-social insects also have overlapping generations [78]. Furthermore, several insect taxa have a social population structure allowing the transmission of socially acquired behaviors over generations, through maternal, paternal and biparental care [79–81]. Maternal and biparental care takes the form of egg and/or offspring guarding, defense, nidification, and/or feeding facilitation or progressive provisioning and underpins the single most widespread form of sociality found in "non-eusocial" insects. These behaviors have been reported for >40 insect families belonging to 12 orders, as well as several non-insect arthropod groups, such as spiders, scorpions, opiliones, mites, chilopodes, and amphipod crustaceans [79]. Moreover, in a diverse array of mainly hemimetabolous arthropods, including treehoppers, true bugs, thrips, cockroaches and social spiders [82], mixed supercolonies of adults and immatures are found. While historically social inheritance has not actively been looked for in most insect taxa to date, the social structure of many insect species provides opportunities for transmission and inheritance of socially acquired behaviors far beyond the few documented cases in well-known, emblematic, social insects.

4. Social Learning of Oviposition-Related Behavior from Con- and Hetero-Specifics

Research on social learning in non-social organisms is becoming a burgeoning field and progressively more evidence is being put forward. We focus on evidence for social learning involved in oviposition behavior (Table 1; but see [30–33] for social learning of foraging, mating, and other behaviors). The first step to show evidence of social learning is that a behavior is modified in response to the perception of a social cue (step 1 in Figure 1). As a large number of studies document the existence of step 1 in various non-social arthropods, we did not include these studies in Table 1 (e.g., [35,83–93]). Historically, most studies on oviposition-related behaviors have focused on parasitoid wasps (Hymenoptera) as model systems, where oviposition takes place in or on the body of another arthropod [94]. These studies were reviewed elsewhere [34] and we only cite a few representative case studies in Table 1. Many wasps use previous experiences with a hetero-specific (i.e., the host) during development or as adults as a social cue leading to a marked change in oviposition behavior compared to naive individuals (Table 1). Table 1 summarizes the evidence of 11 key studies focusing on social learning across 4 taxonomic orders within Arthropoda: the insect orders Hymenoptera (wasps), Diptera (flies), and Coleoptera (beetles) and the arachnid order Trombidiformes (mites). We thus see that modification of oviposition in response to earlier experience of social cues occurs in diverse arthropod orders and we expect many other non-social arthropods to use social learning, with a potential for social transmission and inheritance.

Evidence for social learning of oviposition-related behaviors from con-specifics has been particularly well-documented in *Drosophila* flies (Table 1), where a typical experiment entails comparing fruit substrate preference for oviposition of flies with or without an occasion to observe "trained" congeners displaying a strong preference for a specific oviposition substrate. Training to develop a preference for a specific oviposition substrate (i.e., strawberry) is obtained by associating another substrate (i.e., banana) to an oviposition deterrent, such as quinine. Flies then develop a preference for another, simultaneously available, substrate (i.e., strawberry). Adult female flies further learn to interpret and use a wide variety of cues from con-specifics at different life stages when choosing an oviposition site. Visual cues, such as the presence of con-specific eggs and/or larvae on oviposition substrates, interactions with more experienced female demonstrators, as well olfactory cues produced by con-specifics have been shown to positively influence female oviposition decisions after the original cue has been removed. This suggests that the benefits of con-specific attraction in oviposition site selection may outweigh the costs of competition in the wild [85,86]. In the context of research on social learning in *Drosophila*, the large knowledge-base on cues used for oviposition site selection, as well as the documented evidence for social learning (Table 1), make it an excellent model for testing whether social learning of oviposition sites can be inherited socially.

Acquiring social information from other species can be an efficient way to increase fitness. This is particularly true for non-social insects with limited access to information from con-specifics (such as for early dispersers, insects with small population sizes, and/or species with low con-specific encounter rates. Such species can use information from other species sharing aspects of their ecological niche to make nest choice decisions [95]. An interesting example of hetero-specific social learning can be found in the parasitic wasp *Trichogramma evanescens* [96]. Like its congener *T. brassicae*, this wasp uses the pheromones of its adult host, the butterfly *Pieris brassicae* to identify mated females that will subsequently lay eggs suitable for parasitism by the wasp. By using this information, the wasp will hitch-hike along for the ride to a new oviposition opportunity (i.e., the egg laying site of *P. brassicae*), but unlike *T. brassicae, T. evanescens* needs to learn through an oviposition experience that both host pheromones (to identify adult hosts) and hitch-hiking (towards host eggs) lead to a suitable oviposition site [96]. Several solitary bee species provide another example of social learning from hetero-specifics [95]. The cavity-nesting mason bees, *Osmia caerulescens* and *O. leaiana* examine the nests of another congener, *O. bicornis*, for evidence of brood cell parasites. Though associative learning of nest site quality of congeners (using geometric symbols), *O. caerulescens* and *O. leaiana* preferred to start their own nest at sites associated with healthy nests of *O. bicornis* and rejected sites associated with brood cell parasites. This study is exceptional, because observations and experiments were conducted in the field using wild bees, although the possibility that nest selection behavior is innate and not due to social learning could not be ruled out completely [95].

The value of social information from hetero-specifics has also been studied in *Drosophila*. Particularly noteworthy is the flow of social information in the genus *Drosophila* related to the presence of a parasitoid observed by Kacsoh and co-authors [22]. The divergence in social cues that evolved between different species led to the formation of species-specific communication patterns (referred to as "dialects"). The magnitude of divergence in species-specific communication patterns was found to be correlated with the phylogenetic distance between species. Kacsoh et al. [22] exploited this system to test whether the degree of hetero-specific social information transfer between *Drosophila* species was related to their relative phylogenetic distance, hypothesizing that phylogenetically close species are more successful in sharing social information. Similar to earlier experiments by Kacsoh et al. [21] (Figure 1), *Drosophila* females were presented with visual cues of parasitic wasps that led to a reduction in the number of eggs laid. When the experienced fly belonged to a different species, Kacsoh et al. [22] observed the same decrease in number of eggs laid. While closely related *Drosophila* species were able to efficiently communicate information about the presence of the parasitoid, species that were phylogenetically more distant had limited to no communication abilities. Interestingly, multi-species communities enhanced inter-specific communication, allowing *Drosophila* to learn multiple dialects. This indicates a degree of plasticity in learning abilities that could be adaptive in nature when *Drosophila* species occur in sympatry [22]. This study represents a rare empirical test for socially learned behaviors can be transmitted to others in non-social invertebrates (i.e., up to step 3 in Figure 1).

Evidence for social learning has been based on at least three experimental setups: some studies compare the behavior of individuals before (test a), during (test b) and after (test c) experiencing the social cue. Evidence for social learning becomes apparent when the behaviors observed in tests b and c are similar, but different from the behavior displayed in test a. Another, better design, takes ageing (and its potential confounding effect) into account by comparing groups of naive individuals with experienced individuals (that had an earlier experience with the social cue) of similar age. The behaviors of the naive and experienced groups should differ in the absence of the social cue to show evidence of social learning in the experienced group. A third setup consists of associating a social cue to another cue (that does not need to be social, i.e., color, symbols etc.), and comparing the behavior of a group of naive individuals with a group that experienced the social *and* the associated cue, in the presence of only the associated cue. Evidence for social learning is then based on a significant difference in behavior between the two groups in the presence of the associated (but not the social) cue, for the experienced group. These experimental set-ups, when carefully designed, allow to discriminate beyond any doubt socially learned behaviors from behaviors that are innate or learned as consequence of interactions with abiotic cues. As such, they provide excellent opportunities to study social learning, transmission, and inheritance of oviposition-related (and other) behaviors in a wide range of non-social arthropods. In light of the accumulating evidence for widespread social learning, these experimental designs can greatly contribute to our understanding of the role of social learning in evolution.

Table 1. List of studies on non-social arthropods where social cue perception, social learning, and transmission of socially learned oviposition-related behaviors was quantified. Only studies that document social learning are included (i.e., from step 2 of Figure 1 onwards), as there is a large body of literature covering cue perception (i.e., step 1 of Figure 1). The table includes the species, the order (Diptera = D, Hymenoptera = H, C = Coleoptera, Trombidiformes = T), the type of social cue and the behavior under study, con- (c) or hetero- (h) specific social learning, the steps towards social inheritance (as in Figure 1) and if effects on fitness were quantified in the study. Studies concerned with foraging, mating, host finding and other behaviors, including in non-insect invertebrates, have been discussed elsewhere [30–34].

Species	Order	Social Cue	Behavior	Learning from con- (c) or Hetero- (h) Specifics	Step Towards Social Inheritance	Fitness Tested	Reference
D. melanogaster	D	Experienced females with preferred oviposition site	Site selection	c	1, 2, 3	y	[25]
D. melanogaster	D	Parasitoid presence (i.e., threat to offspring survival)	Clutch size	c	1, 2	y	[21]
Drosophila spp.	D	Parasitoid presence (i.e., threat to offspring survival)	Clutch size	c + h	1, 2, 3	y	[22]
D. melanogaster	D	Mated females	Site selection	c	1, 2	y	[97]
Leptopilina boulardi	H	Host insect	Site selection	h	1, 2	n	[98]
Necremnus tutae	H	Host insect and plant species	Host species preference	h	1, 2	n	[99]
Osmia sp.*	H	Nest site parasitism	Site selection	h	1, 2	n	[95]
Trichogramma evanescens	H	Host adult and eggs	Phoresy to oviposition substrate	h	1, 2	n	[96]
Anisopteromalus calandrae	H	Host insect	Host preference + host-finding + parasitism rates	h	1, 2	y	[100]
Phratora vulgatissima	C	Adult females	Distance between clutches	c	1, 2	y	[101]
Tetranychus urticae, T. kanzawai	T	Predator	Site selection (leaf surface vs web)	h	1, 2	n	[102]

* Tested under field conditions.

Evidence for hetero-specific social learning has also been found for behaviors other than oviposition. Social learning in non-social arthropods was first reported in a cricket, *Nemobius sylvestris*, that changed its predator avoidance behavior based on observations, and memory of such observations, of either predator presence (spiders) or of congener crickets that had already experienced the presence of spiders [26]. Hetero-specific social information can thus also be transmitted from experienced to naive crickets [26], which can decrease predation risk. Hetero-specific social information was also found to increase the efficiency of locating food sources [103–106]. Although social information from hetero-specifics is ubiquitous, it can be challenging to decode, for example because the cue may have had a different original meaning or purpose than what is interpreted by the receiving species [107–110].

5. The Adaptive Value of Social Learning

Social learning is an important mechanism in evolution even when transmission of socially acquired behaviors is limited to a few generations within a season, such that social inheritance will not be maintained over long evolutionary times (step 4 in Figure 1). Indeed, we suggest that building expertise during a lifetime by social experiences can increase the adaptation rate of populations that are using and memorizing social information, for example for the spatial location of essential resources, even if every adult individual dies at the end of the reproductive season. This is, for example, because social information allows individuals to avoid unfavorable oviposition sites, to reach an oviposition site earlier or at lower exploratory costs, compared to individuals that explore and spatially navigate without this information. In this regard, most current evidence for social learning, including in non-social insects, concerns behaviors such as foraging and host location, which are based on resources that vary rapidly in space and time notably due to seasonal changes. Related social information is thus of ephemeral relevance as well and it needs to be updated constantly, suppressing the emergence of any form of longer-term social inheritance. Rupture of socially transmitted behaviors can also take place because most representatives of insect populations die seasonally, for example during winter in temperate regions. In the latter case, social information about resources can be acquired and exchanged socially *de novo* at the beginning of the new reproductive season each year, starting from newly emerged naive individuals in spring that learn about resource distribution in their surrounding environment.

The adaptive value of learned behaviors is documented in some vertebrates [4,5], but experimental evidence for the adaptive value of socially learned behaviors in ecologically relevant conditions currently remains unquantified for the vast majority of living taxa [17,111], including non-social insects [112]. Social learning can increase the fitness of individuals and as such be under positive selection in rapidly changing environments. Yet, this is not necessarily the case as negative effects on fitness were documented from partially or incorrectly interpreted social cues that caused increased energy expenditure in basic tasks, such as foraging [108]. The costs associated with social learning, including energetic costs and time constraints, and the environmental parameters under which social learning becomes adaptive, have been explored both experimentally [113] and through modeling work [114,115]. These studies have revealed that social learning is not necessarily adaptive under all conditions and that learning can lead to evolutionary traps under rapidly changing environmental conditions [116].

A study with *D. melanogaster* convincingly suggested that social learning has adaptive value also in the context of oviposition-related behaviors in non-social insects [21]. Here, the authors exposed ovipositing *D. melanogaster* females to a parasitoid wasp that lays eggs inside *D. melanogaster* larvae, which are subsequently consumed from the inside out by the developing parasitoid. Having been faced with a serious threat to the survival of their offspring [117], female *D. melanogaster* reduced the number of eggs laid in the next clutch [118]. When a parasitoid-experienced fly was then observed by a naive female fly, the latter also reduced her clutch size, even though the original social cue, the wasp, was no longer present [21]. Within an ecological context, reducing egg numbers in the face of an immediate threat to offspring survival can have a clear adaptive value, also under natural conditions. Indeed the wasp species used in this study actively searches for host patches in the environment [119,120], where mating, oviposition or feeding *Drosophila* larvae generate perceivable olfactory cues for the wasp [121]. It remains to be tested whether social learning in *D. melanogaster* females can be transmitted from one generation to the next (as was found in [23]).

6. Conclusions

Perception of social cues, social learning and transmission are the stepping stones towards social inheritance. While perception of social cues is now well known to induce behavioral changes in multiple arthropods (e.g., [35,83–93]), we need to increase our understanding of social learning in non-social arthropods and determine its prevalence, both in the laboratory and in the field. Due to its inherent link to fitness, oviposition site selection offers unparalleled opportunities to study social learning and transmission, also in systems other than *Drosophila*. The increasing number of studies on social learning in non-social arthropods (see Table 1) offer promising possibilities for empirical tests of social transmission and inheritance.

Author Contributions: Conceptualization, C.M.N. and B.V.; writing—original draft preparation, C.M.N., M.M., R.V., T.E., B.V.; writing—review and editing, C.M.N., M.M., R.V., T.E., B.V.; visualization, C.M.N., R.V., T.E.; supervision, C.M.N.; project administration, C.M.N.; funding acquisition, C.M.N., T.E. All authors have read and agreed to the published version of the manuscript.

Funding: This research was funded by UCLouvain, the Fédération Wallonie-Bruxelles, and by the Fonds National de Recherche Scientifique (F.R.S.-FNRS). R.V. was supported by the F.R.S.-FNRS grant number T.0169.21 of C.M.N.; M.M. was supported by the "Action de Recherche concertée" grant number 17/22-086 of C.M.N. TE. was funded by the Fyssen foundation.

Institutional Review Board Statement: Not applicable.

Informed Consent Statement: Not applicable.

Data Availability Statement: Not applicable.

Conflicts of Interest: The authors declare no conflict of interest.

References

1. Whiten, A.; Caldwell, C.A.; Mesoudi, A. Cultural diffusion in humans and other animals. *Curr. Opin. Psychol.* **2016**, *8*, 15–21. [CrossRef] [PubMed]
2. Fisher, J.; Hinde, R. Opening of milk bottles by birds. *Br. Birds* **1950**, *42*, 347–357. [CrossRef]
3. Heyes, C.M.; Street, G. Social learning in animals: Categories and mechanisms. *Biol. Rev.* **1994**, *69*, 207–231. [CrossRef] [PubMed]
4. Morand-Ferron, J. Why learn? The adaptive value of associative learning in wild populations. *Curr. Opin. Behav. Sci.* **2017**, *16*, 73–79. [CrossRef]
5. Morand-Ferron, J.; Cole, E.F.; Quinn, J.L. Studying the evolutionary ecology of cognition in the wild: A review of practical and conceptual challenges. *Biol. Rev.* **2016**, *91*, 367–389. [CrossRef]
6. Baldwin, J. A new factor in evolution. *Am. Nat.* **1896**, *30*, 441–451. [CrossRef]
7. Whiten, A. A second inheritance system: The extension of biology through culture. *Interface Focus* **2017**, *7*, 20160142. [CrossRef]
8. Whiten, A. Culture extends the scope of evolutionary biology in the great apes. *Proc. Natl. Acad. Sci. USA* **2017**, *114*, 7790–7797. [CrossRef]
9. Hoppitt, W.; Laland, K. *Social Learning: An Introduction to Mechanisms, Methods, and Models*; Princeton University Press: Princeton, NJ, USA, 2013.
10. Slater, P.J.B. The cultural transmission of bird song. *Trends Ecol. Evol.* **1986**, *1*, 94–97. [CrossRef]
11. Deecke, V.B.; Ford, J.K.B.; Spong, P. Dialect change in resident killer whales: Implications for vocal learning and cultural transmission. *Anim. Behav.* **2000**, *60*, 629–638. [CrossRef]
12. Garland, E.C.; Goldizen, A.W.; Rekdahl, M.L.; Constantine, R.; Garrigue, C.; Hauser, N.D.; Poole, M.M.; Robbins, J.; Noad, M.J. Dynamic horizontal cultural transmission of humpback whale song at the ocean basin scale. *Curr. Biol.* **2011**, *21*, 687–691. [CrossRef]
13. Cantor, M.; Whitehead, H. The interplay between social networks and culture: Theoretically and among whales and dolphins. *Philos. Trans. R. Soc. B Biol. Sci.* **2013**, *368*, 20120340. [CrossRef]

14. Lamon, N.; Neumann, C.; Gruber, T.; Zuberbühler, K. Kin-based cultural transmission of tool use in wild chimpanzees. *Sci. Adv.* **2017**, *3*, 1–10. [CrossRef]
15. Alem, S.; Perry, C.J.; Zhu, X.; Loukola, O.J.; Ingraham, T.; Søvik, E.; Chittka, L. Associative Mechanisms Allow for Social Learning and Cultural Transmission of String Pulling in an Insect. *PLoS Biol.* **2016**, *14*, e1002564. [CrossRef]
16. Nieberding, C.M.; van Alphen, J.J. Culture in bumblebees. *Peer Community Evol. Biol.* **2017**, 2–4. [CrossRef]
17. Grüter, C.; Leadbeater, E. Insights from insects about adaptive social information use. *Trends Ecol. Evol.* **2014**, *29*, 177–184. [CrossRef]
18. Avarguès-Weber, A.; Lihoreau, M.; Isabel, G.; Giurfa, M. Information transfer beyond the waggle dance: Observational learning in bees and flies. *Front. Ecol. Evol.* **2015**, *3*, 1–7. [CrossRef]
19. Worden, B.D.; Papaj, D.R. Flower choice copying in bumblebees. *Biol. Lett.* **2005**, *1*, 504–507. [CrossRef]
20. Jones, P.L.; Agrawal, A.A. Learning in Insect Pollinators and Herbivores. *Annu. Rev. Entomol.* **2017**, *62*, 53–71. [CrossRef]
21. Kacsoh, B.Z.; Bozler, J.; Ramaswami, M.; Bosco, G. Social communication of predator-induced changes in *Drosophila* behavior and germline physiology. *Elife* **2015**, *4*, 1–36. [CrossRef]
22. Kacsoh, B.; Bozler, J.; Bosco, G. *Drosophila* species learn dialects through communal living. *PLoS Genet.* **2018**, *14*, e1007430. [CrossRef] [PubMed]
23. Danchin, É.; Nobel, S.; Pocheville, A.; Dagaeff, A.-C.; Demay, L.; Alphand, M.; Ranty-Roby, S.; van Renssen, L.; Monier, M.; Gazagne, E.; et al. Cultural flies: Conformist social learning in fruitflies predicts long-lasting mate-choice traditions. *Science* **2019**, *366*, 1–7. [CrossRef]
24. Durisko, Z.; Dukas, R. Attraction to and learning from social cues in fruitfly larvae. *Proc. R. Soc. B Biol. Sci.* **2013**, *280*, 1–7. [CrossRef] [PubMed]
25. Battesti, M.; Moreno, C.; Joly, D.; Mery, F. Spread of social information and dynamics of social transmission within *Drosophila* groups. *Curr. Biol.* **2012**, *22*, 309–313. [CrossRef]
26. Coolen, I.; Dangles, O.; Casas, J. Social learning in noncolonial insects? *Curr. Biol.* **2005**, *15*, 1931–1935. [CrossRef]
27. Davies, A.D.; Lewis, Z.; Dougherty, L.R. A meta-analysis of factors influencing the strength of mate-choice copying in animals. *Behav. Ecol.* **2020**, *31*, 1279–1290. [CrossRef]
28. Mery, F.; Varela, S.A.M.; Danchin, É.; Blanchet, S.; Parejo, D.; Coolen, I.; Wagner, R.H. Public versus personal information for mate copying in an invertebrate. *Curr. Biol.* **2009**, *19*, 730–734. [CrossRef]
29. Belkina, E.G.; Shiglik, A.; Sopilko, N.G.; Lysenkov, S.N.; Markov, A.V. Mate choice copying in *Drosophila* is probably less robust than previously suggested. *Anim. Behav.* **2021**, *176*, 175–183. [CrossRef]
30. Dion, E.; Monteiro, A.; Nieberding, C.M. The role of learning on insect and spider sexual behaviors, sexual trait evolution, and speciation. *Front. Ecol. Evol.* **2019**, *6*, 225. [CrossRef]
31. Dukas, R. Evolutionary biology of insect learning. *Annu. Rev. Entomol.* **2008**, *53*, 145–160. [CrossRef]
32. Wright, G.A.; Schiestl, F.P. The evolution of floral scent: The influence of olfactory learning by insect pollinators on the honest signalling of floral rewards. *Funct. Ecol.* **2009**, *23*, 841–851. [CrossRef]
33. Webster, S.J.; Fiorito, G. Socially guided behaviour in non-insect invertebrates. *Anim. Cogn.* **2001**, *4*, 69–79. [CrossRef]
34. Hoedjes, K.M.; Kruidhof, H.M.; Huigens, M.E.; Dicke, M.; Vet, L.E.M.; Smid, H.M. Natural variation in learning rate and memory dynamics in parasitoid wasps: Opportunities for converging ecology and neuroscience. *Proc. R. Soc. B Biol. Sci.* **2011**, *278*, 889–897. [CrossRef]
35. Jones, P.L.; Agrawal, A.A. Beyond preference and performance: Host plant selection by monarch butterflies, *Danaus plexippus*. *Oikos* **2019**, *128*, 1092–1102. [CrossRef]
36. Traynier, R.M.M. Associative learning in the ovipositional behaviour of the cabbage butterfly, *Pieris rapae*. *Physiol. Entomol.* **1984**, *9*, 465–472. [CrossRef]
37. Papaj, D.R. Interpopulation differences in host preference and the evolution of learning in the butterfly, *Battus philenor*. *Evolution* **1986**, *40*, 518–530. [CrossRef]
38. Traynier, R.M.M. Visual learning in assays of sinigrin solution as an oviposition releaser for the cabbage butterfly, *Pieris rapae*. *Entomol. Exp. Appl.* **1986**, *40*, 25–33. [CrossRef]
39. Visser, M.E.; van Alphen, J.J.; Hemerik, L. Adaptive superparasitism and patch time allocation in solitary parasitoids: An ESS model. *J. Anim. Ecol.* **1992**, *61*, 93–101. [CrossRef]
40. Vet, L.E.M.; De Jong, A.G.; Franchi, E.; Papaj, D.R. The effect of complete versus incomplete information on odour discrimination in a parasitic wasp. *Anim. Behav.* **1998**, *55*, 1271–1279. [CrossRef]
41. Mery, F.; Kawecki, T.J. Experimental evolution of learning ability in fruit flies. *Proc. Natl. Acad. Sci. USA* **2002**, *99*, 14274–14279. [CrossRef]
42. Liu, S.S.; Li, Y.H.; Liu, Y.Q.; Zalucki, M.P. Experience-induced preference for oviposition repellents derived from a non-host plant by a specialist herbivore. *Ecol. Lett.* **2005**, *8*, 722–729. [CrossRef]
43. Braem, S.; Turlure, C.; Nieberding, C.; van Dyck, H. Oviposition site selection and learning in a butterfly under niche expansion: An experimental test. *Anim. Behav.* (In press). **2021**. [CrossRef]
44. Kawecki, T.J. Evolutionary ecology of learning: Insights from fruit flies. *Popul. Ecol.* **2010**, *52*, 15–25. [CrossRef]
45. Feldman, M.W.; Laland, K.N. Gene-culture coevolutionary theory. *Trends Ecol. Evol.* **1996**, *11*, 453–457. [CrossRef]

46. Danchin, É.; Wagner, R.H. Inclusive heritability: Combining genetic and non-genetic information to study animal behavior and culture. *Oikos* **2010**, *119*, 210–218. [CrossRef]
47. Mesoudi, A.; Chang, L.; Dall, S.R.X.; Thornton, A. The evolution of individual and cultural variation in social learning. *Trends Ecol. Evol.* **2016**, *31*, 215–225. [CrossRef]
48. Danchin, E.; Pocheville, A.; Rey, O.; Pujol, B.; Blanchet, S. Epigenetically facilitated mutational assimilation: Epigenetics as a hub within the inclusive evolutionary synthesis. *Biol. Rev.* **2019**, *94*, 259–282. [CrossRef]
49. Fitzpatrick, M.J.; Ben-Shahar, Y.; Smid, H.M.; Vet, L.E.M.; Robinson, G.E.; Sokolowski, M.B. Candidate genes for behavioural ecology. *Trends Ecol. Evol.* **2005**, *20*, 96–104. [CrossRef] [PubMed]
50. Reaume, C.J.; Sokolowski, M.B. Conservation of gene function in behaviour. *Philos. Trans. R. Soc. B Biol. Sci.* **2011**, *366*, 2100–2110. [CrossRef] [PubMed]
51. Bengston, S.E.; Dahan, R.A.; Donaldson, Z.; Phelps, S.M.; Van Oers, K.; Sih, A.; Bell, A.M. Genomic tools for behavioural ecologists to understand repeatable individual differences in behaviour. *Nat. Ecol. Evol.* **2018**, *2*, 944–955. [CrossRef] [PubMed]
52. Henriksen, R.; Höglund, A.; Fogelholm, J.; Abbey-Lee, R.; Johnsson, M.; Dingemanse, N.J.; Wright, D. Intra-individual behavioural variability: A trait under genetic control. *Int. J. Mol. Sci.* **2020**, *21*, 8069. [CrossRef]
53. Bubac, C.M.; Miller, J.M.; Coltman, D.W. The genetic basis of animal behavioural diversity in natural populations. *Mol. Ecol.* **2020**, *29*, 1957–1971. [CrossRef]
54. Williams-Simon, P.A.; Posey, C.; Mitchell, S.; Ng'oma, E.; Mrkvicka, J.A.; Zars, T.; King, E.G. Multiple genetic loci affect place learning and memory performance in *Drosophila melanogaster*. *Genes Brain Behav.* **2019**, *18*, 1–16. [CrossRef]
55. Mery, F. Natural variation in learning and memory. *Curr. Opin. Neurobiol.* **2013**, *23*, 52–56. [CrossRef]
56. Hughes, E.; Shymansky, T.; Swinton, E.; Lukowiak, K.S.; Swinton, C.; Sunada, H.; Protheroe, A.; Phillips, I.; Lukowiak, K. Strain-specific differences of the effects of stress on memory in *Lymnaea*. *J. Exp. Biol.* **2017**, *220*, 891–899. [CrossRef]
57. Giunti, G.; Canale, A.; Messing, R.H.; Donati, E.; Stefanini, C.; Michaud, J.P.; Benelli, G. Parasitoid learning: Current knowledge and implications for biological control. *Biol. Control* **2015**, *90*, 208–219. [CrossRef]
58. Liefting, M.; Verwoerd, L.; Dekker, M.L.; Hoedjes, K.M.; Ellers, J. Strain differences rather than species differences contribute to variation in associative learning ability in *Nasonia*. *Anim. Behav.* **2020**, *168*, 25–31. [CrossRef]
59. Osborne, K.A.; Robichon, A.; Burgess, E.; Butland, S.; Shaw, R.A.; Coulthard, A.; Pereira, H.S.; Greenspan, R.J.; Sokolowski, M.B. Natural behavior polymorphism due to a cGMP-dependent protein kinase of *Drosophila*. *Science* **1997**, *277*, 834–836. [CrossRef]
60. Sokolowski, M.B. *Drosophila*: Genetics meets behaviour. *Nat. Rev. Genet.* **2001**, *2*, 879–890. [CrossRef]
61. Wahlberg, N.; Wheat, C.W.; Peña, C. Timing and patterns in the taxonomic diversification of Lepidoptera (butterflies and moths). *PLoS ONE* **2013**, *8*, e80875. [CrossRef]
62. Wheat, C.W. Dispersal genetics: Emerging insights from fruitflies, butterflies, and beyond. In *Dispersal Ecology and Evolution*; Clobert, J., Baguette, M., Benton, T., Bullock, J., Eds.; Oxford University Press: Oxford, UK, 2012; p. 498.
63. Fitzpatrick, M.J.; Sokolowski, M.B. In search of food: Exploring the evolutionary link between cGMP-dependent protein kinase (PKG) and behaviour. *Integr. Comp. Biol.* **2004**, *44*, 28–36. [CrossRef] [PubMed]
64. Gapp, K.; Jawaid, A.; Sarkies, P.; Bohacek, J.; Pelczar, P.; Prados, J.; Farinelli, L.; Miska, E.; Mansuy, I.M. Implication of sperm RNAs in transgenerational inheritance of the effects of early trauma in mice. *Nat. Neurosci.* **2014**, *17*, 667–669. [CrossRef] [PubMed]
65. Charlesworth, A.G.; Seroussi, U.; Claycomb, J.M. Next-Gen Learning: The *C. elegans* Approach. *Cell* **2019**, *177*, 1674–1676. [CrossRef] [PubMed]
66. Ledón-Rettig, C.C.; Richards, C.L.; Martin, L.B. Epigenetics for behavioral ecologists. *Behav. Ecol.* **2013**, *24*, 311–324. [CrossRef]
67. Dias, B.G.; Ressler, K.J. Parental olfactory experience influences behavior and neural structure in subsequent generations. *Nat. Neurosci.* **2014**, *17*, 89–96. [CrossRef]
68. Gowri, V.; Dion, E.; Viswanath, A.; Piel, F.M.; Monteiro, A. Transgenerational inheritance of learned preferences for novel host plant odors in *Bicyclus anynana* butterflies. *Evolution* **2019**, *73*, 2401–2414. [CrossRef]
69. Zhang, Y.; Lu, H.; Bargmann, C.I. Pathogenic bacteria induce aversive olfactory learning in *Caenorhabditis elegans*. *Nature* **2005**, *438*, 179–184. [CrossRef]
70. Moore, R.S.; Kaletsky, R.; Murphy, C.T. Piwi/PRG-1 Argonaute and TGF-β mediate transgenerational learned pathogenic avoidance. *Cell* **2019**, *177*, 1827–1841.e12. [CrossRef]
71. Posner, R.; Toker, I.A.; Antonova, O.; Star, E.; Anava, S.; Azmon, E.; Hendricks, M.; Bracha, S.; Gingold, H.; Rechavi, O. Neuronal small RNAs control behavior transgenerationally. *Cell* **2019**, *177*, 1814–1826.e15. [CrossRef]
72. Rösvik, A.; Lhomme, P.; Khallaf, M.A.; Anderson, P. Plant-induced transgenerational plasticity affecting performance but not preference in a polyphagous moth. *Front. Ecol. Evol.* **2020**, *8*, 1–9. [CrossRef]
73. Barrett, L.P.; Stanton, L.A.; Benson-Amram, S. The cognition of 'nuisance' species. *Anim. Behav.* **2019**, *147*, 167–177. [CrossRef]
74. Rausher, M.D. Search image for leaf shape in a butterfly. *Science* **1978**, *200*, 1071–1073. [CrossRef]
75. Williams, K.S.; Gilbert, L.E. Insects as selective agents on plant vegetative morphology: Egg mimicry reduces egg laying by butterflies. *Science* **1981**, *212*, 467–469. [CrossRef]
76. Bar-On, Y.M.; Phillips, R.; Milo, R. The biomass distribution on Earth. *Proc. Natl. Acad. Sci. USA* **2018**, *115*, 6506. [CrossRef]
77. Mora, C.; Tittensor, D.P.; Adl, S.; Simpson, A.G.B.; Worm, B. How many species are there on earth and in the ocean? *PLoS Biol.* **2011**, *9*, e1001127. [CrossRef]

78. Ellner, S.; Hairston, N.G., Jr. Role of overlapping generations in maintaining genetic variation in a fluctuating environment. *Am. Nat.* **1994**, *143*, 403–417. [CrossRef]
79. Choe, J.; Crespi, B. *The Evolution of Social Behavior in Insects and Arachnids*; Cambridge University Press: Cambridge, UK, 1997.
80. Costa, J. *The Other Insect Societies*; Harvard University Press: Cambridge, MA, USA, 2006.
81. Costa, J. Social evolution in "other" insects and arachnids. In *Encyclopedia of Animal Behavior*; Breed, M., Moore, J., Eds.; Academic Press: Cambridge, MA, USA, 2016.
82. Costa, J.T. The other insect societies: Overview and new directions. *Curr. Opin. Insect Sci.* **2018**, *28*, 40–49. [CrossRef]
83. Aluja, M.; Díaz-Fleischer, F. Foraging behavior of *Anastrepha ludens*, *A. obliqua*, and *A. serpentina* in response to feces extracts containing host marking pheromone. *J. Chem. Ecol.* **2006**, *32*, 367–389. [CrossRef]
84. Decker, A.; D'elia, B.; Kuhl, A.; Rosen, S.; Disney, A.; Dial, C.; Linietsky, M.; Taylor-Lilquist, J.; Taylor-Lilquist, B.; Kim, E.; et al. Acoustic stimulus influences ovipositioning in *Drosophila melanogaster*. *Bull. Insectol.* **2020**, *73*, 103–109.
85. Corbet, S.A. Mandibular gland secretion of larvae of the flour moth, Anagasta kuehniella, contains an epideictic pheromone and elicits oviposition movements in a hymenopteran parasite. *Nature* **1971**, *232*, 481–484. [CrossRef]
86. Otake, R.; Dobata, S. Copy if dissatisfied, innovate if not: Contrasting egg-laying decision making in an insect. *Anim. Cogn.* **2018**, *21*, 805–812. [CrossRef]
87. Malek, H.L.; Long, T.A.F. On the use of private versus social information in oviposition site choice decisions by *Drosophila melanogaster* females. *Behav. Ecol.* **2020**, *31*, 739–749. [CrossRef]
88. Battesti, M.; Moreno, C.; Joly, D.; Mery, F. Biased social transmission in *Drosophila* oviposition choice. *Behav. Ecol. Sociobiol.* **2015**, *69*, 83–87. [CrossRef]
89. Battesti, M.; Pasquaretta, C.; Moreno, C.; Teseo, S.; Joly, D.; Klensch, E.; Petit, O.; Sueur, C.; Mery, F. Ecology of information: Social transmission dynamics within groups of non-social insects. *Proc. R. Soc. B Biol. Sci.* **2015**, *282*, 20142480. [CrossRef]
90. Elsensohn, J.E.; Aly, M.F.K.; Schal, C.; Burrack, H.J. Social signals mediate oviposition site selection in *Drosophila suzukii*. *Sci. Rep.* **2021**, *11*, 1–10. [CrossRef]
91. Stelinski, L.L.; Rodriguez-Saona, C.; Meyer, W.L. Recognition of foreign oviposition-marking pheromone in a multi-trophic context. *Naturwissenschaften* **2009**, *96*, 585–592. [CrossRef]
92. Pasqualone, A.A.; Davis, J.M. The use of conspecific phenotypic states as information during reproductive decisions. *Anim. Behav.* **2011**, *82*, 281–284. [CrossRef]
93. Yadav, P.; Desireddy, S.; Kasinathan, S.; Bessière, J.M.; Borges, R.M. History matters: Oviposition resource acceptance in an exploiter of a nursery pollination mutualism. *J. Chem. Ecol.* **2018**, *44*, 18–28. [CrossRef]
94. Godfray, H.C.J. *Parasitoids: Behavioural and Evolutionary Ecology*; Princeton University Press: West Sussex, NJ, USA, 1994.
95. Loukola, O.J.; Gatto, E.; Híjar-Islas, A.C.; Chittka, L. Selective interspecific information use in the nest choice of solitary bees. *Anim. Biol.* **2020**, *70*, 215–225. [CrossRef]
96. Huigens, M.E.; Pashalidou, F.G.; Qian, M.H.; Bukovinszky, T.; Smid, H.M.; Van Loon, J.J.A.; Dicke, M.; Fatouros, N.E. Hitch-hiking parasitic wasp learns to exploit butterfly antiaphrodisiac. *Proc. Natl. Acad. Sci. USA* **2009**, *106*, 820–825. [CrossRef]
97. Sarin, S.; Dukas, R. Social learning about egg-laying substrates in fruitflies. *Proc. R. Soc. B Biol. Sci.* **2009**, *276*, 4323–4328. [CrossRef] [PubMed]
98. Couty, A.; Kaiser, L.; Huet, D.; Pham-Delegue, M.H. The attractiveness of different odour sources from the fruit-host complex on Leptopilina boulardi, a larval parasitoid of frugivorous *Drosophila* spp. *Physiol. Entomol.* **1999**, *24*, 76–82. [CrossRef]
99. Bodino, N.; Ferracini, C.; Tavella, L. Is host selection influenced by natal and adult experience in the parasitoid *Necremnus tutae* (Hymenoptera: Eulophidae)? *Anim. Behav.* **2016**, *112*, 221–228. [CrossRef]
100. Ghimire, M.N.; Phillips, T.W. Effects of prior experience on host selection and host utilization by two populations of *Anisopteromalus calandrae* (Hymenoptera: Pteromalidae). *Environ. Entomol.* **2008**, *37*, 1300–1306. [CrossRef]
101. Stephan, J.G.; Stenberg, J.A.; Björkman, C. How far away is the next basket of eggs? Spatial memory and perceived cues shape aggregation patterns in a leaf beetle. *Ecology* **2015**, *96*, 908–914. [CrossRef]
102. Murase, A.; Fujita, K.; Yano, S. Behavioural flexibility in spider mites: Oviposition site shifts based on past and present stimuli from conspecifics and predators. *R. Soc. Open Sci.* **2017**, *4*, 170328. [CrossRef]
103. Kujtan, L.; Dukas, R. Learning magnifies individual variation in heterospecific mating propensity. *Anim. Behav.* **2009**, *78*, 549–554. [CrossRef]
104. Mair, M.M.; Seifert, N.; Ruther, J. Previous interspecific courtship impairs female receptivity to conspecifics in the parasitoid wasp *Nasonia longicornis* but not in *N. vitripennis*. *Insects* **2018**, *9*, 112. [CrossRef]
105. Hostachy, C.; Couzi, P.; Portemer, G.; Hanafi-Portier, M.; Murmu, M.; Deisig, N.; Dacher, M. Exposure to conspecific and heterospecific sex-pheromones modulates gustatory habituation in the moth *Agrotis ipsilon*. *Front. Physiol.* **2019**, *10*, 1–8. [CrossRef]
106. Romano, D.; Benelli, G.; Stefanini, C. Opposite valence social information provided by bio-robotic demonstrators shapes selection processes in the green bottle fly. *J. R. Soc. Interface* **2021**, *18*, 20210056. [CrossRef]
107. Verzijden, M.N.; ten Cate, C.; Servedio, M.R.; Kozak, G.M.; Boughman, J.W.; Svensson, E. The impact of learning on sexual selection and speciation. *Trends Ecol. Evol.* **2012**, *27*, 511–519. [CrossRef]
108. Vosteen, I.; van den Meiracker, N.; Poelman, E.H. Getting confused: Learning reduces parasitoid foraging efficiency in some environments with non-host-infested plants. *Oecologia* **2019**, *189*, 919–930. [CrossRef]

109. Magrath, R.D.; Haff, T.M.; Fallow, P.M.; Radford, A.N. Eavesdropping on heterospecific alarm calls: From mechanisms to consequences. *Biol. Rev.* **2015**, *90*, 560–586. [CrossRef]
110. Muramatsu, D. Sand-bubbler crabs distinguish fiddler crab signals to predict intruders. *Behav. Ecol. Sociobiol.* **2021**, *75*, 1–11. [CrossRef]
111. Rieucau, G.; Giraldeau, L.A. Exploring the costs and benefits of social information use: An appraisal of current experimental evidence. *Philos. Trans. R. Soc. B Biol. Sci.* **2011**, *366*, 949–957. [CrossRef]
112. Nieberding, C.M.; Van Dyck, H.; Chittka, L. Adaptive learning in non-social insects: From theory to field work, and back. *Curr. Opin. Insect Sci.* **2018**, *27*, 75–81. [CrossRef]
113. Costa, T.M.; Hebets, E.A.; Melo, D.; Willemart, R.H. Costly learning: Preference for familiar food persists despite negative impact on survival. *Biol. Lett.* **2016**, *12*, 20160256. [CrossRef]
114. Botero, C.A.; Weissing, F.J.; Wright, J.; Rubenstein, D.R. Evolutionary tipping points in the capacity to adapt to environmental change. *Proc. Natl. Acad. Sci. USA* **2015**, *112*, 184–189. [CrossRef]
115. Dechaume-Moncharmont, F.X.; Dornhaus, A.; Houston, A.I.; McNamara, J.M.; Collins, E.J.; Franks, N.R. The hidden cost of information in collective foraging. *Proc. R. Soc. B Biol. Sci.* **2005**, *272*, 1689–1695. [CrossRef]
116. Greggor, A.L.; Trimmer, P.C.; Barrett, B.J.; Sih, A. Challenges of Learning to Escape Evolutionary Traps. *Front. Ecol. Evol.* **2019**, *7*, 408. [CrossRef]
117. Fleury, F.; Gibert, P.; Ris, N.; Allemand, R. Ecology and life history evolution of frugivorous Drosophila parasitoids. *Adv. Parasitol.* **2009**, *70*, 3–44. [CrossRef] [PubMed]
118. Lefèvre, T.; De Roode, J.C.; Kacsoh, B.Z.; Schlenke, T.A. Defence strategies against a parasitoid wasp in *Drosophila*: Fight or flight? *Biol. Lett.* **2012**, *8*, 230–233. [CrossRef] [PubMed]
119. van Lenteren, J.C.; Bakker, K. Behavioural aspects of the functional responses of a parasite (*Pseudocoila bochei*) to its host (*Drosophila melanogaster*). *Netherlands J. Zool.* **1978**, *28*, 213–233. [CrossRef]
120. Vet, L.E.; Papaj, D. Effects of experience on parasitoid movement in odour plumes. *Physiol. Entomol.* **1992**, *17*, 90–96. [CrossRef]
121. Wertheim, B.; Vet, L.E.M.; Dicke, M. Increased risk of parasitism as ecological costs of using aggregation pheromones: Laboratory and field study of *Drosophila-Leptopilina* interaction. *Oikos* **2003**, *100*, 269–282. [CrossRef]

genes

MDPI

Article

Evolution of Oxidative Phosphorylation (OXPHOS) Genes Reflecting the Evolutionary and Life Histories of Fig Wasps (Hymenoptera, Chalcidoidea)

Yi Zhou, Dawei Huang, Zhaozhe Xin and Jinhua Xiao *

Institute of Entomology, College of Life Sciences, Nankai University, Tianjin 300071, China; 1120170366@mail.nankai.edu.cn (Y.Z.); huangdw@nankai.edu.cn (D.H.); 1120180392@mail.nankai.edu.cn (Z.X.)
* Correspondence: xiaojh@nankai.edu.cn; Tel.: +86-185-2245-2108

Received: 25 October 2020; Accepted: 13 November 2020; Published: 15 November 2020

Abstract: Fig wasps are a peculiar group of insects which, for millions of years, have inhabited the enclosed syconia of fig trees. Considering the relatively closed and dark environment of fig syconia, we hypothesize that the fig wasps' oxidative phosphorylation (OXPHOS) pathway, which is the main oxygen consumption and adenosine triphosphate (ATP) production system, may have adaptively evolved. In this study, we manually annotated the OXPHOS genes of 11 species of fig wasps, and compared the evolutionary patterns of OXPHOS genes for six pollinators and five non-pollinators. Thirteen mitochondrial protein-coding genes and 30 nuclear-coding single-copy orthologous genes were used to analyze the amino acid substitution rate and natural selection. The results showed high amino acid substitution rates of both mitochondrial and nuclear OXPHOS genes in fig wasps, implying the co-evolution of mitochondrial and nuclear genes. Our results further revealed that the OXPHOS-related genes evolved significantly faster in pollinators than in non-pollinators, and five genes had significant positive selection signals in the pollinator lineage, indicating that OXPHOS genes play an important role in the adaptation of pollinators. This study can help us understand the relationship between gene evolution and environmental adaptation.

Keywords: oxidative phosphorylation; positive selection; fig wasps

1. Introduction

The symbiotic relationship between fig trees (Moraceae, *Ficus*) and their pollinating fig wasps (or abbreviated as pollinators) (Chalcidoidea, Agaonidae) can be traced back to 75 million years ago and represents extreme obligate mutualism [1]. Fig trees have enclosed inflorescences (syconia), which have a narrow opening (ostiole) connecting the inside and outside world. Therefore, the internal environment of fig syconia should be closed and lightless, and the oxygen content in the fig syconia may also be different from that of the outside atmosphere. The figs also provide habitats for other wasps (Hymenoptera, Chalcidoidea), which do not pollinate figs (that is, non-pollinators). Pollinators and non-pollinators have different lifestyles related to figs, as a mature female pollinator leaves the natal fig syconium to enter a new receptive syconium for pollination and oviposition, while non-pollinators do not pollinate, and most of them only lay eggs outside syconia through the fig wall to the fig ovaries [2]. For pollinators, the process of drilling into the fig syconia is not easy, and they have flat heads and a row of teeth of mandibular appendages, which can help in pushing bracts of the ostiole [3]. Furthermore, previous studies have shown that the pollinators, which emerged during the Eocene, are older than non-pollinators, which emerged during the Oligocene or Miocene [4]. Therefore, the fig pollinators and non-pollinators display different life and evolutionary histories related to the peculiar living environment within fig syconia, implying that they may have different evolutionary patterns in terms of the oxygen utilization strategy and energy consumption.

The oxidative phosphorylation (OXPHOS) pathway is the main oxygen consumption and adenosine triphosphate (ATP) production system in eukaryotic cells [5]. It is composed of five complexes. Except for complex II, in which the proteins are all encoded by nuclear genes, the remaining four complexes (complex I, III, IV, and V) are composed of subunits encoded by both mitochondrial and nuclear genes [6]. Therefore, maintenance of the function of OXPHOS may involve the co-evolution of mitochondrial genes and nuclear genes. Generally, mitochondrial genes evolve faster than most nuclear genes, so nuclear genes related to OXPHOS may also show faster evolution rates than the nuclear genes which are not involved in OXPHOS [7,8]. Many studies have been carried out to detect the evolutionary patterns of the OXPHOS system in other insects, such as fruit flies [9]; parasitic wasps *Nasonia* [10]; and a comparison of Hymenoptera, Lepidoptera, Coleoptera, and Diptera [11]. The articles on the OXPHOS genes of fig wasps are mostly limited to mitochondrial-coding genes [12,13], and there is only one work on mitochondrial and nuclear OXPHOS genes based on unpublished data from our laboratory [14]. In fact, considering the special living environment of fig wasps related to figs, the evolution of genes related to the OXPHOS pathway is a topic worth exploring.

In this study, we used mitochondrial and nuclear genomic data to explore the evolutionary pattern of OXPHOS-related genes in fig wasps, in order to detect whether OXPHOS genes display different evolutionary patterns in the adaptation of pollinators and non-pollinators to living within fig syconia.

2. Materials and Methods

2.1. Sequence Determination

We used mitochondrial genomes, nuclear genomes, and transcriptome data from 11 fig wasps, including six pollinators and five non-pollinators (Table S1). The species classification was provided by the website (http://www.figweb.org/Fig_wasps/Classification/index.htm).

The nuclear-encoded OXPHOS genes for *Drosophila melanogaster* (dme00190), *Nasonia vitripennis* (nvi00190), and *Apis mellifera* (ame00190) were downloaded from the KEGG database [15]. We performed TBLASTN searches of the whole genome sequences of fig wasps using the amino acid sequences of *D. melanogaster*, *N. vitripennis*, and *A. mellifera* as queries. If the BLAST E-scores were lower than 10^{-4}, the sequences were mapped to the genome to identify the positions of each exon and modified with the actual transcripts shown in the transcriptomic data, using IGV v2.4.14 [16,17]. If the quality of transcriptome data was poor and there were no clear exons on IGV, we used the Softberry website (www.softberry.com/berry.phtml?topic=fgenes_plus&group=programs&subgroup=gfs) to predict the complete coding sequences of genes. Then, the genes were extracted from genomes and translated to protein sequences using BioEdit v7.0.9.0. Using these protein sequences as queries, we searched for the conserved domains on NCBI CD-search tool [18].

The complete mitogenomes were annotated using the MITOS webserver [19] and NCBI ORF Finder (https://www.ncbi.nlm.nih.gov/orffinder). When the annotation results from both webservers were different, we compared them with the 13 protein-coding genes of the closely related species *N. vitripennis* (EU746609.1 and EU746613.1) to evaluate the final length of the reading frames.

In order to obtain a nuclear non-OXPHOS gene set, we used OrthoMCL v2.0.9 [20] to screen the single-copy orthologous genes of 11 fig wasp species (the genome accessions in Table S1). Then, the OXPHOS genes were removed from the single-copy gene set. Finally, a set of 4622 single-copy non-OXPHOS genes was obtained.

2.2. Phylogenetic Reconstruction

The phylogenetic tree was built based on nuclear single-copy orthologous genes and 13 mitochondrial protein-coding genes. OrthoMCL v2.0.9 [20] was used to identify single-copy orthologous genes in the genomes of 11 fig wasps and *A. mellifera* (GCF_000002195). The amino acids of each gene were aligned using MAFFT v7.313 [21] implemented by PhyloSuite v1.1.16 [22]. Gblocks 0.91b [23] was used to remove poor alignments. These alignment sequences were concatenated by PhyloSuite v1.1.16 [22]. The phylogenetic tree was reconstructed using IQ-TREE v1.6.1 [24] based on the maximum likelihood method with the best model of JTT+F+I+G4 and 5000 ultrafast bootstrap replicates. *A. mellifera* was used as the outgroup. The tree topology (Figure 1a) was used for subsequent analysis.

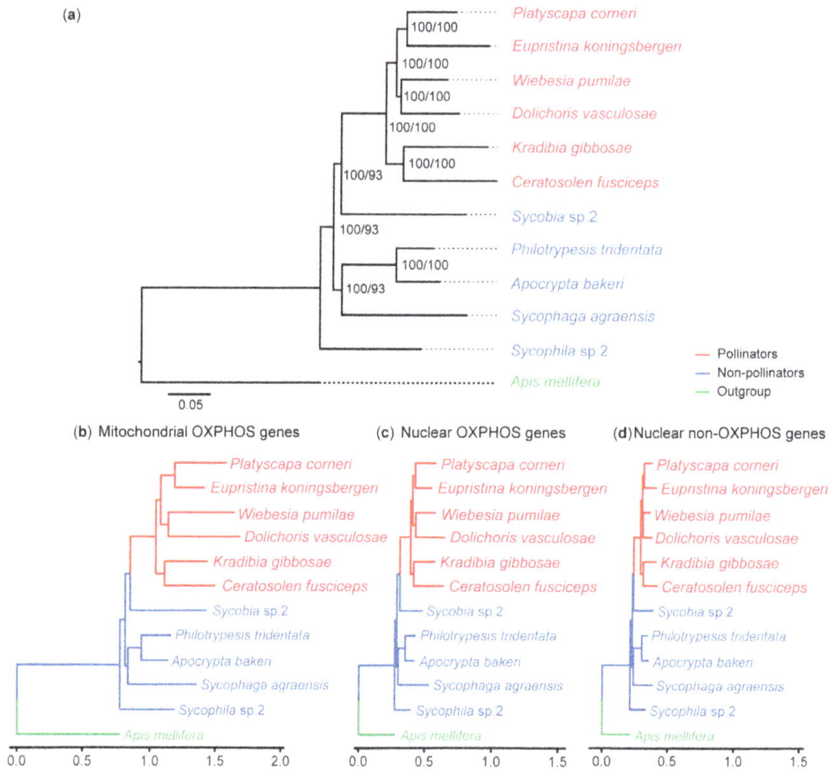

Figure 1. The phylogenetic trees used in this study. Red species represent pollinators, blue species represent non-pollinators, and green species represent the outgroup. (**a**) Phylogenetic tree based on mitochondrial-coding genes and nuclear single-copy orthologous genes. Number on the node represents SH-aLRT%/UFBoot%. Phylogenetic trees based on concatenated amino acid alignments of (**b**) mitochondrial OXPHOS genes, (**c**) nuclear OXPHOS genes, and (**d**) nuclear non-OXPHOS genes. Branch lengths represent the average number of amino acid substitutions per site.

2.3. Estimation of the Amino Acid Substitution Rate

We used the branch length from the outgroup (*A. mellifera*) to each ingroup as the estimation of the amino acid substitution rate, based on the maximum likelihood trees [11].

In the mitochondrial-encoded OXPHOS gene category, the protein sequence of each gene was aligned using MAFFT v7.313 [21] and concatenated using PhyloSuite v1.1.16 [22]. The poor blocks were removed using Gblocks 0.91b [23]. Then, the maximum likelihood tree was reconstructed using IQ-TREE v1.6.1 [24] with the automatically selecting model and the tree topology of Figure 1a. The same procedures were carried out for the nuclear OXOPHS gene category and nuclear non-OXPHOS gene category. These phylogenetic trees were visualized and colored with the R package "ggtree" [25].

For each individual gene in every category, the protein sequence was aligned using MAFFT v7.313 [21]. Then, each gene tree was reconstructed based on the topology from Figure 1a using IQ-TREE v1.6.1 [24] with the parameter automatically selecting the best fitting model.

The branch length from each ingroup species to outgroup species (*A. mellifera*) was retrieved using Newick Utilities v1.6 [26], which was used as the amino acid substitution rate of the gene [11]. The Wilcoxon rank sum test in R software was employed to test the amino acid substitution rate difference between the pollinators and non-pollinators. The pairwise Wilcoxon rank sum test was used to test the difference between mitochondrial OXPHOS, nuclear OXPHOS, and nuclear non-OXPHOS genes. The p values for multiple comparisons were adjusted by Holm correction [27].

2.4. Natural Selection Analysis

We used codeml implemented in PAML v4.9f [28] to test the natural selection for each gene in mitochondrial and nuclear OXPHOS gene categories. The nonsynonymous (dN) to synonymous (dS) rate ($\omega = d\text{N}/d\text{S}$) ratios represent the changes of selective pressures ($\omega = 1$, neutral evolution; $\omega < 1$, purifying selection; $\omega > 1$, positive selection). The tree topology (Figure 1a) without *A. mellifera* was used for the selection analysis.

The one-ratio models were used to estimate the single ω ratio for all branches in the phylogenetic tree. The free-ratio models [29,30] were used to obtain an independent ω ratio for each branch. We compared the ω ratio of pollinators and non-pollinators in each gene using the Wilcoxon rank sum test in R software. The boxplots of ω ratios were made using "ggpubr" in the R package.

Positive selection may act in short episodes, affecting only a few sites along particular lineages. The modified branch-site model A [31] and its null model (branch-site model A with $\omega_2 = 1$ fixed) were used to test for the positive selection of sites along foreground branches (pollinator lineages). The likelihood ratio test was used to test whether the alternative model was significantly better than the null model and Bayes empirical Bayes (BEB) [32] analysis was employed to identify positive sites. When the likelihood ratio test was significant and the posterior probability was greater than 95%, we considered this gene to be a positively selected gene.

3. Results

3.1. OXPHOS Gene Annotation

We manually annotated OXPHOS genes for 11 fig wasp species by using the genomic and transcriptomic data. A total of 72 genes related to OXPHOS were found, but the copy numbers were different in different species (Table S2). Finally, we found that 30 genes were single-copy orthologs with completed conserved domains for every species, and 13 orthologous genes had multiple copies in some species (Table S2). In particular, the COX4 gene was the most special in that it had a single copy in each pollinator species, but two copies in each non-pollinator (Table S2). The gene tree of COX4 showed that the lineage of one copy (namely COX4.2) in non-pollinators exhibited a basal position and another copy (namely COX4.1) in non-pollinators was clustered with COX4 in pollinators (Figure S1). The 30 single-copy genes were used in subsequent analyses as the gene category of "nuclear OXPHOS genes".

3.2. Amino Acid Substitution Rate Analysis

We compared the amino acid substitution rate of mitochondrial-encoded OXPHOS genes, nuclear OXPHOS genes, and nuclear non-OXPHOS genes, based on the maximum likelihood trees (Figure 1b–d). The results of three concatenated sequences showed that the highest substitution rate was found in mitochondrial genes, followed by nuclear OXPHOS genes, and finally nuclear non-OXPHOS genes (Figure 2a and Table S3).

We also compared the amino acid substitution rates of the genes for pollinators and non-pollinators, and the results showed that the mitochondrial genes of the pollinators exhibited a higher amino acid substitution rate than those of the non-pollinators ($p < 0.01$); nuclear OXPHOS genes displayed the same difference pattern ($p < 0.01$) (Figure 2b and Table S3). However, differences of nuclear non-OXPHOS genes between pollinators and non-pollinators were not significant ($p > 0.05$) (Figure 2b and Table S3). In order to be more specific about the substitution rate of each gene, we compared the substitution rate of each individual gene for pollinators and non-pollinators, and declared that 25 out of the 43 mitochondrial and nuclear OXPHOS genes showed higher amino acid substitution rates in pollinators than in non-pollinators ($p < 0.05$) (Figure 3 and Table S4).

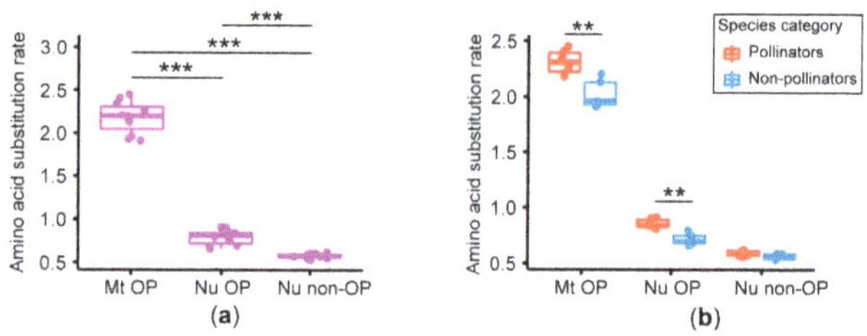

Figure 2. The amino acid substitution rate estimation based on concatenated mitochondrial OXPHOS genes, concatenated nuclear OXPHOS genes, and concatenated nuclear non-OXPHOS genes. **, $p < 0.01$ and ***, $p < 0.001$. Abbreviations: Mt OP, mitochondrial OXPHOS genes; Nu OP, nuclear OXPHOS genes; Nu non-OP, nuclear non-OXPHOS genes. (**a**) The comparisons of the amino acid substitution rate for mitochondrial OXPHOS, nuclear OXPHOS, and nuclear non-OXPHOS genes. The p values were determined by the pairwise Wilcoxon rank sum test and adjusted by Holm correction. (**b**) Comparisons of the amino acid substitution rate for pollinators and non-pollinators in three gene categories. The p values were determined by the Wilcoxon rank sum test. Red boxplots represent pollinators and blue boxplots represent non-pollinators.

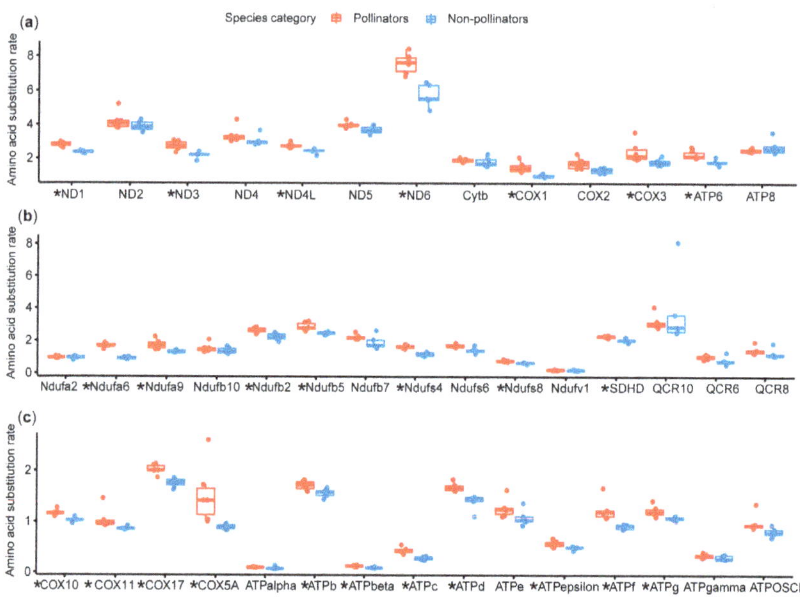

Figure 3. Amino acid substitution rate estimation of each OXPHOS-related gene. A comparison of pollinators and non-pollinators based on 13 mitochondrial genes (**a**); nuclear-encoded genes in OXPHOS complex I, II, and III (**b**); and nuclear-encoded genes in OXPHOS complex IV and V (**c**). Red boxplots represent pollinators and blue boxplots represent non-pollinators. Genes with a significant difference in rates are marked with an asterisk.

3.3. Natural Selection Analysis on the OXPHOS Genes

We tested for selective pressure on OXPHOS genes by comparing the ratio of non-synonymous to synonymous substitutions. The one-ratio model results showed that each gene had a ω ratio far smaller than one (Table S5), indicating that all tested genes were under purifying selection. We conducted a comparison of ω ratios for pollinators and non-pollinators on mitochondrial and nuclear OXPHOS genes using the free-ratio model results (Figure 4 and Table S6). The results demonstrated that the median ω ratio of pollinators was higher than that of non-pollinators in nuclear-encoded OXPHOS genes ($p < 0.05$), but there was no significant difference in mitochondrial OXPHOS genes (Figure 4a). At the level of individual genes, pollinators exhibited higher ω ratios than non-pollinators in most genes; however, the differences were only significant in seven genes (Figure 4b–d and Table S6).

Considering that the higher ω ratios of the OXPHOS genes in the pollinators may indicate positive selection or relaxed selection, we used the branch-site model to test for positive selection in pollinators (Table S7). The results showed that five nuclear-encoded OXPHOS genes were under positive selection in the branch of pollinators compared with non-pollinators (Table 1). BEB analysis suggested that eight sites were well-supported (posterior probability > 95%) in the five genes, which encode proteins on complex I (Ndufb5 and Ndufb10), complex IV (COX11), and complex V (ATPd and ATPOSCP), respectively (Table 1).

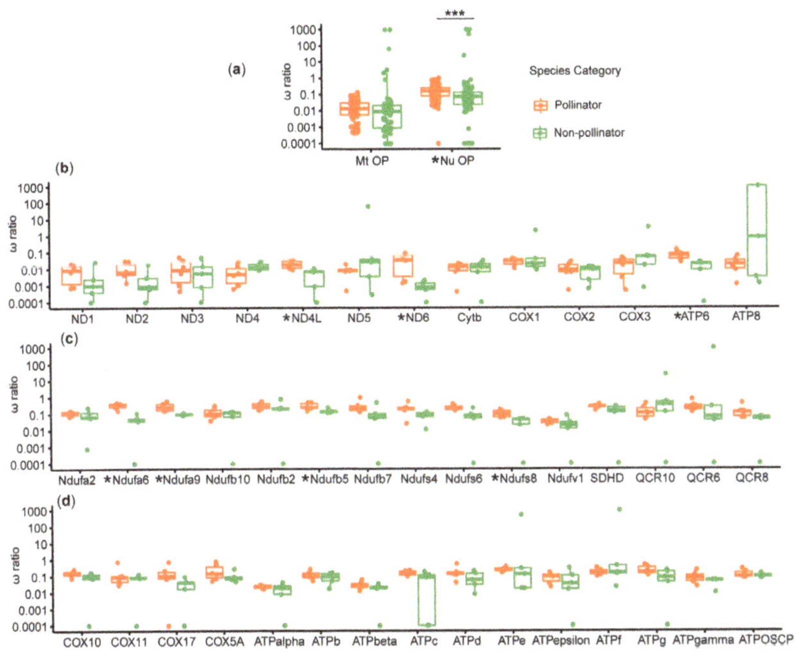

Figure 4. The ω ratios of mitochondrial and nuclear OXPHOS genes. Orange boxplots represent the ω ratio distribution of pollinator species and green boxplots represent the ω ratio distribution of non-pollinators. Genes with a significant difference in ω ratios are marked with an asterisk. ***, $p < 0.001$. Abbreviations: Mt OP, mitochondrial OXPHOS genes; Nu OP, nuclear OXPHOS genes. A comparison of the ω ratios for pollinators and non-pollinators based on (**a**) all mitochondrial and nuclear OXPHOS genes; (**b**) each mitochondrial gene; (**c**) genes involved in nuclear OXPHOS complex I, II, and III; and (**d**) genes related to OXPHOS complex IV and V.

Table 1. Positively selected genes and sites detected in branch-site model tests.

Gene	lnL (Alter)	lnL (Null)	2ΔlnL	*p* Value	Positive Sites
Ndufb5	−5095.22	−5100.57	10.69	0.001	75 G **
Ndufb10	−3997.50	−4001.73	8.45	0.004	118 N **
COX11	−4689.52	−4691.67	4.28	0.039	92 I *, 171 M **
ATPd	−4605.99	−4608.06	4.13	0.042	47 S *, 129 Q *
ATPOSCP	−4263.97	−4266.42	4.89	0.027	92 G **, 93 T **

*, posterior probability > 95% and **, posterior probability > 99%. Only the sites with the significant likelihood ratio test and posterior probability are shown and other sites are detailed in supplementary Table S7.

4. Discussion

In this study, we manually annotated the OXPHOS genes in 11 fig wasp species and conducted a thorough molecular evolution analysis, in order to understand whether they present different evolutionary patterns in fig pollinators and non-pollinators, due to their different life history and evolutionary history related to fig syconia.

Based on the comparative analysis of gene members related to OXPHOS, we found that most orthologous OXPHOS genes are single-copy in fig wasps, but some orthologous genes have multiple copies in several fig wasp species. An interesting result that caught our attention is that COX4, encoding a key subunit of cytochrome c oxidase, has single copies in pollinators and two paralogous copies in non-pollinators. According to the phylogenetic tree of COX4, we can infer that the ancestral species of fig wasps had two copies of COX4, and the fact that the pollinators have only one copy may have been caused by gene loss. There are also two paralogs in vertebrates, except birds [33], and functionally, the genes in mammals are hypoxia-responsive [34], and in fish, they expressed in specific tissues [35]. These previous studies indicate that COX4 genes may play important roles in the adaptive evolution of organisms, enabling them to adapt to different oxygen environments, and our results thus lead us to infer that pollinators and non-pollinators may present different evolutionary patterns in the function of COX4 in their divergent evolutionary histories with figs. Additionally, the proteins encoded by the two paralogs may have been differentiated in non-pollinating fig wasps.

The level of the amino acid substitution rate can often well-reflect the rate of protein evolution [36]. In this study, the amino acid substitution rate of OXPHOS genes was higher than non-OXPHOS genes, and these results are consistent with previous findings on other insects in Hymenoptera [11]. In view of the rapid evolution of mitochondria in fig wasps [12], the rapid evolution of nuclear-coding OXPHOS may represent compensatory evolution for maintaining function, which supports the viewpoint of the coevolution of mitochondria and the nucleus [7]. This is consistent with previous estimations of OXPHOS evolution patterns among different insect orders [11]. We also found that the concatenated mitochondrial OXPHOS genes and concatenated nuclear OXPHOS genes display higher amino acid substitution rates in pollinators than in non-pollinators, respectively. In the results of individual genes, in 58.1% of mitochondrial and nuclear OXPHOS genes, the amino acid substitution rate of pollinators is significantly higher than that of non-pollinators. This indicates that the evolution of OXPHOS genes in pollinators occurred more rapidly than in non-pollinators.

Genes with a fast evolution rate often have a high number of nucleotide substitutions which are driven by positive selection or neutral mutation [37]. In view of the fact that the OXPHOS-related genes of the pollinator fig wasps may have evolved faster than those of the non-pollinators, we performed an evolutionary selection pressure analysis on these genes. The results of the natural selection analysis based on the branch model in PAML show that the overall ω ratio pattern of pollinators is significantly higher than that of non-pollinators in nuclear OXPHOS genes, suggesting that these genes in pollinators have undergone relaxed purifying selective constraint or positive selection [38]. We used branch-site models and found that five genes are positively selected in pollinators among 43 OXPHOS-related genes. Ndufb5 and Ndufb10 encode accessory subunits, which are not directly involved in catalysis and may be required for stabilizing complex I (NADH: ubiquinone oxidoreductase) [39]. COX11 is a copper-binding protein, and it is essential for complex IV (cytochrome c oxidase) assembly and function [40]. ATPd and ATPOSCP are parts of the peripheral stalk and play vital roles in maintaining the structural stability of complex V (F1Fo ATP synthase) [41,42]. Therefore, complex I, IV, and V of the mitochondrial respiratory chain may play important roles in the adaption of pollinators.

Three genes (Ndufb5, COX11, and ATPd) have significantly higher amino acid substitution rates in pollinators than those in non-pollinators and they are positively selected in the pollinator lineage. Therefore, we speculate that the high substitution rates of these genes have been driven by positive selection. However, there are other genes that are not positively selected, but show high substitution rates, which may be caused by other factors, such as neutral mutations.

Positive selection may drive OXPHOS genes to better adapt to the high energy requirements and hypoxia [43–45]. Therefore, we infer that pollinators expend more energy than non-pollinators, which may be needed in the life behaviors of drilling into syconia and pollinating flowers. In the future, we will study the causes of fig–wasp mutualism affecting OXPHOS genes of pollinators and attempt to obtain experimental evidence for positively selected genes affecting the function of OXPHOS.

5. Conclusions

In the compact environment of fig syconia, where fig wasps have lived for tens of millions of years, the fig pollinators not only have a longer coevolutionary history with fig trees, but also a longer lifetime inside syconia in their adult life history than the non-pollinators, which may indicate their different evolutionary patterns in the genes of OXPHOS pathway. Our molecular evolution analysis results based on the OXPHOS genes in fig wasps showed that the nuclear-encoded OXPHOS genes had a faster evolutionary rate than the nuclear-encoded non-OXPHOS genes, which may be due to co-evolution with the fast evolution of mitochondrial genes. Our results also found that both mitochondrial and nuclear OXPHOS genes in pollinators were evolving faster than the genes in non-pollinators, which may be driven by positive selection on the pollinators. This study provides us with some evidence on the adaptation of insects living in an enclosed environment from the evolution of OXPHOS genes.

Supplementary Materials: The following are available online at http://www.mdpi.com/2073-4425/11/11/1353/s1: Figure S1: The phylogenetic tree of COX4 genes; Table S1: List of fig wasp species used in this study; Table S2: The number of nuclear OXPHOS genes of each species in fig wasps; Table S3: The amino acid substitution rate estimation based on concatenated genes; Table S4: The details of the amino acid substitution rate estimation of each OXPHOS-related gene; Table S5: The one-ratio model results of mitochondrial and nuclear OXPHOS genes; Table S6: The free-ratio model results of mitochondrial and nuclear OXPHOS genes; Table S7: The branch-site model results of mitochondrial and nuclear OXPHOS genes.

Author Contributions: Methodology, formal analysis, investigation, data curation, writing—original Draft, and writing—review and editing, Y.Z.; writing—review and editing, Z.X.; conceptualization, writing—review and editing, supervision, project administration, and funding acquisition, J.X. and D.H. All authors have read and agreed to the published version of the manuscript.

Funding: This research was funded by the National Natural Science Foundation of China (Nos. 31830084, 31970440 and 31672336), and also supported by the construction funds for the "Double First-Class" initiative for Nankai University (Nos. 96172158, 96173250, and 91822294).

Conflicts of Interest: The authors declare no conflict of interest.

References

1. Cruaud, A.; Ronsted, N.; Chantarasuwan, B.; Chou, L.S.; Clement, W.L.; Couloux, A.; Cousins, B.; Genson, G.; Harrison, R.D.; Hanson, P.E. An extreme case of plant-insect codiversification: Figs and fig-pollinating wasps. *Syst. Biol.* **2012**, *61*, 1029–1047. [CrossRef]
2. Borges, R.M. How to be a fig wasp parasite on the fig-fig wasp mutualism. *Curr. Opin. Insect Sci.* **2015**, *8*, 34–40. [CrossRef] [PubMed]
3. Weiblen, G.D. How to be a fig wasp. *Annu. Rev. Entomol.* **2002**, *47*, 299–330. [CrossRef] [PubMed]

4. Peters, R.S.; Niehuis, O.; Gunkel, S.; Bläser, M.; Mayer, C.; Podsiadlowski, L.; Kozlov, A.; Donath, A.; van Noort, S.; Liu, S.; et al. Transcriptome sequence-based phylogeny of chalcidoid wasps (Hymenoptera: Chalcidoidea) reveals a history of rapid radiations, convergence, and evolutionary success. *Mol. Phylogenet. Evol.* **2018**, *120*, 286–296. [CrossRef] [PubMed]
5. Sunnucks, P.; Morales, H.E.; Lamb, A.M.; Pavlova, A.; Greening, C. Integrative approaches for studying mitochondrial and nuclear genome co-evolution in oxidative phosphorylation. *Front. Genet.* **2017**, *8*. [CrossRef]
6. Signes, A.; Fernandez-Vizarra, E. Assembly of mammalian oxidative phosphorylation complexes I–V and supercomplexes. *Essays Biochem.* **2018**, *62*, 255–270. [CrossRef]
7. Hill, G.E. Mitonuclear ecology. *Mol. Biol. Evol.* **2015**, *32*, 1917–1927. [CrossRef]
8. Zhang, F.; Broughton, R.E. Mitochondrial-nuclear interactions: Compensatory evolution or variable functional constraint among vertebrate oxidative phosphorylation genes? *Genome Biol. Evol.* **2013**, *5*, 1781–1791. [CrossRef]
9. Tripoli, G.; D'Elia, D.; Barsanti, P.; Caggese, C. Comparison of the oxidative phosphorylation (OXPHOS) nuclear genes in the genomes of *Drosophila melanogaster*, *Drosophila pseudoobscura* and *Anopheles gambiae*. *Genome Biol.* **2005**, *6*, R11. [CrossRef]
10. Gibson, J.D.; Niehuis, O.; Verrelli, B.C.; Gadau, J. Contrasting patterns of selective constraints in nuclear-encoded genes of the oxidative phosphorylation pathway in holometabolous insects and their possible role in hybrid breakdown in *Nasonia*. *Heredity* **2010**, *104*, 310–317. [CrossRef]
11. Li, Y.; Zhang, R.; Liu, S.; Donath, A.; Peters, R.S.; Ware, J.; Misof, B.; Niehuis, O.; Pfrender, M.E.; Zhou, X. The molecular evolutionary dynamics of oxidative phosphorylation (OXPHOS) genes in Hymenoptera. *BMC Evol. Biol.* **2017**, *17*, 269. [CrossRef] [PubMed]
12. Xiao, J.H.; Jia, J.G.; Murphy, R.W.; Huang, D.W. Rapid evolution of the mitochondrial genome in Chalcidoid wasps (Hymenoptera: Chalcidoidea) driven by parasitic lifestyles. *PLoS ONE* **2011**, *6*, e26645. [CrossRef] [PubMed]
13. Xiao, J.H.; Wang, N.X.; Murphy, R.W.; Cook, J.; Jia, L.Y.; Huang, D.W. *Wolbachia* infection and dramatic intraspecific mitochondrial DNA divergence in a fig wasp. *Evolution* **2012**, *66*, 1907–1916. [CrossRef] [PubMed]
14. Li, Z.Z. The Association between *Wolbachia* and the Evolution of OXPHOS Genes in a Fig Wasp (*Ceratosolen solmsi*). Ph.D. Thesis, University of Chinese Academy of Sciences, Beijing, China, 2014.
15. Kanehisa, M.; Goto, S. KEGG: Kyoto encyclopedia of genes and genomes. *Nucleic Acids Res.* **2000**, *28*, 27–30. [CrossRef]
16. Robinson, J.T.; Thorvaldsdóttir, H.; Winckler, W.; Guttman, M.; Lander, E.S.; Getz, G.; Mesirov, J.P. Integrative genomics viewer. *Nat. Biotechnol.* **2011**, *29*, 24–26. [CrossRef]
17. Thorvaldsdóttir, H.; Robinson, J.T.; Mesirov, J.P. Integrative Genomics Viewer (IGV): High-performance genomics data visualization and exploration. *Brief. Bioinform.* **2013**, *14*, 178–192. [CrossRef]
18. Marchler-Bauer, A.; Bryant, S.H. CD-Search: Protein domain annotations on the fly. *Nucleic Acids Res.* **2004**, *32*, W327–W331. [CrossRef]
19. Bernt, M.; Donath, A.; Jühling, F.; Externbrink, F.; Florentz, C.; Fritzsch, G.; Pütz, J.; Middendorf, M.; Stadler, P.F. MITOS: Improved de novo metazoan mitochondrial genome annotation. *Mol. Phylogenet. Evol.* **2013**, *69*, 313–319. [CrossRef]
20. Li, L.; Stoeckert, C.J., Jr.; Roos, D.S. OrthoMCL: Identification of ortholog groups for eukaryotic genomes. *Genome Res.* **2003**, *13*, 2178–2189. [CrossRef]
21. Katoh, K.; Standley, D.M. MAFFT multiple sequence alignment software version 7: Improvements in performance and usability. *Mol. Biol. Evol.* **2013**, *30*, 772–780. [CrossRef]
22. Zhang, D.; Gao, F.; Jakovlić, I.; Zou, H.; Zhang, J.; Li, W.X.; Wang, G.T. PhyloSuite: An integrated and scalable desktop platform for streamlined molecular sequence data management and evolutionary phylogenetics studies. *Mol. Ecol. Resour.* **2020**, *20*, 348–355. [CrossRef] [PubMed]
23. Talavera, G.; Castresana, J. Improvement of phylogenies after removing divergent and ambiguously aligned blocks from protein sequence alignments. *Syst. Biol.* **2007**, *56*, 564–577. [CrossRef] [PubMed]
24. Nguyen, L.-T.; Schmidt, H.A.; von Haeseler, A.; Minh, B.Q. IQ-TREE: A fast and effective stochastic algorithm for estimating maximum-likelihood phylogenies. *Mol. Biol. Evol.* **2015**, *32*, 268–274. [CrossRef] [PubMed]

25. Yu, G.; Smith, D.K.; Zhu, H.; Guan, Y.; Lam, T.T.-Y. ggtree: An R package for visualization and annotation of phylogenetic trees with their covariates and other associated data. *Methods Ecol. Evol.* **2017**, *8*, 28–36. [CrossRef]
26. Junier, T.; Zdobnov, E.M. The Newick utilities: High-throughput phylogenetic tree processing in the UNIX shell. *Bioinformatics* **2010**, *26*, 1669–1670. [CrossRef]
27. Holm, S. A simple sequentially rejective multiple test procedure. *Scand. J. Stat.* **1979**, *6*, 65–70.
28. Yang, Z. PAML 4: Phylogenetic analysis by maximum likelihood. *Mol. Biol. Evol.* **2007**, *24*, 1586–1591. [CrossRef]
29. Yang, Z. Likelihood ratio tests for detecting positive selection and application to primate lysozyme evolution. *Mol. Biol. Evol.* **1998**, *15*, 568–573. [CrossRef]
30. Yang, Z.; Nielsen, R. Synonymous and nonsynonymous rate variation in nuclear genes of mammals. *J. Mol. Evol.* **1998**, *46*, 409–418. [CrossRef]
31. Zhang, J.; Nielsen, R.; Yang, Z. Evaluation of an improved branch-site likelihood method for detecting positive selection at the molecular level. *Mol. Biol. Evol.* **2005**, *22*, 2472–2479. [CrossRef]
32. Yang, Z.; Wong, W.S.W.; Nielsen, R. Bayes empirical bayes inference of amino acid sites under positive selection. *Mol. Biol. Evol.* **2005**, *22*, 1107–1118. [CrossRef] [PubMed]
33. Little, A.G.; Kocha, K.M.; Lougheed, S.C.; Moyes, C.D. Evolution of the nuclear-encoded cytochrome oxidase subunits in vertebrates. *Physiol. Genom.* **2010**, *42*, 76–84. [CrossRef] [PubMed]
34. Fukuda, R.; Zhang, H.; Kim, J.W.; Shimoda, L.; Dang, C.V.; Semenza, G.L. HIF-1 regulates cytochrome oxidase subunits to optimize efficiency of respiration in hypoxic cells. *Cell* **2007**, *129*, 111–122. [CrossRef] [PubMed]
35. Porplycia, D.; Lau, G.Y.; McDonald, J.; Chen, Z.; Richards, J.G.; Moyes, C.D. Subfunctionalization of COX4 paralogs in fish. *Am. J. Physiol. Regul. Integr. C* **2017**, *312*, 671–680. [CrossRef]
36. Kimura, M. Evolutionary rate at the molecular level. *Nature* **1968**, *217*, 624–626. [CrossRef]
37. Nei, M. The new mutation theory of phenotypic evolution. *Proc. Natl. Acad. Sci. USA* **2007**, *104*, 12235–12242. [CrossRef]
38. Yang, Z. *Computational Molecular Evolution*; Oxford University Press: New York, NY, USA, 2006; p. 271.
39. Rhooms, S.-K.; Murari, A.; Goparaju, N.S.V.; Vilanueva, M.; Owusu-Ansah, E. Insights from *Drosophila* on mitochondrial complex I. *Cell. Mol. Life Sci.* **2020**, *77*, 607–618. [CrossRef]
40. Carr, H.S.; George, G.N.; Winge, D.R. Yeast Cox11, a protein essential for cytochrome c oxidase assembly, is a Cu(I)-binding protein. *J. Biol. Chem.* **2002**, *277*, 31237–31242. [CrossRef]
41. He, J.; Ford, H.C.; Carroll, J.; Douglas, C.; Gonzales, E.; Ding, S.; Fearnley, I.M.; Walker, J.E. Assembly of the membrane domain of ATP synthase in human mitochondria. *Proc. Natl. Acad. Sci. USA* **2018**, *115*, 2988–2993. [CrossRef]
42. Gauba, E.; Chen, H.; Guo, L.; Du, H. Cyclophilin D deficiency attenuates mitochondrial F1FO-ATP synthase dysfunction via OSCP in Alzheimer's disease. *Neurobiol. Dis.* **2019**, *121*, 138–147. [CrossRef]
43. Zhang, Z.Y.; Chen, B.; Zhao, D.J.; Kang, L. Functional modulation of mitochondrial cytochrome c oxidase underlies adaptation to high-altitude hypoxia in a Tibetan migratory locust. *Proc. R. Soc. B Biol. Sci.* **2013**, *280*, 20122758. [CrossRef] [PubMed]
44. Zhang, F.F.; Broughton, R.E. Heterogeneous natural selection on oxidative phosphorylation genes among fishes with extreme high and low aerobic performance. *BMC Evol. Biol.* **2015**, *15*, 173. [CrossRef] [PubMed]
45. Li, X.-D.; Jiang, G.-F.; Yan, L.-Y.; Li, R.; Mu, Y.; Deng, W.-A. Positive selection drove the adaptation of mitochondrial genes to the demands of flight and high-altitude environments in grasshoppers. *Front. Genet.* **2018**, *9*, 605. [CrossRef] [PubMed]

genes

Article

The Ribosomal Protein RpL22 Interacts In Vitro with 5′-UTR Sequences Found in Some *Drosophila melanogaster* Transposons

Crescenzio Francesco Minervini [1], Maria Francesca Berloco [2], René Massimiliano Marsano [2,3,*,†] and Luigi Viggiano [2,*,†]

1 Department of Emergency and Organ Transplantation (D.E.T.O.), Hematology and Stem Cell Transplantation Unit, University of Bari "Aldo Moro", 70124 Bari, Italy; crescenziofrancesco.minervini@uniba.it
2 Department of Biology, Università degli Studi di Bari "Aldo Moro", 70125 Bari, Italy; mariafrancesca.berloco@uniba.it
3 Department of Genetics Anthropology Evolution, University of Parma, Parco Area delle Scienze 11/A, 43124 Parma, Italy
* Correspondence: renemassimiliano.marsano@uniba.it (R.M.M.); luigi.viggiano@uniba.it (L.V.)
† These authors contributed equally to this work.

Abstract: Mobility of eukaryotic transposable elements (TEs) are finely regulated to avoid an excessive mutational load caused by their movement. The transposition of retrotransposons is usually regulated through the interaction of host- and TE-encoded proteins, with non-coding regions (LTR and 5′-UTR) of the transposon. Examples of new potent cis-acting sequences, identified and characterized in the non-coding regions of retrotransposons, include the insulator of *gypsy* and Idefix, and the enhancer of *ZAM* of *Drosophila melanogaster*. Recently we have shown that in the 5′-UTR of the LTR-retrotransposon *ZAM* there is a sequence structured in tandem-repeat capable of operating as an insulator both in *Drosophila* ($S2R^+$) and human cells (HEK293). Here, we test the hypothesis that tandem repeated 5′-UTR of a different LTR-retrotransposon could accommodate similar regulatory elements. The comparison of the 5′-UTR of some LTR-transposons allowed us to identify a shared motif of 13 bp, called Transposable Element Redundant Motif (TERM). Surprisingly, we demonstrated, by Yeast One-Hybrid assay, that TERM interacts with the *D. melanogaster* ribosomal protein RpL22. The *Drosophila* RpL22 has additional Ala-, Lys- and Pro-rich sequences at the amino terminus, which resembles the carboxy-terminal portion of histone H1 and histone H5. For this reason, it has been hypothesized that RpL22 might have two functions, namely the role in organizing the ribosome, and a potential regulatory role involving DNA-binding similar to histone H1, which represses transcription in *Drosophila*. In this paper, we show, by two independent sets of experiments, that DmRpL22 is able to directly and specifically bind DNA of *Drosophila melanogaster*.

Keywords: ribosomal protein; Rpl22; *Drosophila*; DNA-protein interaction; transposable elements; histone 1-like

check for updates

Citation: Minervini, C.F.; Berloco, M.F.; Marsano, R.M.; Viggiano, L. The Ribosomal Protein RpL22 Interacts In Vitro with 5′-UTR Sequences Found in Some *Drosophila melanogaster* Transposons. *Genes* **2022**, *13*, 305. https://doi.org/10.3390/genes13020305

Academic Editors: Maciej Wnuk and Przemyslaw Szafranski

Received: 6 November 2021
Accepted: 1 February 2022
Published: 5 February 2022

1. Introduction

Transposable elements (TE) are DNA sequences which are able to move throughout the host genome. These elements were first identified more than 60 years ago by the geneticist Barbara McClintock [1].

TEs constitute a large fraction of the eukaryotic genome (i.e., up to 45% of the human genome and at least 50% of the maize genome [2,3]). The activity of these elements has been linked to more than 75 human diseases including hemophilia A, breast cancer, colorectal cancer, amyotrophic lateral sclerosis, and frontotemporal lobar degeneration [4–8]. In addition, TEs contribute to both neurodevelopment and neurological diseases and disorders [9,10]. Thus, it is important to understand how TEs transpose and how their mobilization is regulated in eukaryotic organisms. While most TEs in the human genome are completely inactive, the thirty percent of the elements in the *Drosophila melanogaster* genome are intact and active [11,12]. As such, *D. melanogaster* has always been considered a model organism for the study of eukaryotic TEs.

TEs are divided into two major classes based on their mechanism of transposition: DNA transposons and retrotransposons.

The elements of class I, also known as retrotransposons, are mobilized through a "copy and paste" mechanism according to which an intermediate of RNA is reverse transcribed into a cDNA copy and it is integrated elsewhere in the genome [13].

Retrotransposons include Long Terminal Repeat (LTR) retrotransposons, non-LTR retrotransposons (LINEs and LINE-like elements), and short interspersed nuclear elements (SINEs) [14].

The genome may be viewed as an ecosystem inhabited by diverse communities of TEs, which seek to propagate and multiply through sophisticated interactions with each other and with other components of the cell [15].

The transposition of class I TEs and its control take place thanks to the interactions between specific non-coding regions of TEs, tRNAs, self-encoded, and host-encoded molecules, including tRNAs and proteins [16].

These non-coding regions are able to control the transcription of the ORFs present in the transposon, and this determines the regulation of their life cycle itself.

Along with these regulatory elements, mostly located in the LTRs, new classes of functional elements have been identified and characterized, the most important of which is called "insulator". The characterization of TE-related regulatory sequences can also boost the development of new biotechnological tools [17,18].

One of the first TEs where a potent regulatory element has been characterized was *gypsy*, which harbors an insulator in their 5′-UTR. In a previous article, we have shown that also the 5′-UTR of the LTR-retrotransposon *ZAM* acts as an insulator both in *Drosophila* (S2-R$^+$) and human cells (HEK293) [19].

Notably, *ZAM*'s insulator has the same tandemly repeated structure and the same localization (5′-UTR) like *gypsy*'s insulator.

These observations led us to formulate the hypothesis that the tandem repeat regions present in the 5′-UTR of some other retrotransposons (RTs) could accommodate similar regulatory elements. In a previous paper, we grouped the *D. melanogaster* LTR-retrotransposons into three distinct subsets, based on the presence and the complexity of the repeats in the 5′-UTR [20]. Among the retrotransposons with complex repeats in the 5′-UTR, *Tirant* [21,22], *accord* [12], and *ZAM* [23] were selected due to the greater linguistic complexity and lower AT/GC ratio in their tandem-repeat sequences. In the tandemly repeated region of *ZAM*'s 5′-UTR, we had previously identified for the first time the DNA binding site of the HP1 protein [20]. The binding of HP1 to the 5′-UTR of *ZAM* could have a repressive role, inhibiting the retrotransposition of *ZAM*, possibly by recruiting chromodomain-containing proteins, such as protein of the Polycomb group, and thus burying the TE in a heterochromatic domain. The ability to bind the 5′-UTR of a retrotransposon could then represent a generalized defense mechanism of the genome to keep certain species of retrotransposons under control. With the aim to test this hypothesis, we have identified in the 5′-UTR of *ZAM*, *accord*, and *Tirant* a shared motif of 13 bp that we have called TERM (Transposable Element Redundant Motif).

"In vivo" and "in vitro" experiments demonstrated that TERM specifically interacts with the RpL22 protein. We demonstrate here that the peculiar H1/H5-like N-terminal domain of RpL22 of *D. melanogaster* [24,25] is responsible for binding to the TERM motif. We propose that the nuclear localization of RpL22, demonstrated by immunofluorescence experiments, could be supportive of a possible role of Rpl22 as a candidate for the role of controller of the activity of a restricted group of retrotransposons carrying TERM.

2. Materials and Methods

2.1. In Silico Analysis

Multiple alignments were performed using Multalin [26]. The TERM motif has been detected by using DNA pattern discovery programs which use either enumerative algorithms to examine all oligomers of a given length, reporting those that occur more often than expected as output, or alignment methods to identify unknown signals by local multiple alignment of submitted sequences. We used both approaches to analyze 5′-UTRs of the RTs *accord*, *Tirant*, and *ZAM* using the programs oligo-analysis [27], and MEME [28]. The analyses resulted in similar patterns that can be suitably described by the TERM position weight matrix.

Pattern search. DNA pattern search programs are based on a positional weight matrix (PWM) description of the pattern to be searched. The weight score associated with each examined DNA segment represents a measure of its similarity to the collection of sequences that constitute the PWM—the more a given DNA segment matches the PWM, the higher its weight score. We used the Matrix-scan program [29] to scan the comparable random sequences, generated using *D. melanogaster* as background model with TERM PWM. Analyses were performed with a weight score threshold of 5.29, established as the lower value that is associated with a conserved TERM element in the 5′-UTR of *ZAM*, *accord*, and *Tirant*.

2.2. Plasmid Construction and Sequencing

The yeast integration and reporter vector used to produce the one-hybrid reporter plasmid was pHISi-1 (Clontech, Palo Alto, CA, USA). The reporter plasmid (pTERM3-HISi-1) was constructed by cloning the couple of annealed anticomplementary primers, containing TERM3, into EcoRI/XbaI sites of pHISi-1.

TERM3-F→5′-aattcATCAAtcgctgaTATCAAtcgctgaTATCAAtcgctgaTg-3′

TERM3-R→3′-gTAGTTAGCGACTATAGTTAGCGACTATAGTTAGCGACTA agatc-5′

Plasmid's map and corresponding nucleotide sequence are available from the authors.

2.3. Yeast One-Hybrid Assay

One-hybrid selection was performed according to the manufacturer's protocol (Clontech's MATCHMAKER One Hybrid System) as already depicted in Minervini et al. 2007 [20].

2.4. Expression and Purification of RpL22 Protein and RpL22/H5 RpL22/L22 Polypeptides

The full-length cDNA of RpL22 gene was amplified by High-Fidelity PCR using primer pair pETup/pETlow from one of the Yeast One-Hybrid assay positive clones (Figure 1).

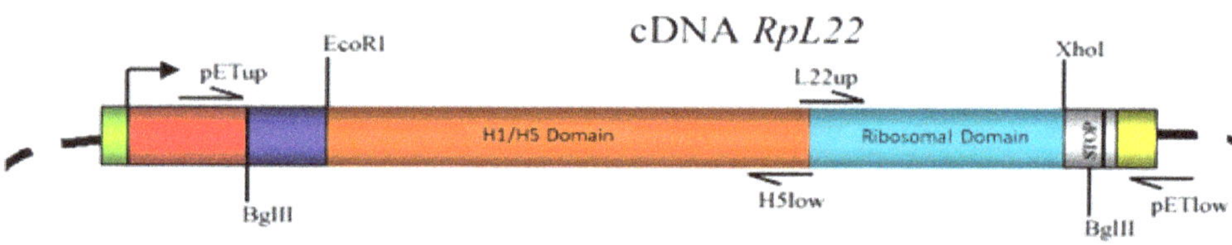

Figure 1. Full-length RpL22 cDNA clone. Graphical representation of the RpL22 cDNA clone which was used to construct vectors expressing RpL22 and its sub-domains (ribosomal and histone-like). Arrows indicate the name and position of the PCR oligo-primers.

The purified PCR product was cloned into the pET-200 expression vector (Invitrogen, USA) to obtain the pET-200/RpL22 plasmid. This plasmid was transformed into the *E. coli* BL21 Star™ expression host. RpL22 gene was expressed in BL21 Star ™ following the manufacturer's instructions (Invitrogen, Waltham, MA, USA). His6-Rpl22 protein was purified from harvested cells using the Ni-NTA Fast Start kit (Qiagen, Hilden, Germany) under native conditions following the manufacturer's instructions. The molecular mass of the protein was determined by SDS-PAGE (12% (w/v) after staining with Coomassie brilliant blue R-250 The concentrations of the purified protein were determined by the Bradford method [30].

The Histone-like domain and the Ribosomal domain of the RpL22 gene were amplified by high-fidelity PCR using, respectively, the primer pair pETup/H5low and pETlow/L22up and cloned the PCR products into pET200 vector to obtain the plasmid pET-200/RpL22_H5, and pET-200/RpL22_L22. Next, the pET plasmids were transformed into *E. coli* BL21 Star™. To obtain the purified RpL22/H5 and RpL22/L22 polypeptides we followed the same procedures described above.

pETup 5′-CACCATGGCTTACCCATA-3′
pETlow 5′-ATAAAAGAAGGCAAAACGATG-3′
H5low 5′-CTAACGCAGCACGTTCTTCTT-3′
L22up 5′-CACCAAGGTGGTCAAGAAGAA-3′

2.5. DNA-Binding Assays

Gel mobility shift assays were performed essentially as previously described [31]. Unspecific λ-DNA was sonicated to obtain DNA of average fragments size comparable to that of TERM3.

2.6. Production of Antibody Anti RpL22/H5

Detection of RpL22/H5 polypeptides in Sodium Dodecyl Sulfate-Polyacrylamide Gels was performed as previously described [32].

Briefly, the preparation of RpL22/H5 polypeptide from BL21 lysates has been performed by using sodium dodecyl sulfate-polyacrylamide gel electrophoresis (SDS-PAGE). After electrophoresis, the band of RpL22/H5 polypeptide has been located in the gel by light staining of the gel with 0.05% Coomassie Brilliant Blue R-250 prepared in water. After 10 min of staining the gel has been washed with numerous changes of water over the next few hours. Once the appropriate band was visible, it has been excised with a scalpel.

Preparation of RpL22/H5 polypeptides from Sodium Dodecyl Sulfate-Polyacrylamide Gels for Immunization was performed as previously described [33].

Briefly, after removing the RpL22/H5 band from a polyacrylamide gel, it needs to be fragmented into small pieces before being injected into animals, making it more easily phagocytized and presented to cells of the immune system. We removed the plungers from the barrels of two 5-mL syringes and place the gel fragment into one of the barrels. Afterwards we replaced the plunger and place the syringe outlet in the barrel of the second syringe. Using firm and rapid pressure on the plunger, we pushed the gel into the second syringe. We repeated the process five times, passing the gel fragments back and forth between the two syringes. Finally, we placed 21-gauge needles onto the outlet of the syringes and repeat the process a couple of times. After preparing the antigen (RpL22/H5 polypeptide) we sent it to Invitrogen Custom Polyclonal Antibody Service to obtain the production and purification of the Anti-RpL22/H5 antibody from rabbit.

2.7. Immunofluorescence and Immunocytochemistry

Immunofluorescence (IF) and immunocytochemistry (ICC) were performed as follows. Cells were fixed with 4% paraformaldehyde (PFA) for 10 min at room temperature and permeabilized with 0.2% Triton X-100 before immunostaining. The cells were washed in PBS and blocked for 1 h in blocking buffer (10% goat serum in PBS). Samples were incubated with Anti-Fibrillarin antibody (G-8) (sc-374022_Santa Cruz Biotechnology), Anti-Ribosomal ProteinL28 (A-16) (sc-14151_Santa Cruz Biotechnology), Anti-Histone H1 (AE-4) (sc-8030_Santa Cruz Biotechnology), and our Anti RpL22/H5 for 1 h at r.t., washed three times in PBS and incubated with Alexa Fluor 488 goat anti-rabbit secondary antibody (Life Technologies Carlsbad, CA, USA, 1:200 dilution) and Alexa Fluor 488 goat anti-mouse secondary antibody (Life Technologies, Carlsbad, CA, USA 1:200 dilution) for 1 h at r.t. for detection. Counterstaining was done with DAPI. Images were acquired using a Leica IL

MD LED inverted fluorescence microscope. To ensure the validity and specificity of the anti RpL22/H5 antibody, we conducted IF and ICC experiments using the pre-immune serum under the same experimental conditions described above without obtaining any signal on the tested cells or tissues.

3. Results

3.1. Search for Shared Motifs

In a previous work, we published the comparative analysis of the 5′-UTR of the known LTR-retrotransposons of *D. melanogaster* [19]. This analysis revealed that 19 out of 49 5′-UTRs tested (39%) have a tandemly repeated organization and that it was possible to cluster them based on their linguistic complexity and A/T content. It was possible to cluster *accord*, *Tirant*, and *ZAM* into the same group, that share very complex and extended tandemly repeated regions in the 5′-UTRs.

Using the software oligo-analysis, available in the RSATools package [27] (available online: http://rsat.ulb.ac.be/rsat; accessed on 25 April 2018) we performed an analysis to find shared motifs among all tandem repeats identified in the 5′-UTR of the retrotransposons analyzed.

The output of oligo-analysis describes the consensus sequence of the motifs, their position, and its score value representing the statistical significance of the motif and a graphic representation (Figure 2A).

A 13 bp-long motif shared by the LTR-retrotransposons *ZAM*, *accord*, and *Tirant* shows a very high score value. The consensus sequence of the motif is the following: ATCCATCGCTGAT.

The analysis was repeated using an alternative tool (MEME 26), available in the MEME Suite (available at: http://meme-suite.org/; last accessed 25 April 2018) which gave the same. This motif has been called "TERM" (Transposable Element Redundant Motif).

To exclude that TERM was a pattern emerged by chance, we repeated the same analyzes on a group of comparable random sequences, generated using *D. melanogaster* as background model. The sequence generator used is available in RSATools. The TERM matrix was used to scan the 5′-UTR-comparable random sequences using RSATools Matrix-scan program [29]. This analysis did not produce any statistically significant results. Furthermore, we have scanned the 5′-UTRs of all retrotransposons and we found TERM just in *ZAM*, *accord* e *Tirant*. This data acquires further importance considering that in *D. melanogaster* there are several hundred copies of the TERM motif, as highlighted by genome to matrix-scan analysis of the genome. So, while it is a relatively common motif, it is only present in the 5′-UTRs of *ZAM*, *accord*, and *Tirant*.

3.2. Identification of Proteins Able to Interact with TERM

One way to determine if a short DNA sequence may have a function is to identify protein(s) interacting with it.

For this purpose, we performed a Yeast One-Hybrid assay using the TERM motif as bait (see Material and Methods).

To carry out the One-Hybrid assay with TERM, we designed a pair of complementary oligonucleotides in which the TERM element was repeated three times ($TERM^3$). The double stranded fragment obtained from the annealing of the two oligonucleotides was cloned into the One-Hybrid vectors.

His^- yeast mutant strain (YM427), bearing $pTERM^3$-HISi-1, the reporter plasmid carrying three TERM tandem-repeat copies cloned upstream of the HIS3 selectable marker, was transformed with a cDNA recombinant plasmid library obtained from 0–21 h embryos of *D. melanogaster* fused with yeast GAL4 activation domain.

The screening of approximately 8.3×10^5 yeast transformed cells led to the identification of 51 full-length cDNA clones that reproducibly shown to restore the His+ phenotype in yeast bearing the $TERM^3$ reporter plasmid, whereas they failed to transform the yeast cells bearing the parental reporter plasmid lacking $TERM^3$ to His+ phenotype.

A

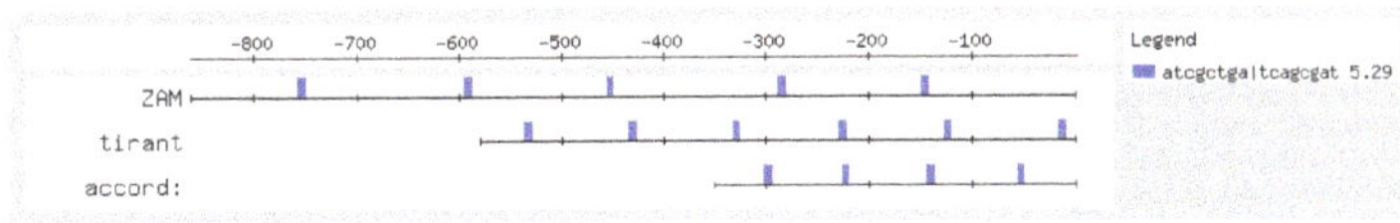

B

A	15	0	0	9	15	0	0	0	0	0	0	15	3
C	0	0	15	0	0	0	15	0	15	0	0	0	0
G	0	0	0	0	0	0	0	15	0	0	15	0	0
T	0	15	0	6	0	15	0	0	0	15	0	0	12

C

D

```
>ZAM
acgcatataaaaaaaaaaaaaaaaaaaaaagaaaaaaaaaaacccaaaacttaaaaatgccgtaaccg
cgaaacatgatatgcgttgtacttgtgtgaaATCAATCGCTGATagtcactgccgaagtttattaaggc
caagtaccatatcattactttcatgtttacatacatatatatgccccacaattaaaacaacatacacac
acacaaatatttcaaatgcaaaaaaaaaaaaaagaatgtagtgtacctgcgtggcATCAATCGCTGATa
aaccactgccgaaatattaaaggcccggtactacatcacaaaacacgtatatatgcaacaaaaatatac
acaacaaaaccatatatacaaacgtgtatgagtgacgtgtaatgtacttgtgtgaaATCAATCGCTGAT
aatcactgccgaagcttagtaaggccaagtaccacatcattactaacatgtgtacatatatatatatgc
aaaacaattaaaacaacatacacacacacaaatatttcaaatgcaaaaaaaaaaaaaaagaggaaatg
ttgtgtacctgcgtggcATCAATCGCTGATaaaccactgccgaaatattaaaggcccggtactacatca
caaaacacgtatatatgcaacaaaaatatacacaacaaaaccatatatacaaacgtgtatgagtgacgt
gtaatgtacttgtgtgaaATCAATCGCTGATaatcactgccgaagcttagtaaggccaagtaccacatc
attactaacatgtgtacatatatatatgcaaaccaccaaaacaaatacatatacacatacaaacactcc
aaaaaaaaaaacaaataatactatatgaacggc

>tirant
aaaaaaagtgcaagtgtaccgtactgccgcgctgacgtggaATCTATCGCTGATcatcacgccatcggt
atgtccatactctgccgaacgtcataatttttttaaaaaagtgcaagtgtaccgtactgccgcactgac
gtggaATCTATCGCTGATcatcacgccatcggtatgtccatactctgccgaacgtcataattttttataa
aaaagtgcaagtgtaccgtactgccgcgctgacgtggaATCTATCGCTGATcatcacgccatcggtatg
tccatactctgccaaacgtcataattttttataaaaaagtgcaagtgtaccgtactgccgcgctgacgtg
gaATCTATCGCTGATcatcacgccatcggtatgtccatactctgccaaacgtcataattttttataaaaa
aagtgcaagtgtaccgtactgctgcgctgacgtggaATCTATCGCTGATcaccacgccatcggtatgtc
catactctgccaaacgtcataagttttttataaaaaaaaaagagtgcaagtgtaccgtactgccgcgctg
acgtggaATCTATCGCTGATcatcacgtc

>Accord
gacatacagtataattacaaactattgtgaccgaatatgtgcacatcaATCGCTGAAagggcaccggga
tcacacataaaaaaaatgtctttacattcaaagtgaccgaatatgtgcacatcaATCGCTGAAagggca
ccgggatcacttatgacatacataaaaagttaacatcattcagtgctatggccaaatatgtgtacatca
ATCGCTGATgagacaccgggaccatacataaataagcaaaacaaataaatctagttcaatgttgtggcc
gaatatgtgtacatcaATCGCTGAAaaggcaccgggaccacaacattgaaagaaattaaacaataaatc
agta
```

Figure 2. Bioinformatics analysis of the 5′-UTRs of *ZAM*, *Tirant*, and Accord. (**A**) Feature map of over-represented TERM in the 5-UTRs of the indicated RTs. The scale bar provides coordinates relative to the first ORF (GAG) start of the retrotransposons. Note the regularity of the TERM motif in the 5′-UTRs; (**B**) display of the logos of the TERM motif. The graphic representation was created using WebLogo. Sequence logos are a graphical representation of an alignment of multiple nucleic acid sequences (PWM) developed by Tom Schneider and Mike Stephens [34]. Each logo is made up of stacks of symbols, one stack for each position in the sequence. The overall height of the stack indicates the conservation of the sequence at that location, while the height of the symbols within the stack indicates the relative frequency of each nucleic acid at that location; (**C**) positional weight Matrix of TERM motif; (**D**) sequence of the tandem repeats present in the 5′-UTR of the RTEs under examination. The single tandem repeats are in blue and red, while the TERM motifs are in uppercase underscored.

All the positive clones were then sequenced, and the identity of each insert was obtained by a BLAST search of the *D. melanogaster* predicted genes databases.

The results showed that 35 independent clones (69% of the positive clones) correspond to the gene CG7434 that encodes the Ribosomal protein L22 (RpL22) (see Table 1).

Table 1. Yeast One-Hybrid assay results.

Clone's Name	BLAST Results	BLASTX Results	Notes
L1	mitochondrial sequence	#	#
L2	Grn	GRN	GATA trascription factor
L3	CG7434	RpL22	Ribosomal protein
L4	CG7434	RpL22	Ribosomal protein
L5	Csn6	CSN6	Signalosome
L6	CG7434	RpL22	Ribosomal protein
L7	Mis	MIS	body pigmentation
pTERM3lig01	CG4314	st	eye pigment precursor transport
pTERM3lig02	GS1	GS1	glutammina sintetasi
pTERM3lig03	CG1883	CG1883	Rps7-like
pTERM3lig05	CG7434	RpL22	Ribosomal protein
pTERM3lig06	CG7434	RpL22	Ribosomal protein
pTERM3lig07	Hrb27C	Hrb27C	RNA binding protein
pTERM3lig08	CG7434	RpL22	Ribosomal protein
pTERM3lig09	CG7434	RpL22	Ribosomal protein
pTERM3lig10	CG7434	RpL22	Ribosomal protein
pTERM3lig11	CG9253	CG9253	RNA helicase activity
pTERM3lig12	Mod(mdg4)	Mod(mdg4)	FLYWCH domain
pTERM3lig13	CG7434	RpL22	Ribosomal protein
pTERM3lig14	CG7434	RpL22	Ribosomal protein
pTERM3lig15	RpS16	RpS16	Ribosomal protein
pTERM3lig16	CG7434	RpL22	Ribosomal protein
pTERM3lig17	CG7434	RpL22	Ribosomal protein
pTERM3lig19	CG7434	RpL22	Ribosomal protein
pTERM3lig20	CG7434	RpL22	Ribosomal protein
pTERM3lig21	CG7434	RpL22	Ribosomal protein
pTERM3lig25	CG7434	RpL22	Ribosomal protein
pTERM3lig26	CG7434	RpL22	Ribosomal protein
pTERM3lig27	CG6007	GatA	serine hydrolase activity
pTERM3lig29	CG7434	RpL22	Ribosomal protein
pTERM3lig31	CG7434	RpL22	Ribosomal protein
pTERM3lig32	CG30389	CG30389	actin filament binding activity
pTERM3lig33	CG7434	RpL22	Ribosomal protein
pTERM3lig34	CG7434	RpL22	Ribosomal protein
pTERM3lig35	CG9415	CG9415	trascription factor
pTERM3lig36	CG9277	CG9277	beta tubulina
pTERM3lig38	CG7434	RpL22	Ribosomal protein
pTERM3lig39	CG7434	RpL22	Ribosomal protein
pTERM3lig41	CG7434	RpL22	Ribosomal protein
pTERM3lig43	CG7434	RpL22	Ribosomal protein
pTERM3lig44	CG7434	RpL22	Ribosomal protein
pTERM3lig45	CG7434	RpL22	Ribosomal protein
pTERM3lig46	CG7434	RpL22	Ribosomal protein
pTERM3lig47	CG7434	RpL22	Ribosomal protein
pTERM3lig48	CG7434	RpL22	Ribosomal protein
pTERM3lig49	CG7434	RpL22	Ribosomal protein
pTERM3lig50	CG7434	RpL22	Ribosomal protein
pTERM3lig51	CG17326	luna	Zinc finger C2H2-type

Table 1. *Cont.*

Clone's Name	BLAST Results	BLASTX Results	Notes
pTERM3lig52	CG7434	RpL22	Ribosomal protein
pTERM3lig53	CG7434	RpL22	Ribosomal protein
pTERM3lig54	CG7434	RpL22	Ribosomal protein

Fifty-nine percent (35 out of 51) of the positive clones isolated in the One-Hybrid assay correspond to gene CG7434 (in green) encoding the ribosomal protein RpL22. The remaining 16 clones correspond to non-coding mitochondrial sequences (L1), structural or enzymatic proteins (L5, L7, pTERM3lig01, pTERM3lig02, pTERM3lig36, pTERM3lig27), other ribosomal protein (pTERM3lig15), and some transcription factors and/or other DNA binding proteins (L2, pTERM3lig12, pTERM3lig32, pTERM3lig35, pTERM3lig51). It is important to underline that, except for the CG7434 gene, all clones screened with the One-Hybrid assay are represented only once.

3.3. Analysis of RpL22/TERM Interaction

RpL22 encodes for a ribosomal protein of the major subunit of the ribosome in *D. melanogaster*. It encodes a 299 aa-long protein consisting of two domains: a ribosomal L22e C-terminal domain (L22e) which accounts for 1/3 of the protein and a N-terminal H1/H5-like domain (H1/H5) which occupies the remaining 2/3 of the protein [24,25].

The latter domain is a highly basic domain and is characteristic of the H1, and H5 linker histones, and is responsible for their DNA-binding proprieties.

There is no evidence in the literature that RpL22 is able to directly interact with DNA. However, the presence of the H1/H5-like domain, and the result of our Yeast One-Hybrid assay suggest that Rpl22 is a DNA-binding protein.

Gel mobility shift assays were performed to confirm the ability of RpL22 to bind DNA, and especially the direct interaction between RpL22 and TERM.

$TERM^3$ was terminally radiolabeled and used as substrate. The purified recombinant RpL22 protein was incubated with $TERM^3$ and the DNA-protein complexes were resolved by native PAGE.

The shifted signals indicated that RpL22, when incubated with $TERM^3$, was able to produce a slower migrating complex (Figure 3A).

Several EMSA competition experiments were performed to investigate the specificity of this interaction. Purified RpL22 was incubated with a constant amount of radiolabeled $TERM^3$ and in presence of increasing amounts of unlabeled DNA competitors.

The $TERM^3$-RpL22 interaction appears to be specific since using up to 500-fold molar excess of unlabeled non-specific competitor (sonicated λ-DNA) did not affect the shift of the $TERM^3$-RpL22 complex (Figure 3A). Specific competition experiments performed with an increasing amount of unlabeled $TERM^3$ fragments showed that the $TERM^3$-RpL22 complex was easily destroyed by adding small quantities (5x) of specific competitor (Figure 3A).

The EMSA results confirmed our assumption that RpL22 may be able to interact specifically with $TERM^3$. To understand which domain was responsible for this interaction, we separately cloned the two domains L22e and H1/H5 in the pET expression vector. The two domains were purified and used in further EMSA experiments (Figure 3B). As shown in the figure, only the H1/H5-like domain is able to bind $TERM^3$, while the ribosomal domain (C-term domain) does not interact with $TERM^3$. This means that the ability of RpL22 to bind TERM3 is conferred by its H1/H5-like domain.

It is interesting to note that the size of the shift, obtained using the entire protein, is lower than that obtained using the H1/H5 domain alone. Apparently, this seems to be inconsistent, as we would expect an opposite behavior, given the reduction in the size of the protein. We believe that the lack of the L22 domain causes a change in the charge density of the polypeptide, which becomes more positive and this slows the run of Rpl22/H5 polypeptide.

3.4. RpL22 Sub-Cellular Localization

From what we have shown so far, it emerges that the RpL22 protein has a bivalent nature with an additional non-ribosomal function. It is indeed composed of a Histone-like portion (N-term) and a ribosomal portion (C-term). Consistent with the presence of the histone portion, RpL22 is able to interact with DNA both "in vivo" (Yeast One-Hybrid assay) and "in vitro" (EMSA).

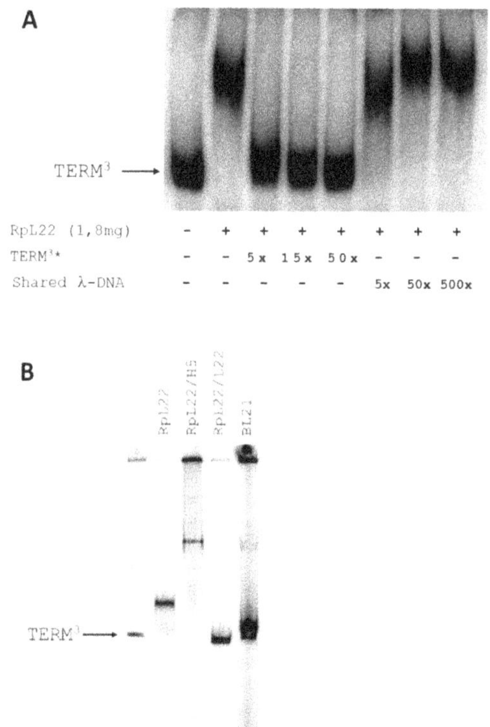

Figure 3. Rpl22 binds the TERM^3 in vitro. Each lane contains an identical amount of input labeled TERM^3 DNA (2 ng) incubated with recombinant purified Rpl22 protein. (**A**) TERM3-Rpl22 complex formation has shown in the lane 2, whereas the remaining lanes are committed to specific and not-specific competition experiments: specific competitor (unlabeled TERM^{3*}) or a large excess of non-specific competitor (shared λ-DNA) were used as shown in figure; (**B**) identification of which domain of RpL22 is responsible for binding with TERM^3: we used purified Rpl22 (1.8 μg), RpL22/H5 (1.2 μg), and RpL22/L22 (0.6 μg). We used different amounts of the proteins to maintain the same stoichiometric ratio. The experiment suggests that only the RpL22/H5 polypeptide is able to bind TERM^3.

To try to uncover in vivo in *D. melanogaster* the extra-ribosomal role of Rpl22, we analyzed its cellular location.

Mageeney and colleagues showed that RpL22, in the testes, displays a punctate nuclear pattern, probably in the nucleoli [35].

Given the nature of the tissue examined, together with the fact that the expression of the RpL22 gene is not ubiquitous in all testicular cell subtypes, we wanted to verify the protein sub-cellular localization in a more tractable experimental cellular system, such as the *D. melanogaster* S2R$^+$ cell line.

We produced a polyclonal antibody from the H1/H5-like domain to be used in immunofluorescence experiments on S2R$^+$ cells. To allow a more precise sub-cellular localization of the RpL22 protein, we performed both immunofluorescence and co-immunofluorescence experiments. We used anti-RpL22, anti-Rpl28, anti-H1, and anti-fibrillarin antibodies (Figure 4). Immunofluorescence staining experiments using formaldehyde-fixed S2R$^+$ cells showed that RpL22 has the expected ribosomal distribution pattern, namely a cytoplasmic and nucleolar localization (Panel A). Notably, the nucleolar localization of RpL22 roughly corresponds to DAPI staining loss in the nucleus. The pattern of RpL22 corresponds to the pattern of other ribosomal proteins (such as RpL28) (Panel B). We also performed co-immunofluorescence staining experiments using anti-RpL22 and anti-H1 antibodies. As expected, the two proteins have very different and almost specular localization, while RpL22 is localized in the cytoplasm and the nucleolus, H1 is positioned only in the nucleus with a staining free region that corresponds to the volume occupied by RpL22 in the nucleolus (Panel B).

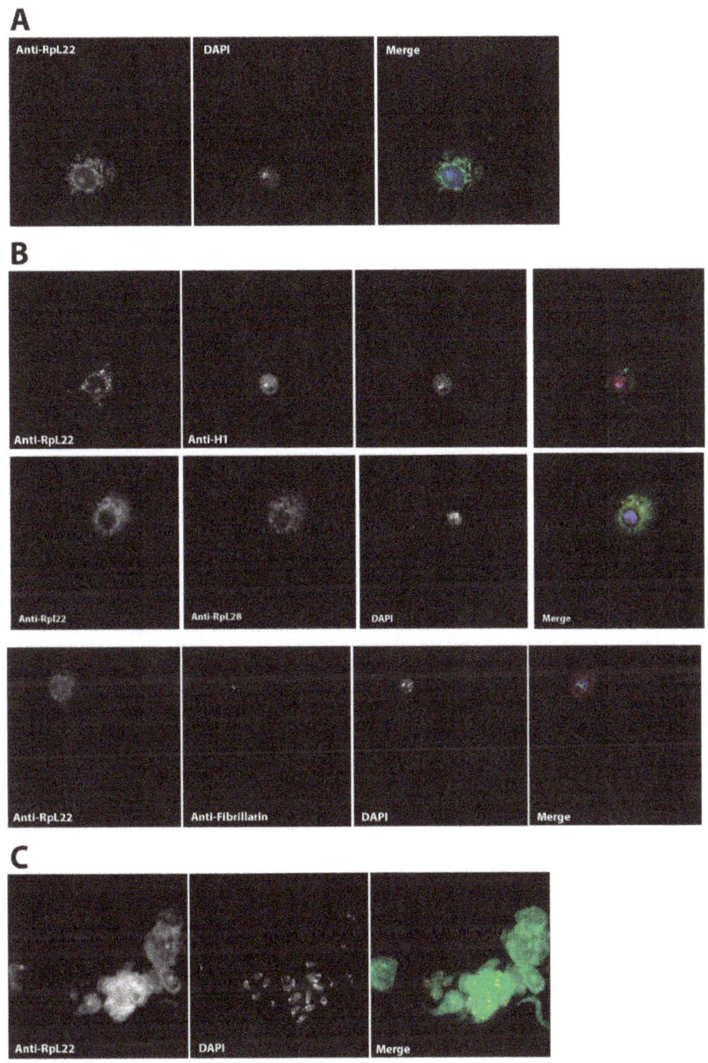

Figure 4. RpL22 localization in *Drosophila* cell line S2R$^+$ and in the brain cells. (**A**) Rpl22 localizes both in cytoplasm and nucleolus in S2R$^+$ cell. (**B**) To highlight the "ribosomal" behavior of RpL22, co-immunofluorescence experiments were performed both with the anti-H1 antibody and anti-RpL28 antibody, finally, as further confirmation of the nucleolar localization, RpL22 co-localizes with the nucleolar marker of fibrillarin. (**C**) The same localization pattern occurs (cytoplasm and nucleolus) also in neurons.

To confirm the cellular and subcellular localization of RpL22 also in tissues other than cultured cells and germline tissues, we performed immunofluorescence experiments also on *Drosophila* brain tissue, confirming the same cytoplasmic and nucleolar localization of RpL22 already highlighted in S2R$^+$ cells (panel C).

4. Discussion

The relationship between retrotransposons and the host genome has a double nature. On the one hand, RTs are a source of genetic variability that has been exploited during the evolution to develop both metabolic and physiological innovations such as the placental syncytins [36], or they are involved in genomic stress response and adaptation, mainly through rewiring of transcriptional networks [16,37,38]. On the other hand, retrotransposons constitute a hazard both to the structural integrity of the genome and its functionality. To balance these two opposite conditions, the eukaryotic genome has evolved mechanisms to control transposition and exploit RT genes and their regulatory regions to develop new gene functions (gain of function). Eukaryotic cells have developed a series of transcriptional repression mechanisms to tame retrotransposons, basically based on small RNAs. *gypsy* is controlled by a *Drosophila* locus called *flamenco*, which maintains the retrovirus in a repressed state. The *flamenco* locus acts as a source of piRNAs in the ovary, while in somatic tissues it acts as a source of endo-siRNAs [39–41].

Alternative mechanisms for controlling RTs activity have also been proposed, for example the transposon homing [42]. It has been hypothesized that the interaction between a motif present in the 5′-UTR of *ZAM* and the HP1 protein allows cells to direct the insertion of *ZAM* in a biased manner into the genome. This would force the new transposed copies of *ZAM* to converge in the heterochromatic regions of the genome where they would remain inactive [19].

However, the host-RT co-evolution has sometimes resulted in the development of potent regulatory sequences within the non-coding regions of RTs, due to their physical interaction with host factors. This phenomenon has been used by the host to create additional variability associated with the transcriptional rewiring of gene networks.

The 5′-UTR regions of some RTs are of particular interest from this standpoint. The 5′-UTR of RTs like *gypsy*, *ZAM*, *Tirant*, and *Idefix* have a polymorphic repeated organization. In some cases, the repeated structure has been associated with enhancer (*ZAM*), silencer (*gypsy*), or insulator (*gypsy*, *ZAM* and *Idefix*) functions. These functions are fulfilled through the physical interaction of one or more host proteins with the 5′-UTR of the RT, as reported for *gypsy* [43,44] and *ZAM* [20]. We have therefore tested the hypothesis that RTs harboring a structured 5′-UTR could be bounded by host proteins that could either regulate their transposition or confer them new functions.

In this work, we first identified a shared motif (TERM) in the repeat-containing 5′-UTRs of some RTs using a comparative analysis approach, and subsequently found the RpL22 protein as the main interacting protein of the TER motif, using the Yeast One-Hybrid assay. We have finally confirmed the Rpl22/TERM interaction in vitro and mapped the DNA binding domain to the NH-terminal portion of the protein. Although our IF experiments are not informative of the Rpl22/DNA interaction in vivo, this connection cannot be excluded, as discussed below.

So, what is the biological relevance of our findings? RpL22 is a ribosomal protein mainly localized in the cytoplasm. Nevertheless, other studies have highlighted its role in establishing a state of generalized transcriptional repression [45], as already demonstrated for histone H1 [45]. The results of our experiments show that not only RpL22 can interact directly with DNA, but also that this interaction is sequence-specific (TERM motif). The fact that RpL22 possesses a histone H1/H5-like domain capable of binding DNA leads us to hypothesize that the RpL22 protein could act, through the binding to the TERM motif, as a transcriptional repressor, especially on *ZAM*, *accord*, and *Tirant* where several copies of TERM are clustered (Figure 2A,D).

Although we have found that the Histone-like region is responsible for the binding of RpL22 to TERM both in vivo (Yeast One-Hybrid assay) and in vitro (EMSA), unlike Ni and colleagues [45], we were not able to pinpoint a chromosomal localization of RpL22 by IF and ICC experiments except for the nucleolus region (Figure 4). This may be due to the cell type (S2R$^+$), or tissue (*Drosophila* brain) used in our study.

The behavior of RpL22 may depend on the cell type. Some post-translational modifications of RpL22 (SUMOylation and phosphorylation) are known [46], and they may be able to modify the localization and/or function of RpL22 in a tissue- and/or developmental stage-dependent manner. It could also be hypothesized that, in S2R$^+$, neuron, and salivary gland cells [47], the putative chromatin-associated function of Rpl22 could be dispensable, while it could be essential in other tissues not investigated in this study. Imaginal discs are tissues experiencing profound changes in the transcriptional program and Rpl22 is one of the very few ribosomal genes active during metamorphosis (Figure 5) [48]. Therefore, Rpl22 might exert its role in controlling TEs, during the metamorphosis.

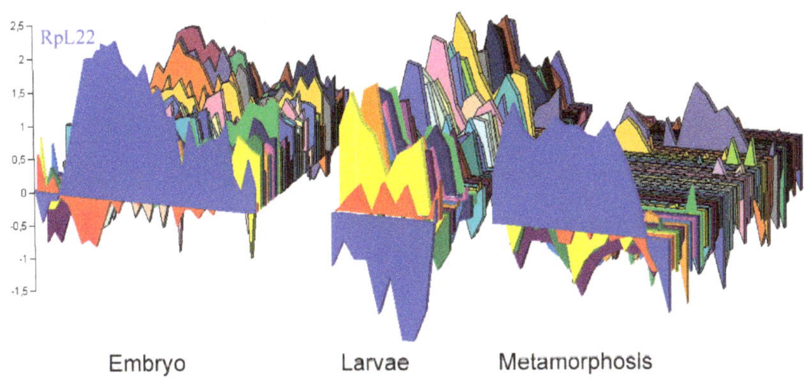

Figure 5. Comparison of the expression profile of RpL22 (in blue) with the other ribosomal proteins during the development of *Drosophila melanogaster*. Microarray data of ribosomal *D. melanogaster* gene expression during development was downloaded from the FLYMINE database [49] (available at: https://www.flymine.org/flymine; last accessed 15 December 2021). These data were used to construct the graph. Y axis: fold change. Reference sample is a pooled mRNA representing all stages of the life cycle as reported in Arbeitman et al. [48].

In a parallel study [47], we have also demonstrated that the *Doc5* transposon, which is located exquisitely in the heterochromatin of *D. melanogaster*, is also a binding site for Rpl22. Being a LINE-like transposon, *Doc5* has not been included in this study. At least six TERM-like motifs can be found in the *Doc5* sequence (Figure S1), which suggests that Rpl22 exhibits sequence specificity. Moreover, the study by Berloco et al. [47] confirms the connection between Rpl22 and transposable elements.

Additional studies, aimed at the identification of the Rpl22/DNA interaction in vivo are required to support our current hypotheses. However, studies aimed at revealing Rpl22 as a chromatin component require antibody optimization and the development of a transgenic line that expresses efficiently the Rpl22 protein. The only transgenic line allowing the overexpression of Rpl22 available to date [50] does not allow for an efficient testing of our hypothesis. Furthermore, it is possible that the Rpl22 binding to chromosomes could be only highlighted in vivo under particular physiological conditions (such as development, tissue, or stress specific conditions), making it difficult to uncover the role of Rpl22 in chromatin dynamics.

In conclusion, our results show that the *D. melanogaster* Rpl22 protein specifically interacts in vitro with DNA sequences related to TEs. While our findings open up the possibility for RpL22 to participate in controlling TEs of *D. melanogaster*, such interactions need an experimental validation in vivo using specific approaches such as Chip-seq or similar methods.

Supplementary Materials: The following supporting information can be downloaded at: https://www.mdpi.com/article/10.3390/genes13020305/s1, Figure S1: Doc5 transposon fragment harbors at least 6 TERM-like elements

Author Contributions: C.F.M., L.V. performed One Hybrid Assays; L.V., M.F.B. performed Immunofluorescence experiments; C.F.M. performed in silico analyses; L.V., R.M.M. supervised the project; L.V., R.M.M. drafted the manuscript. All authors have read and agreed to the published version of the manuscript.

Funding: This research received no external funding.

Institutional Review Board Statement: Not applicable.

Informed Consent Statement: Not applicable.

Data Availability Statement: Not applicable.

Conflicts of Interest: The authors declare no conflict of interest. The funders had no role in the design of the study; in the collection, analyses, or interpretation of data; in the writing of the manuscript, or in the decision to publish the results.

References

1. McClintock, B. Induction of Instability at Selected Loci in Maize. *Genetics* **1953**, *38*, 579–599. [CrossRef]
2. Lander, E.S.; Linton, L.M.; Birren, B.; Nusbaum, C.; Zody, M.C.; Baldwin, J.; Devon, K.; Dewar, K.; Doyle, M.; FitzHugh, W.; et al. Initial sequencing and analysis of the human genome. *Nature* **2001**, *409*, 860–921. [CrossRef]
3. SanMiguel, P.; Tikhonov, A.; Jin, Y.K.; Motchoulskaia, N.; Zakharov, D.; Melake-Berhan, A.; Springer, P.S.; Edwards, K.J.; Lee, M.; Avramova, Z.; et al. Nested retrotransposons in the intergenic regions of the maize genome. *Science* **1996**, *274*, 765–768. [CrossRef] [PubMed]
4. Batzer, M.A.; Deininger, P.L. Alu repeats and human genomic diversity. *Nat. Rev. Genet.* **2002**, *3*, 370–379. [CrossRef] [PubMed]
5. Chen, L.; Dahlstrom, J.E.; Chandra, A.; Board, P.; Rangasamy, D. Prognostic value of LINE-1 retrotransposon expression and its subcellular localization in breast cancer. *Breast Cancer Res. Treat.* **2012**, *136*, 129–142. [CrossRef]
6. Lannoy, N.; Hermans, C. Principles of genetic variations and molecular diseases: Applications in hemophilia A. *Crit. Rev. Oncol./Hematol.* **2016**, *104*, 1–8. [CrossRef] [PubMed]
7. Li, W.; Jin, Y.; Prazak, L.; Hammell, M.; Dubnau, J. Transposable elements in TDP-43-mediated neurodegenerative disorders. *PLoS ONE* **2012**, *7*, e44099. [CrossRef]
8. Solyom, S.; Ewing, A.D.; Rahrmann, E.P.; Doucet, T.; Nelson, H.H.; Burns, M.B.; Harris, R.S.; Sigmon, D.F.; Casella, A.; Erlanger, B.; et al. Extensive somatic L1 retrotransposition in colorectal tumors. *Genome Res.* **2012**, *22*, 2328–2338. [CrossRef] [PubMed]
9. Baillie, J.K.; Barnett, M.W.; Upton, K.R.; Gerhardt, D.J.; Richmond, T.A.; De Sapio, F.; Brennan, P.M.; Rizzu, P.; Smith, S.; Fell, M.; et al. Somatic retrotransposition alters the genetic landscape of the human brain. *Nature* **2011**, *479*, 534–537. [CrossRef] [PubMed]
10. Reilly, M.T.; Faulkner, G.J.; Dubnau, J.; Ponomarev, I.; Gage, F.H. The role of transposable elements in health and diseases of the central nervous system. *J. Neurosci.* **2013**, *33*, 17577–17586. [CrossRef]
11. Barrón, M.G.; Fiston-Lavier, A.-S.; Petrov, D.A.; González, J. Population Genomics of Transposable Elements in Drosophila. *Annu. Rev. Genet.* **2014**, *48*, 561–581. [CrossRef] [PubMed]
12. Kaminker, J.S.; Bergman, C.M.; Kronmiller, B.; Carlson, J.; Svirskas, R.; Patel, S.; Frise, E.; Wheeler, D.A.; Lewis, S.E.; Rubin, G.M.; et al. The transposable elements of the Drosophila melanogaster euchromatin: A genomics perspective. *Genome Biol.* **2002**, *3*, RESEARCH0084. [CrossRef]
13. Boeke, J.D.; Garfinkel, D.J.; Styles, C.A.; Fink, G.R. Ty elements transpose through an RNA intermediate. *Cell* **1985**, *40*, 491–500. [CrossRef]
14. McCullers, T.J.; Steiniger, M. Transposable elements in Drosophila. *Mob. Genet. Elem.* **2017**, *7*, 1–18. [CrossRef]
15. Venner, S.; Feschotte, C.; Biemont, C. Dynamics of transposable elements: Towards a community ecology of the genome. *Trends Genet.* **2009**, *25*, 317–323. [CrossRef] [PubMed]
16. Chuong, E.B.; Elde, N.C.; Feschotte, C. Regulatory activities of transposable elements: From conflicts to benefits. *Nat. Rev. Genet.* **2017**, *18*, 71–86. [CrossRef] [PubMed]
17. Kawakami, K.; Largaespada, D.A.; Ivics, Z. Transposons As Tools for Functional Genomics in Vertebrate Models. *Trends Genet.* **2017**, *33*, 784–801. [CrossRef]

18. Palazzo, A.; Marsano, R.M. Transposable elements: A jump toward the future of expression vectors. *Crit. Rev. Biotechnol.* **2021**, 1–27. [CrossRef]
19. Minervini, C.F.; Ruggieri, S.; Traversa, M.; D'Aiuto, L.; Marsano, R.M.; Leronni, D.; Centomani, I.; De Giovanni, C.; Viggiano, L. Evidences for insulator activity of the 5′UTR of the Drosophila melanogaster LTR-retrotransposon ZAM. *Mol. Genet. Genom.* **2010**, *283*, 503–509. [CrossRef]
20. Minervini, C.F.; Marsano, R.M.; Casieri, P.; Fanti, L.; Caizzi, R.; Pimpinelli, S.; Rocchi, M.; Viggiano, L. Heterochromatin protein 1 interacts with 5′UTR of transposable element ZAM in a sequence-specific fashion. *Gene* **2007**, *393*, 1–10. [CrossRef]
21. Viggiano, L.; Caggese, C.; Barsanti, P.; Caizzi, R. Cloning and characterization of a copy of Tirant transposable element in Drosophila melanogaster. *Gene* **1997**, *197*, 29–35. [CrossRef]
22. Marsano, R.M.; Moschetti, R.; Caggese, C.; Lanave, C.; Barsanti, P.; Caizzi, R. The complete Tirant transposable element in Drosophila melanogaster shows a structural relationship with retrovirus-like retrotransposons. *Gene* **2000**, *247*, 87–95. [CrossRef]
23. Desset, S.; Conte, C.; Dimitri, P.; Calco, V.; Dastugue, B.; Vaury, C. Mobilization of two retroelements, ZAM and Idefix, in a novel unstable line of Drosophila melanogaster. *Mol. Biol. Evol.* **1999**, *16*, 54–66. [CrossRef] [PubMed]
24. Zhao, W.; Bidwai, A.P.; Glover, C.V.C. Interaction of casein kinase II with ribosomal protein L22 of Drosophila melanogaster. *Biochem. Biophys. Res. Commun.* **2002**, *298*, 60–66. [CrossRef]
25. Koyama, Y.; Katagiri, S.; Hanai, S.; Uchida, K.; Miwa, M. Poly(ADP-ribose) polymerase interacts with novel Drosophila ribosomal proteins, L22 and L23a, with unique histone-like amino-terminal extensions. *Gene* **1999**, *226*, 339–345. [CrossRef]
26. Corpet, F. Multiple sequence alignment with hierarchical clustering. *Nucleic Acids Res.* **1988**, *16*, 10881–10890. [CrossRef]
27. Van Helden, J.; André, B.; Collado-Vides, J. Extracting regulatory sites from the upstream region of yeast genes by computational analysis of oligonucleotide frequencies11Edited by G. von Heijne. *J. Mol. Biol.* **1998**, *281*, 827–842. [CrossRef]
28. Bailey, T.L.; Johnson, J.; Grant, C.E.; Noble, W.S. The MEME Suite. *Nucleic Acids Res.* **2015**, *43*, W39–W49. [CrossRef]
29. Turatsinze, J.-V.; Thomas-Chollier, M.; Defrance, M.; van Helden, J. Using RSAT to scan genome sequences for transcription factor binding sites and cis-regulatory modules. *Nat. Protoc.* **2008**, *3*, 1578–1588. [CrossRef]
30. Bradford, M.M. A rapid and sensitive method for the quantitation of microgram quantities of protein utilizing the principle of protein-dye binding. *Anal. Biochem.* **1976**, *72*, 248–254. [CrossRef]
31. Garner, M.M.; Revzin, A. A gel electrophoresis method for quantifying the binding of proteins to specific DNA regions: Application to components of the Escherichia coli lactose operon regulatory system. *Nucleic Acids Res.* **1981**, *9*, 3047–3060. [CrossRef]
32. Greenfield, E.A.; DeCaprio, J.; Brahmandam, M. Detecting Protein Antigens in Sodium Dodecyl Sulfate-Polyacrylamide Gels. *Cold Spring Harbor Protoc.* **2019**, *2019*, pdb.prot099994. [CrossRef] [PubMed]
33. Greenfield, E.A.; DeCaprio, J.; Brahmandam, M. Preparing Protein Antigens from Sodium Dodecyl Sulfate-Polyacrylamide Gels for Immunization. *Cold Spring Harbor Protoc.* **2019**, *2019*, pdb.prot100008. [CrossRef] [PubMed]
34. Schneider, T.D.; Stephens, R.M. Sequence logos: A new way to display consensus sequences. *Nucleic Acids Res.* **1990**, *18*, 6097–6100. [CrossRef] [PubMed]
35. Mageeney, C.M.; Ware, V.C. Specialized eRpL22 paralogue-specific ribosomes regulate specific mRNA translation in spermatogenesis in Drosophila melanogaster. *Mol. Biol. Cell* **2019**, *30*, 2240–2253. [CrossRef]
36. Chuong, E.B. The placenta goes viral: Retroviruses control gene expression in pregnancy. *PLoS Biol.* **2018**, *16*, e3000028. [CrossRef]
37. Sundaram, V.; Wysocka, J. Transposable elements as a potent source of diverse cis-regulatory sequences in mammalian genomes. *Philos. Trans. R. Soc. Lond. B Biol. Sci.* **2020**, *375*, 20190347. [CrossRef]
38. Moschetti, R.; Palazzo, A.; Lorusso, P.; Viggiano, L.; Marsano, R.M. "What You Need, Baby, I Got It": Transposable Elements as Suppliers of Cis-Operating Sequences in Drosophila. *Biology* **2020**, *9*, 25. [CrossRef]
39. Brennecke, J.; Aravin, A.A.; Stark, A.; Dus, M.; Kellis, M.; Sachidanandam, R.; Hannon, G.J. Discrete small RNA-generating loci as master regulators of transposon activity in Drosophila. *Cell* **2007**, *128*, 1089–1103. [CrossRef]
40. Ghildiyal, M.; Seitz, H.; Horwich, M.D.; Li, C.; Du, T.; Lee, S.; Xu, J.; Kittler, E.L.; Zapp, M.L.; Weng, Z.; et al. Endogenous siRNAs derived from transposons and mRNAs in Drosophila somatic cells. *Science* **2008**, *320*, 1077–1081. [CrossRef]
41. Mével-Ninio, M.; Pelisson, A.; Kinder, J.; Campos, A.R.; Bucheton, A. The flamenco locus controls the gypsy and ZAM retroviruses and is required for Drosophila oogenesis. *Genetics* **2007**, *175*, 1615–1624. [CrossRef]
42. Kassis, J.A. 14—Pairing-Sensitive Silencing, Polycomb Group Response Elements, and Transposon Homing in Drosophila. In *Advertisment Genet*; Dunlap, J.C., Wu, C.T., Eds.; Academic Press: Cambridge, MA, USA, 2002; Volume 46, pp. 421–438.
43. Georgiev, P.; Kozycina, M. Interaction between mutations in the suppressor of Hairy wing and modifier of mdg4 genes of Drosophila melanogaster affecting the phenotype of gypsy-induced mutations. *Genetics* **1996**, *142*, 425–436. [CrossRef]
44. Pai, C.Y.; Lei, E.P.; Ghosh, D.; Corces, V.G. The centrosomal protein CP190 is a component of the gypsy chromatin insulator. *Mol. Cell* **2004**, *16*, 737–748. [CrossRef] [PubMed]
45. Ni, J.-Q.; Liu, L.-P.; Hess, D.; Rietdorf, J.; Sun, F.-L. Drosophila ribosomal proteins are associated with linker histone H1 and suppress gene transcription. *Genes Dev.* **2006**, *20*, 1959–1973. [CrossRef]
46. Kearse, M.G.; Ireland, J.A.; Prem, S.M.; Chen, A.S.; Ware, V.C. RpL22e, but not RpL22e-like-PA, is SUMOylated and localizes to the nucleoplasm of Drosophila meiotic spermatocytes. *Nucleus* **2013**, *4*, 241–258. [CrossRef] [PubMed]

47. Berloco, M.F.; Minervini, C.F.; Moschetti, R.; Palazzo, A.; Viggiano, L.; Marsano, R.M. Evidence of the Physical Interaction between Rpl22 and the Transposable Element Doc5, a Heterochromatic Transposon of Drosophila melanogaster. *Genes* **2021**, *12*, 1997. [CrossRef] [PubMed]
48. Arbeitman Michelle, N.; Furlong Eileen, E.M.; Imam, F.; Johnson, E.; Null Brian, H.; Baker Bruce, S.; Krasnow Mark, A.; Scott Matthew, P.; Davis Ronald, W.; White Kevin, P. Gene Expression During the Life Cycle of Drosophila melanogaster. *Science* **2002**, *297*, 2270–2275. [CrossRef]
49. Lyne, R.; Smith, R.; Rutherford, K.; Wakeling, M.; Varley, A.; Guillier, F.; Janssens, H.; Ji, W.; McLaren, P.; North, P.; et al. FlyMine: An integrated database for Drosophila and Anopheles genomics. *Genome Biol.* **2007**, *8*, R129. [CrossRef]
50. Bischof, J.; Björklund, M.; Furger, E.; Schertel, C.; Taipale, J.; Basler, K. A versatile platform for creating a comprehensive UAS-ORFeome library in Drosophila. *Development* **2013**, *140*, 2434–2442. [CrossRef] [PubMed]

genes

Article

Evidence of the Physical Interaction between Rpl22 and the Transposable Element *Doc5*, a Heterochromatic Transposon of *Drosophila melanogaster*

Maria Francesca Berloco [1,†], Crescenzio Francesco Minervini [2,†], Roberta Moschetti [1], Antonio Palazzo [1], Luigi Viggiano [1,*,‡] and René Massimiliano Marsano [1,*,‡,§]

1 Department of Biology, University of Bari "Aldo Moro", 70126 Bari, Italy; mariafrancesca.berloco@uniba.it (M.F.B.); roberta.moschetti@uniba.it (R.M.); antonio.palazzo@uniba.it (A.P.)
2 Department of Emergency and Organ Transplantation (D.E.T.O.), Hematology and Stem Cell Transplantation Unit, University of Bari "Aldo Moro", 70124 Bari, Italy; crescenziofrancesco.minervini@uniba.it
* Correspondence: luigi.viggiano@uniba.it (L.V.); renemassimiliano.marsano@uniba.it (R.M.M.)
† joint first authors.
‡ joint corresponding authors.
§ Former affiliation: Department of Genetics Anthropology Evolution, University of Parma, Parco Area delle Scienze 11/A, 43124 Parma, Italy.

Abstract: Chromatin is a highly dynamic biological entity that allows for both the control of gene expression and the stabilization of chromosomal domains. Given the high degree of plasticity observed in model and non-model organisms, it is not surprising that new chromatin components are frequently described. In this work, we tested the hypothesis that the remnants of the *Doc5* transposable element, which retains a heterochromatin insertion pattern in the melanogaster species complex, can be bound by chromatin proteins, and thus be involved in the organization of heterochromatic domains. Using the Yeast One Hybrid approach, we found Rpl22 as a potential interacting protein of *Doc5*. We further tested in vitro the observed interaction through Electrophoretic Mobility Shift Assay, uncovering that the *N*-terminal portion of the protein is sufficient to interact with *Doc5*. However, in situ localization of the native protein failed to detect Rpl22 association with chromatin. The results obtained are discussed in the light of the current knowledge on the extra-ribosomal role of ribosomal protein in eukaryotes, which suggests a possible role of Rpl22 in the determination of the heterochromatin in *Drosophila*.

Keywords: ribosomal protein; Rpl22; *Drosophila*; DNA–protein interaction; transposable elements; heterochromatin; *Doc5/Porto1*

Citation: Berloco, M.F.; Minervini, C.F.; Moschetti, R.; Palazzo, A.; Viggiano, L.; Marsano, R.M. Evidence of the Physical Interaction between Rpl22 and the Transposable Element *Doc5*, a Heterochromatic Transposon of *Drosophila melanogaster*. *Genes* **2021**, *12*, 1997. https://doi.org/10.3390/genes12121997

Academic Editor: Miroslav Plohl

Received: 6 November 2021
Accepted: 12 December 2021
Published: 16 December 2021

1. Introduction

Chromatin [1] is a nucleoprotein complex that plays a key role in controlling cell behavior and chromosomal structure [2,3]. Its regulation is important in the control of cellular events, including genome packaging, replication, recombination, DNA repair, and transcription. The nucleosome, which comprises the four core histones (H2A, H2B, H3, H4), wrapped around with 168 bp of DNA, and the linker histones H1 or H5 form the chromatosome, the structural unit of the chromatin [4].

Chromatin is found in two fundamental states during the cell cycle, the loosely condensed euchromatin and the highly compacted heterochromatin. A huge number of DNA–protein and protein–protein interactions contribute to the maintenance of these two structures, the plasticity of which is tightly regulated at the epigenetic level.

Many proteins act as structural components or regulators of the chromatin state, and post-translational modifications of many chromatin components play a fundamental role in maintaining the dynamic state of different chromatin domains. The ongoing EN-

CODE projects [5,6] aim to determine the nature of the epigenetic code and to what extent chromatin remodeling could influence the phenotypes.

Several pieces of observation suggest that ribosomal proteins (RPs) could have an active role in chromatin dynamics. First, RNA-mediated processes have a functional role in regulating chromatin structure and gene expression through the action of non-coding RNA molecules [7–9]. Second, a large fraction of the expressed lncRNA interacts with ribosomes in humans and mice (roughly 39% and 48%, respectively) [10].

Third, the presence of RPs in the nucleus is well-recognized since RPs are imported into the nucleus and assembled into pre-ribosomes in the nucleolus [11].

Therefore, a subset of RPs could be co-opted as chromatin components to perform additional functions under either physiological or exceptional conditions.

Heterochromatin is a partition of the eukaryotic genome, often regarded as useless and functionless. This concept is due to its low gene density and the consequent low impact of mutational load in this compartment on viability and fertility. The massive presence of satellite DNA and transposons in the constitutive heterochromatin has further reinforced this idea. However, since heterochromatin is associated with important functions and structures of the eukaryotic chromosomes, its role has been recently re-evaluated, both in model and non-model organisms. In *D. melanogaster*, several hundreds of genes have been mapped in the constitutive heterochromatin, thus demonstrating its importance in the physiology of cells, tissues, and organs in the fly [12], an observation largely supported by classic and modern genetics studies.

Several additional features make heterochromatin a fascinating genomic compartment. These include the massive presence of repeats and transposable elements, whose structural and functional roles remain elusive, despite decades of studies. In this respect, noteworthy examples still come from *D. melanogaster*. Several repeated loci have been characterized so far in the heterochromatin of *D. melanogaster*, and some of them play an extremely important role in determining critical phenotypes [13–15]. One of these relevant loci lies in the h39 region, a Hoechst-bright chromosomal band adjacent to the centromere of the second chromosome. Two well-studied satellite DNA sequences are clustered in the h39 region, the Responder locus (*Rsp*), and the *Bari1* repeat. The *Rsp* locus, in combination with the *Sd* euchromatic locus, constitutes the key components of one of the best-known segregation distortion systems [13]. The *Bari1* cluster is an array of roughly 80 copies of the *Bari1* transposon [16–18], depending on the fly strain [18,19] of the *Bari1* transposon. Elements of the *Bari1* family are *Tc1*-like transposons that have colonized the genome of several *Drosophila* species [17] and are active in the respective genomes [20,21].

The characterization of the *Bari1* copy number variation in several *D. melanogaster* populations [16,18,19] revealed that this is an extremely static array if compared to the closely linked *Rsp* locus [22,23]. Considering that the *Bari1* cluster origin probably dates back to the split of the *melanogaster* and simulans species, approximately 5 Mya, and that it is only present in *D. melanogaster* [16,17,24], it has been speculated that either it could be functionally connected to the *Rsp* locus or to other structural features of the h39 region. However, the presence of a small *Bari1* cluster on the X chromosome [17,25] and an additional small *Rsp* repeat on the third chromosome [26] and the highly repetitive nature of the h39 region complicate the molecular and genetic investigations of this chromosomal region.

Many transposon relics map at both sides of the *Bari1* cluster in the h39 region of the mitotic chromosomes of *D. melanogaster* [27]. A direct duplication of a 596 bp sequence identified upstream and downstream of the *Bari1* cluster is of particular interest. We hypothesized that this short duplication could be the signature of the transposition-mediated origin of the *Bari1* cluster (as also hypothesized for the minor X-linked *Bari1* cluster [25]). Alternatively, it could have a functional role in the h39 locus or in the heterochromatin [27]. Since the first hypothesis seems unreliable (due to the size incompatibility and outcome of the transposition event), in this work we tested the hypothesis that the above-described 596 bp sequence could have acquired a new function in the heterochromatin through the binding of a chromatin protein. Here, we present evidence that the ribosomal protein Rpl22 binds DNA in vitro, which suggests the possibility that it could be recruited as chromatin protein. The CG7434 gene, which encodes RpL22 protein, maps on the X chromosome. Three additional genes are present within the coding region of RpL22, two encoding snoRNAs (CR34590 and CR33918) and one encoding a ncoRNA (CR42491). This structure complicates the genetic analysis of the locus, and, in fact, no genetic studies have been performed focusing on this gene. At least two post-translational modification events have been characterized, involving phosphorylation of the Ser 289 and Ser290 residues of the RpL22 in *Drosophila* [28]. Among RPs, some members of the RpL22e family have unique structural features and several, apparently unrelated, possible functions. The *Drosophila* Rpl22 has additional Ala-, Lys- and Pro-rich sequences at the amino terminus, which resembles the carboxyl-terminal portion of histone H1 and histone H5 that have been demonstrated to be important in genome stability [29]. For this reason, it has been already hypothesized that *Drosophila* L22 might have two functions, namely, the role of DNA-binding similar to histone H1 and the role of organizing the ribosome [30]. Moreover, as hypothesized in previous works, any potential biological difference between Rpl22 and Rpl22-like proteins should be ascribed to the presence of the extra N-terminal domain of Rpl22, which can be the target of post-translational modifications [31].

We also have evidence that Rpl22 enters into the nucleus of different cell types, in addition to what was demonstrated previously in the male germline cells [32]. The possible implications in the stability of a specific heterochromatin region are discussed.

2. Materials and Methods

2.1. Plasmids Construction

The *Doc5* fragment flanking the *Bari1* cluster was PCR-amplified from the purified DNA of the BACR16M08 clone (described in [25]) using specific primers containing EcoRI adapters at the 5′ end. The PCR fragment was cloned into the EcoRI site of the pGEM-T vector (Promega) and verified by Sanger sequencing.

2.2. PCR Amplification

Primers used for PCR amplification are reported in Table 1.

Table 1. List of primers used in this study.

Primer	Sequence	Usage
ADread	5′-CTATTCGATGATGAAGAT-3′	sequencing
pACT2seq	5′-TACCACTACAATGGATG-3′	sequencing
pACT2 up	5′-CTATTCGATGATGAAGATACCCCACCAAACCC-3′	Amplification/cloning
pACT2 low	5′-GTGAACTTGCGGGGTTTTTCAGTATCTACGAT-3′	Amplification/cloning
His1_up	5′-GAGGCCCTTTCGTCTTCAA-3′	Amplification/cloning
His1_low	5′-CTAGGGCTTTCTGCTCTGTCATCT-3′	Amplification/cloning
Doc5_up	5′-ACGGCTATTATTGTTTCTTATTGCT-3′	Amplification/cloning
Doc5_low	5′-TTATCCTCATCCCTTATCCTATGT-3′	Amplification/cloning
pETup	5′-CACCATGGCTTACCCATA-3′	Amplification/cloning
pETlow	5′-ATAAAAGAAGGCAAAACGATG-3′	Amplification/cloning
H5low	5′-CTAACGCAGCACGTTCTTCTT-3′	Amplification/cloning
L22up	5′-CACCAAGGTGGTCAAGAAGAA-3′	Amplification/cloning

2.3. One Hybrid Screening

The one hybrid screening was performed using the Matchmaker One-Hybrid System (Clontech, Kyoto, Japan) following the manufacturer recommendations.

A *Drosophila* embryonic cDNA library (cDNA pool from 0–21 h embryos of the Canton-s strain) in the pACT2 vector (Clontech) was used for the yeast one-hybrid screens.

The *Doc5* sequence was subcloned into the pHISi-1 vector at the EcoRI site and into the pLacZi vector. Both plasmids were linearized using either BamHI (pHISi-1) or NcoI (pLacZi) and transformed in the YM4271 *S. cerevisiae* strain using the TRAFO system [33]. Recombinant colonies, carrying the integrated constructs, were selected onto selective SD medium lacking either histidine (pHISi-1 vector) or uracil (pLacZi vector).

The background expression of the LacZ reporter was determined by a standard β-galactosidase assay. Colonies were transferred to Whatman filter paper discs and lysed with liquid nitrogen. Filters were then exposed to Z-buffer ($Na_2HPO_4 \cdot 7H_2O$ 60 mM, $NaH_2PO_4 \cdot H_2O$ 40 mM, KCl 10 mM, MgSO4 1 mM, β–mercaptoethanol 50 mM, pH 7) containing X-gal (5-bromo-4-chloro-indolyl-b-D-galactopyranoside 0.33 mg/mL). Only clones without LacZ basal expression in 8 h were selected for further analyses. These clones were further transformed to integrate the linearized pHisi-1 vector. Background expression of the His cassette was found to be inhibited by 15 mM 3-AT.

2.4. Protein Expression and Purification

The plasmid sets used to express proteins in *E. coli* (pET/RpL22 for the full-length protein expression; pET/H5 for the H1-H5 domain expression; pET/L22 for the ribosomal domain expression) were constructed by PCR amplification of either the full-length, the 5′-terminal or the 3′-terminal part of the cDNA and subsequent cloning into the pET-200 vector.

Plasmids were transformed in chemically competent *E. coli* (BL21-DE3), and the cultures were induced with 1 mM IPTG at a cell density equivalent to 0.5 OD_{600} and maintained for 2.5 h at 37 °C. Cells were sonicated in 25 mM HEPES (pH 7.5), 1 M NaCl, 15% glycerol, 0.25% Tween 20, 2 mM β-mercaptoethanol, and 1 mM PMSF. A total of 10 mM imidazole (pH 8.0) was added to the soluble fraction before it was mixed with Ni-NTA resin (Qiagen, Hilden, Germany) according to the manufacturer's recommendations. The resin was washed with sonication buffer containing 30% glycerol and 50 mM imidazole. Bounded proteins were eluted with sonication buffer containing 300 mM imidazole and dialyzed overnight against sonication buffer without imidazole. Purified proteins were analyzed on 12% SDS-polyacrylamide gel. Protein concentration was determined using the Protein Assay ESL Kit (Roche Basel, Switzerland).

2.5. Electrophoretic Mobility Shift Assay (EMSA)

In total, 5 µg of the pT/Doc5 plasmid was EcoRI-digested and the released fragment was gel-purified using the QIAquick Gel Extraction Kit (Qiagen). A filling-in reaction was performed to end-label the target DNA. A total of 50 ng of the eluted fragment was incubated with [α^{32}P]ATP (Perkin Elmer, Waltham, MA, USA), 1X Klenow reaction buffer and 2U of Klenow fragment (Roche, Basel, Switzerland).

Labeled fragments were purified using Sephadex G50 exclusion chromatography columns.

A total of 2 ng of the labeled fragment was incubated with the appropriate protein (either the full-length Rpl22, the H1-H5 domain or the ribosomal domain) in binding buffer as described in [34] (25 mM HEPES, pH 7.6, 50 mM NaCl, 1 mM EDTA, 1 mM DTT, 0.1 mg/mL BSA, 2.5 mM spermidine, 10% glycerol, and 0.1 mg/mL poly (dI-dC). Competition experiments were performed using either linear pUC19 (SmaI linearized) or sonicated λ phage DNA (200–1000 bp size range enrichment). The binding reaction was started by adding the protein extract and incubated for 20 min at 25 °C, then loaded directly onto 5% polyacrylamide (75:1 acrylamide:bisacrylamide) pre-run gel in 40 mM Tris–acetate, 2.5 mM EDTA (pH 7.8). Gels were run for 4.5 h at 4 °C at 10 V/cm and dehydrated using a gel-dryer. DNA–protein complexes were visualized by autoradiography using a STORM phosphorimager (Molecular Dynamics).

2.6. Fluorescence In Situ Hybridization and Immunofluorescence on Polytene Chromosomes

Fluorescence in situ hybridization experiments on polytene chromosomes were performed as described in [35]. Polytene chromosomes were prepared from third instar larvae of *D. simulans* and *D. sechellia*, reared on standard cornmeal medium at 18 °C. Salivary glands were dissected in PBS using a pair of dissection needles, fixed in 40% acetic acid, and squashed onto microscopy slides. Probes were labeled using the nick translation method with Cy3-dUTP, hybridized overnight at 37 °C.

Digital images were obtained using an Olympus epifluorescence microscope equipped with a cooled CCD camera. Gray scale images, recording Cy3 and DAPI fluorescence, were obtained separately using specific filters and were pseudo colored and merged to obtain the final image using the Adobe Photoshop software.

Immunodetection experiments of Rpl22 and fibrillarin on polytene chromosomes of the Oregon-R (wild type) were performed according to James et al. [36] using the polyclonal primary anti-Rpl22 antibody (diluted 1:50) raised in rabbit (Invitrogen Carlsbad, CA, USA, Minervini et al. submitted) and the monoclonal (G-8sc-374022 Santa Cruz Biotechnology Inc., Dallas, TX, USA) anti-fibrillarin antibody raised in mouse. An FITC (fluorescein isothiocyanate)-conjugated anti-rabbit Ig (whole antibody) raised in sheep (diluted 1:20) and the Alexa Fluor 488 goat anti-mouse antibody (Life Technologies, Carlsbad, CA, USA, 1:200 dilution) were used as secondary antibodies. Following incubation, the slides were washed three times in PBS, stained with DAPI (4,6-diamidino-2-phenilindole) at 0.01 μg/mL and mounted in anti-fading medium. Immunodetection on S2R+ cells were performed as previously described in [20,21] using the above-described antibodies.

2.7. Other Methods

Sequencing of the cloned fragments was performed at the BMR Genomics sequencing facility (Padova, Italy).

Global alignments were performed using DNA Strider [37]. Local alignments were performed using BLAST at the NCBI website.

NLS signals were searched with cNLS Mapper (http://nls-mapper.iab.keio.ac.jp/cgi-bin/NLS_Mapper_form.cgi (accessed on 1 March 2021)) [38] using a cutoff score = 7 in the entire protein sequence, and with Nucpred (https://nucpred.bioinfo.se/cgi-bin/single.cgi (accessed on 2 March 2021)) [39].

3. Results

We have previously identified a 596 bp DNA sequence duplication (formerly named DRM8) at both sides of the *Bari1* cluster in the heterochromatin of 2R chromosome of *D. melanogaster* [27]. Specifically, this repetitive sequence maps in the h39 region, and it has been proven lately to be a remnant of the *Doc5/Porto1* element, a highly repeated non-LTR retrotransposon in the heterochromatin of *D. melanogaster* [40]. The similarity between the DRM8 sequence and the reference *Doc5/Porto1* element is shown in Figure 1. Hereafter, we will refer to this sequence as *Doc5*.

Several copies of the *Doc5* can be found in the reference genome of *D. melanogaster* (see Table 2). In silico analyses reveal that *Doc5* maps exclusively in the constitutive heterochromatin of the two major autosomes of *D. melanogaster*, including the centromere, as well as at the eu-heterochromatin transition.

The heterochromatic localization of the *Doc5* element is also a conserved feature in closely related species of the *melanogaster* complex, such as *D. simulans* and *D. sechellia*, as demonstrated by the results of FISH experiments on polytene chromosomes (Figure 2).

suggested that the preservation of a repetitive non-coding DNA sequence, especially in the heterochromatin, could be promoted with the aid of stabilizing binding proteins [41], such as chromatin proteins. To test this hypothesis, we performed a One-Hybrid System assay aimed at the identification of proteins that potentially interact with the *Doc5* fragment.

The double selection method (i.e., His prototrophy and positivity to the β-galactosidase test) applied to identify positive clones ensures that the false positive rate is minimized.

Twenty-four positive clones, selected on selective media lacking histidine, were further tested with the β-galactosidase activity (Table S1). Many of these clones turned rapidly blue upon β-galactosidase testing (3–5 h). However, a large fraction (46%) of such clones matched to Rpl22 transcripts after Sanger sequencing and BLASTN analysis. Based on the fastness of color turning (the smallest the better) and the relative abundance of positive clones, we chose Rpl22 as the candidate for further investigations.

The positive clones obtained from the first round of screening were further validated by independent transformation of the isolated plasmids into the bait-containing yeast strains (i.e., yeast strains containing the *Doc5*-His and *Doc5*-lacZ cassette, data not shown) to confirm the bait–pray interaction.

To further solidify this result, we assayed the Rpl22–*Doc5* interaction in vitro. The Rpl22 protein was expressed and purified in *E. coli* and used for in vitro binding assays in order to test its ability to bind an end-labeled *Doc5* fragment (see Materials and Methods).

As can be observed in Figure 3 (lanes 3–5), increasing amounts of purified protein led to a slower migration in a polyacrylamide gel, suggesting the formation of progressively slower DNA–protein complexes. In our hands, 3 μg of the protein extract led to the formation of the slowest DNA–protein complex. This pattern could be explained by either the presence of multiple binding sites in the target or by the possible formation of multimeric protein complexes that bind the target fragment.

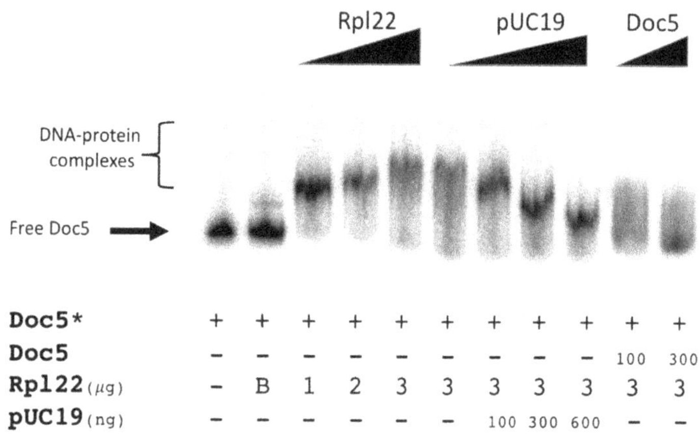

Figure 3. The binding of Rpl22 to *Doc5*. The amount of labeled fragment (Doc5 *, Figure 3) in each lane is 3 ng. The amounts of unlabeled specific competitor (ng of Doc5), Rpl22 (μg), and unlabeled non-specific competitor (ng of linearized pUC19) are indicated in the figure legend under the respective lanes. Increasing amounts of purified Rpl22 protein (lanes 3–5) and non-specific (lanes 6–9) and specific (lanes 10–11) competitors are indicated on the top by triangles. A negative control (lane 2) was performed following the incubation of the Doc5-labeled probe with 3 μg of non-induced *E. coli* (BL21 strain) lysate (indicated with B). The labeled fragments are indicated with an asterisk (*).

The observed protein binding is specific and reversible, as demonstrated by the competition assays in Figure 3. While a 200-fold amount of unspecific competitor is not sufficient to disrupt the Rpl22–*Doc5* interaction (Figure 3, lanes 6–9), a 30-fold amount of target fragment completely disrupts the observed DNA–protein binding (Figure 3, lanes 10–11). Additional controls to assess the specificity of the binding were performed using either an unrelated DNA fragment, or using a different non-specific competitor DNA (Figure S1).

We next investigated whether the two domains of Rpl22 could differentially contribute to the observed DNA–protein interaction. The H1-H5 domain and the ribosomal domain were independently tested in EMSA assays for their ability to interact with *Doc5*. As can be observed in Figure 4, only the H1-H5 domain retains the ability to bind the *Doc5* fragment tested (Figure 4, lane 3), whereas the ribosomal domain does not (Figure 4, lane 2) if compared to the binding observed for the wild-type Rpl22 protein (Figure 4, lane 4). Similar to what observed for the wild-type protein (Figure 3, lanes 3–5), the H1–H5 domain interacts with the *Doc5* sequence in a dose-dependent manner (Figure 4B).

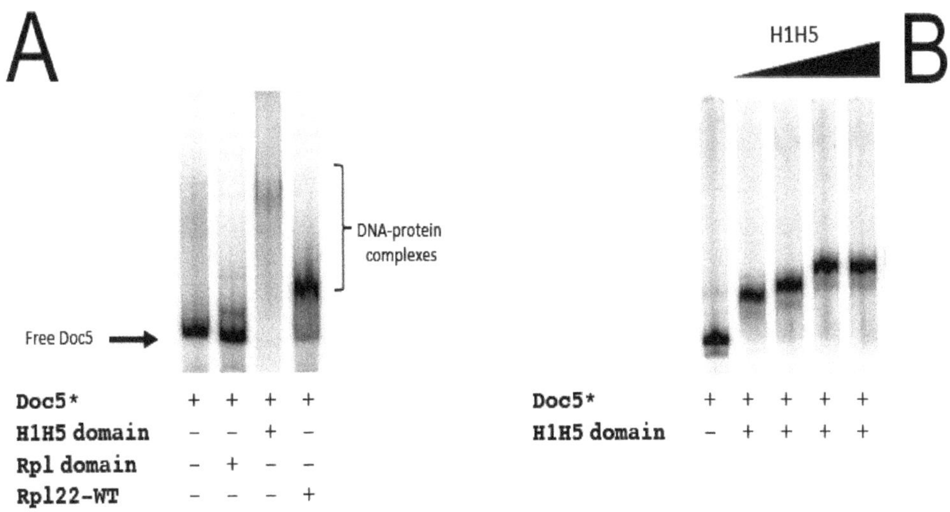

Figure 4. Dissection of the DNA-binding domain of Rpl22 in vitro. Labeled fragments are indicated with an asterisk (*). (**A**) EMSA analysis of the ribosomal and the histone-like domains of Rpl22. (**B**) EMSA analysis of the histone-like domain. A total of 3 µg of the Rpl22 (WT) and 1.5 µg of the H1-H5 and ribosomal domains were used to maintain the unaltered DNA:protein molar ratio. A schematic representation of the two main domains of Rpl22 protein is depicted at the top of the figure. Asterisk indicates that the fragment is labelled.

To further investigate the possible role of Rpl22 in the chromatin dynamics, we tested the Rpl22 protein localization in both *D. melanogaster* cultured cells, in order to check whether the protein co-localizes with chromosomes. We performed immunofluorescence localization of the native Rpl22 protein on polytene chromosomes of the Oregon-R wild-type strain and inn cultured S2R+ cells, using a polyclonal antibody raised against the Rpl22 protein.

The results obtained (Figure 5) clearly show that Rpl22 localizes to the nuclei, with a marked nucleolar localization that has been further confirmed by co-localization with the nucleolar marker fibrillarin, (Figure S2) both in salivary gland cells (Figure 5A) and in cultured cells (Figure 5B), without any additional evidence of localization to chromatin.

In silico prediction of the nuclear localization of Rpl22 using cNLS Mapper [38] suggests its nuclear localization, with the best scoring NLS signal (score 7/7) mapped at position 234. A similar search, using NucPred [39] as an alternative algorithm, returned the sequence GKGQKKKK (position 181, score 0.28; a 0.30 threshold corresponds to 77% sensitivity and 55% specificity).

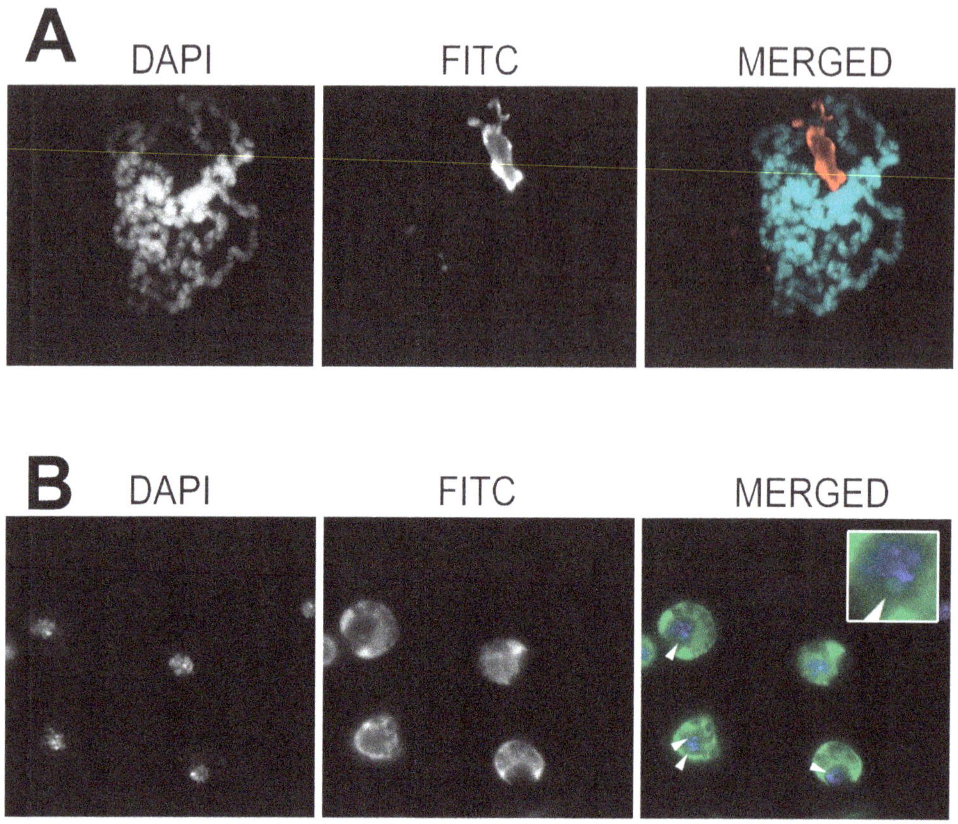

Figure 5. Pattern of subcellular immunolocalization of Rpl22 in *D. melanogaster* salivary gland nuclei (**A**) and in cultured S2R+ cells (**B**). White arrowheads point to nucleoli. A magnified detail of the nucleolar co-localization is reported in the inset. Additional details on the localization of Rpl22 to nucleoli are given in Figure S2.

In the absence of additional experimental evidence, the possible role of Rpl22 in the heterochromatin can be inferred from interactomic data obtained in previously published works. Out of the ninety-one RpL22-interacting proteins that are annotated in FlyBase, 13 are non-RPs. Notably, 12 out of the 13 interacting proteins are not directly linked with the translational machinery.

Rpl22 interacts with protein involved in heterochromatin organization (vig and vig2 [42,43]), piRNA biogenesis (Fmr1 and its associated miRNA, bantam [44]), and transcriptional repression (Ago1 and Ago2 [42]) (reported in Table 3). Such interactions further suggest the involvement of Rpl22 in chromatin determination and transcriptional pathways, supporting our hypothesis.

Table 3. Rpl22 interacting proteins involved in heterochromatin functions. Information retrieved from Flybase (last accessed August 2021).

Gene Name	FlyBase ID	Function	Inferred by	Reference
vig	FBgn0024183	Heterochromatin organization	Co-IP	[42]
AGO1	FBgn0262739	transcriptional repression	Co-IP	[42]
AGO2	FBgn0087035	transcriptional repression	Co-IP	[42]

Table 3. *Cont.*

Gene Name	FlyBase ID	Function	Inferred by	Reference
vig2	FBgn0046214	Heterochromatin organization	Mass-spec	[43]
Fmr1	FBgn0028734	piRNA biogenesis	Co-IP	[42]
ban	FBgn0262451	piRNA biogenesis	Co-IP	[42]
esi2	FBgn0285992	Unknown	Co-IP	[42]
smt3	FBgn0264922	mitosis	Co-IP	[31]

4. Discussion

The stabilization of large chromosomal domains containing extended repeat blocks essentially depends on the chromatin architecture that wraps these loci. Both the Encode [5] and modEncode [45] projects have had a leading role in the determination of the genome-wide chromatin status in *H. sapiens* and model organisms, respectively. The outcome of these huge projects led to the birth of epigenomics that aims to link cell-type-specific gene expression to chromatin structure. The specific features of chromatin domains are also of critical importance for genome evolution since the propensity of certain loci to be converted and relocated from the euchromatin to the heterochromatin is probably determined by the ancestral epigenetic marks [46]. For this reason, profound knowledge of chromatin dynamics is fundamental in the determination of the evolutionary trajectories that chromosomes follow.

However, chromatin is a highly dynamic biological entity, and for this reason, it is difficult to provide a definitive and exhaustive description. Unbiased approaches, i.e., not focused on a particular developmental stage or specific tissue, allow for a near-to-complete characterization of chromatin-associated proteins. It follows that the elucidation of the changing state of chromatin in the most diverse cellular types is of particular importance toward the complete understanding of physiological and pathological conditions [47].

Here, we report that a ribosomal protein binds the *Doc5* transposon, a non-autonomous TE family enriched in the heterochromatin of *D. melanogaster* and closely related species [48], providing in vitro experimental evidence for a functional interaction of Rpl22 with DNA, and possibly to chromosome and chromatin. In *Drosophila*, the direct binding of protein to TEs, especially involving retrotransposons, has been previously reported [49–51]. In a yeast one-hybrid assay, we probed a *D. melanogaster* expression library with *Doc5* as bait and found Rpl22 as the best candidate interacting protein. We have further validated the DNA–protein interaction with a series of EMSA experiments that confirmed the results of the experiments in yeast. We further demonstrated that the NH-terminal domain (H1–H5 domain) of the protein is both necessary and sufficient to bind DNA. Furthermore, the assays performed in vitro show that the *Doc5*–Rpl22 interaction depends on the amount of protein input. We cannot dismiss the hypothesis that this behavior could depend both on the presence of multiple binding sites on the target (which we have not investigated), and on the ability of Rpl22 to multimerize or to form homogeneous aggregates. In addition, the net charge density of the expressed H1-H5 domain is greater than that of the wild-type Rpl22 protein (27.14/15.8 KDa vs. 36.51/30.6 KDa, respectively, at pH = 7), which can account for the increased shift of the H1-H5/*Doc5* complex if compared to the wild-type Rpl22/*Doc5* complex (Figure 4).

What is the relevance of our findings? Our results let us hypothesize that Rpl22 could have a potential role in the organization of chromatin, possibly in heterochromatin, and this hypothesis is supported by several studies reporting that RPs are linked to biological processes occurring in the nucleus [52]. RPs have been found associated at transcription sites in *Drosophila* polytene chromosomes. This unexpected finding suggested that ribosomal subunits could be associated with nascent mRNAs [53]. An additional study in *Saccharomyces cerevisiae* showed that RPs bind to noncoding RNA genes, suggesting that the RPs–RNA association might be independent of the translatability of the transcript and might involve free RPs that are not assembled into ribosomes [54]. Several other examples of RPs with extra ribosomal functions at transcription sites have been reported to date. Some RPs auto-regulate their expression by affecting translation, splicing, or transcription by interacting with their mRNA, or promoter [55–57]. RPs are also able to interact with transcription factors at the promoters of genes. RpL11 binds the oncoprotein c-MYC at the promoter of c-MYC target genes [58,59], RpS3 is a subunit of the NF-κB DNA-binding complex involved in chromatin binding and transcription regulation of specific genes [60]. RpS3 phosphorylation at serine 209 by IKKb is crucial for RPS3 nuclear localization in response to activating stimuli [61].

Rpl22 is a ribosomal protein with a prevailing cytoplasmic localization. Past-published reports claimed that Rpl22 also localizes to the nucleus of *Drosophila* cells. Ni and collaborators [62] demonstrated that Rpl22 expressed at endogenous levels localizes in the nucleus of *Drosophila* Kc (embryo-derived) and cl-8 (derived from imaginal discs) cell cultures, and it is associated with chromatin, resulting in gene suppression. Immunofluorescent staining and chromatin immunoprecipitation (ChIP) analyses demonstrated that RpL22 and H1 are both associated with condensed chromatin. In the same study, it was demonstrated that the overexpression of RpL22 caused the transcriptional repression of two-thirds of the genes suppressed by histone H1. By contrast, RpL22 depletion caused the up-regulation of the transcription of several tested genes, supporting a role for RpL22 as transcriptional repressor [62]. These observations imply the involvement of Rpl22 in global transcriptional processes.

However, Rpl22 has not been previously identified in surveys aimed at the identification of chromatin structure. This can be due to an experimental bias when searching histone modifications [63]. On the other hand, unbiased studies have been focused on euchromatic genomic regions only [64]. Conversely, our approach was based on the search of proteins that interact with a heterochromatic sequence, and our results support a role of Rpl22 in the chromatin. To what extent Rpl22 could participate in the determination of chromatin domains remains to be determined. Another potential implication of our findings concerns the possible role of the non-autonomous *Doc5* transposon in the *D. melanogaster* genome. Non-autonomous TEs often acquire new functions in complex genomes, over evolutionary time. Many examples of evolutionarily inactive TEs that have been co-opted, exapted, or domesticated are described in the scientific literature [65,66].

It has been demonstrated that *Doc5* is under the control of the piRNA pathway [67,68]. Since no potentially active *Doc5* copies are found, these findings suggest that the short RNAs generated from *Doc5* could have alternative roles in the regulation of gene expression, or alternatively in the regulation of the transposition of other, unrelated, TEs.

Alternatively, *Doc5* could mark a chromatin domain with a structural function that prevents the excessive expansion of the *Bari1* cluster. This hypothesis could be extended to other species-specific heterochromatic repeats, since the *Bari1* cluster is unique to the *D. melanogaster* species, while *Doc5* is present in the genome of sibling species [17].

In contrast with previously reported results, we were not able to demonstrate/reproduce a pan-nuclear localization of Rpl22 in S2R+ cells. Our experiments only revealed a nucleolar localization of the protein, without any detectable association to chromatin. This contrasting result could be explained considering the limitations of the immunolocalization technique, which would not allow for the detection of a small amount of chromatin proteins. Moreover, the differences between the cell lines used in our experimental setup (S2R+) and in previous studies (Kc) should be taken into account. Kc are male derived, while S2R+ derive from females. Kc have a plasmatocyte-like phenotype, while S2R+ combines properties of plasmatocytes and crystal cells [69]. Finally, Kc and S2R+ have different ploidy, since Kc are approximately 4n, while S2R+ are 2.5n [70]. We can, therefore, hypothesize that the nuclear localization and the association to chromatin is cell-type dependent. Consequently, we cannot dismiss the fact that, in S2R+ cells, Rpl22 can also enter the nucleus under particular experimental conditions. An additional limitation of our study is the lack of confocal microscopy analyses, which grants a powerful resolution at the subcellular level.

Similarly, we did not detect Rpl22 signals on the polytene chromosome arms, nor in the chromocenter. The limitation in terms of resolution using the polytene chromosome to assess the presence of DNA–protein interactions in heterochromatin is due to the under-replicated nature of the chromocenter [71,72]. Moreover, it has been demonstrated that Rpl22 is subjected to post-translational modifications in testis [31]. SUMOylation, phosphorylation, and possibly other unexplored post-translational modifications could also affect the Rpl22 localization and its ability to be engaged in additional functions, other than translation. Post-translationally modified Rpl22 could potentially exit from the nucleolus and associate to chromatin in particular, unexplored, physiological conditions or in response to environmental stresses. This change in localization could be elicited by protein post-translational modification, as demonstrated in previous studies involving Rpl22 [31].

Despite the lack of localization to chromosomes and chromatin, several additional observations support the hypothesis of an involvement of Rpl22 in chromatin dynamics. There are 12 out of 91 Rpl22-interacting proteins that suggest its involvement in chromatin-related processes. Furthermore, RpL22 has been identified as one of the two hundred genes required for mitotic spindle assembly in *Drosophila* S2 cells in an RNAi screen [73], and the down-regulation of the RpL22 gene also results in aberrantly short, monopolar spindles in S2 cells. These data together with the demonstration of the DNA binding ability of Rpl22 presented in this paper, offer a new perspective of how Rpl22 could participate in chromatin dynamics, at least under specific conditions that has yet to be determined. Additionally, previous genome-wide ChIP-on-chip analysis in the fission yeast *S. pombe* revealed the presence of ribosomal protein complexes at transcription sites with unexpected peaks at centromeres, raising the intriguing hypothesis that RP complexes are involved in tRNA biogenesis and possibly centromere functions [57].

5. Conclusions

We have presented in vitro evidence of the interaction between a typical heterochromatic sequence and a ribosomal protein in *D. melanogaster*. However, experiments in vivo do not confirm the results of experiments in vitro, suggesting that further investigation is needed to reveal the physiological role of Rpl22 in the context of the chromosome structure.

While further studies are needed to understand if the *Doc5* element has been co-opted to absolve further functions in the heterochromatin, several suggestive hypotheses could be proposed. *Doc5* could act as a bidirectional promoter that allows for the transcription of the *Bari1* cluster in order to activate the piRNA-mediated repression of the transposition. In this hypothesis, the ribosomal protein Rpl22 could help in the transcriptional activation from this promoter [74] or in the stabilization of non-coding RNAs [54]. Considering that the *Bari1* elements tested so far are transpositionally active [20,21,75] and are capable of autonomous transcription [76,77], this hypothesis could at least partially explain the low transposition activity observed in *D. melanogaster* laboratory strains [20]. In a companion paper, Minervini et al. (Minervini et al. submitted to *Genes*) also demonstrated that Rpl22 binds in vitro to a transposable element-derived consensus sequence. This observation leads to the hypothesis that ribosomal proteins could be also involved in controlling the activity of TEs in *Drosophila*, not only at the translation level [78] but also at the transcriptional level.

Supplementary Materials: The following are available online at https://www.mdpi.com/article/10.3390/genes12121997/s1, Table S1: List of the yeast clones tested in the β-Galactosidase assay. Supplementary Figure S1: In vitro assays (EMSA) suggesting the specificity of the Doc5/Rpl22 binding. Supplementary Figure S2: (**A**). Rpl22 co-localizes with fibrillarin in S2R+ cells. From the left to the right: DAPI, anti-fibrillarin, anti-Rpl22, merged signals. Signal pseudo-coloring in the merged image is as follows. DAPI: blue; Fibrillarin: red; Rpl22: green. The arrowhead in the merged image point to the nucleolus. A magnified detail of the nucleolar co-localization is reported in the inset. (**B**). Rpl22 co-localizes with fibrillarin in polytene nuclei. From the left to the right: DAPI, anti-fibrillarin, anti-Rpl22, merged signals. Signal pseudo-coloring in the merged image is as follows. DAPI: blue; Fibrillarin: green; Rpl22: red. Arrowheads in the merged image point to nucleoli. Table S1. List of the yeast clones tested in the Galactosidase assay. Clones carrying Rpl22 sequences are reported in bold.

Author Contributions: Conceptualization, R.M.M.; One Hybrid Experiments, C.F.M., R.M.M.; immunofluorescence experiments, M.F.B.; FISH experiments on polytene chromosomes, R.M. and R.M.M.; in silico analyses, A.P.; supervised the project, L.V. and R.M.M.; writing—original draft preparation, C.F.M., M.F.B., L.V., R.M.M.; writing—review and editing, C.F.M., M.F.B., L.V., R.M.M. All authors have read and agreed to the published version of the manuscript.

Funding: This research received no external funding.

Institutional Review Board Statement: Not applicable.

Informed Consent Statement: Not applicable.

Data Availability Statement: Not applicable.

Acknowledgments: A.P. is supported by a grant from Regione Puglia "Research for Innovation (REFIN)"-POR PUGLIA FESR-FSE 2014/2020. Codice Pratica: B39303C8.

Conflicts of Interest: The authors declare no conflict of interest.

References

1. Kornberg, R.D. Chromatin Structure: A Repeating Unit of Histones and DNA. *Science* **1974**, *184*, 868. [CrossRef] [PubMed]
2. Lindsay, S. Chromatin control of gene expression: The simplest model. *Biophys. J.* **2007**, *92*, 1113. [CrossRef]
3. Schmitt, A.D.; Hu, M.; Ren, B. Genome-wide mapping and analysis of chromosome architecture. *Nat. Rev. Mol. Cell Biol.* **2016**, *17*, 743–755. [CrossRef] [PubMed]
4. Simpson, R.T. Structure of the chromatosome, a chromatin particle containing 160 base pairs of DNA and all the histones. *Biochemistry* **1978**, *17*, 5524–5531. [CrossRef]
5. Consortium, E.P. An integrated encyclopedia of DNA elements in the human genome. *Nature* **2012**, *489*, 57–74. [CrossRef]
6. Chen, Z.X.; Sturgill, D.; Qu, J.; Jiang, H.; Park, S.; Boley, N.; Suzuki, A.M.; Fletcher, A.R.; Plachetzki, D.C.; FitzGerald, P.C.; et al. Comparative validation of the D. *melanogaster* modENCODE transcriptome annotation. *Genome Res.* **2014**, *24*, 1209–1223. [CrossRef] [PubMed]
7. Arunkumar, G.; Melters, D.P. Centromeric Transcription: A Conserved Swiss-Army Knife. *Genes* **2020**, *11*, 911. [CrossRef]
8. Tachiwana, H.; Yamamoto, T.; Saitoh, N. Gene regulation by non-coding RNAs in the 3D genome architecture. *Curr. Opin. Genet. Dev.* **2020**, *61*, 69–74. [CrossRef] [PubMed]
9. Barral, A.; Déjardin, J. Telomeric Chromatin and TERRA. *J. Mol. Biol.* **2020**, *432*, 4244–4256. [CrossRef]
10. Zeng, C.; Fukunaga, T.; Hamada, M. Identification and analysis of ribosome-associated lncRNAs using ribosome profiling data. *BMC Genom.* **2018**, *19*, 414. [CrossRef]
11. Mélèse, T.; Xue, Z. The nucleolus: An organelle formed by the act of building a ribosome. *Curr. Opin. Cell Biol.* **1995**, *7*, 319–324. [CrossRef]
12. Marsano, R.M.; Giordano, E.; Messina, G.; Dimitri, P. A New Portrait of Constitutive Heterochromatin: Lessons from *Drosophila melanogaster*. *Trends Genet.* **2019**, *35*, 615–631. [CrossRef] [PubMed]
13. Larracuente, A.M.; Presgraves, D.C. The selfish *Segregation Distorter* gene complex of *Drosophila melanogaster*. *Genetics* **2012**, *192*, 33–53. [CrossRef]
14. Berloco, M.; Fanti, L.; Fau-Breiling, A.; Breiling, A.; Fau-Orlando, V.; Orlando, V.; Fau-Pimpinelli, S.; Pimpinelli, S. The maternal effect gene, abnormal oocyte (abo), of *Drosophila melanogaster* encodes a specific negative regulator of histones. *Proc. Natl. Acad. Sci. USA* **2001**, *98*, 12126–12131. [CrossRef]
15. Tritto, P.; Specchia, V.; Fanti, L.; Berloco, M.; D'Alessandro, R.; Pimpinelli, S.; Palumbo, G.; Pia Bozzetti, M. Structure, Regulation and Evolution of the Crystal–Stellate System of *Drosophila*. *Genetica* **2003**, *117*, 247–257. [CrossRef]
16. Caizzi, R.; Caggese, C.; Pimpinelli, S. Bari-1, a new transposon-like family in *Drosophila melanogaster* with a unique heterochromatic organization. *Genetics* **1993**, *133*, 335–345. [CrossRef]

17. Palazzo, A.; Lovero, D.; D'Addabbo, P.; Caizzi, R.; Marsano, R.M. Identification of Bari Transposons in 23 Sequenced *Drosophila* Genomes Reveals Novel Structural Variants, MITEs and Horizontal Transfer. *PLoS ONE* **2016**, *11*, e0156014. [CrossRef]
18. McGurk, M.P.; Barbash, D.A. Double insertion of transposable elements provides a substrate for the evolution of satellite DNA. *Genome Res.* **2018**, *28*, 714–725. [CrossRef] [PubMed]
19. Caggese, C.; Pimpinelli, S.; Barsanti, P.; Caizzi, R. The distribution of the transposable element Bari-1 in the Drosophila melanogaster and Drosophila *simulans* genomes. *Genetica* **1995**, *96*, 269–283. [CrossRef]
20. Palazzo, A.; Marconi, S.; Specchia, V.; Bozzetti, M.P.; Ivics, Z.; Caizzi, R.; Marsano, R.M. Functional Characterization of the Bari1 Transposition System. *PLoS ONE* **2013**, *8*, e79385. [CrossRef]
21. Palazzo, A.; Moschetti, R.; Caizzi, R.; Marsano, R.M. The *Drosophila* mojavensis Bari3 transposon: Distribution and functional characterization. *Mob. DNA* **2014**, *5*, 21. [CrossRef]
22. Sandler, L.; Hiraizumi, Y.; Fau-Sandler, I.; Sandler, I. Meiotic Drive in Natural Populations of *Drosophila* Melanogaster. I. the Cytogenetic Basis of Segregation-Distortion. *Genetics* **1959**, *44*, 233. [CrossRef] [PubMed]
23. Pimpinelli, S.; Dimitri, P. Cytogenetic analysis of segregation distortion in *Drosophila melanogaster*: The cytological organization of the Responder (Rsp) locus. *Genetics* **1989**, *121*, 765–772. [CrossRef] [PubMed]
24. Moschetti, R.; Palazzo, A.; Lorusso, P.; Viggiano, L.; Marsano, R.M. "What You Need, Baby, I Got It": Transposable Elements as Suppliers of Cis-Operating Sequences in *Drosophila*. *Biology* **2020**, *9*, 25. [CrossRef] [PubMed]
25. Marsano, R.M.; Milano, R.; Minervini, C.; Moschetti, R.; Caggese, C.; Barsanti, P.; Caizzi, R. Organization and possible origin of the Bari-1 cluster in the heterochromatic h39 region of *Drosophila melanogaster*. *Genetica* **2003**, *117*, 281–289. [CrossRef]
26. Moschetti, R.; Caizzi, R.; Pimpinelli, S. Segregation distortion in *Drosophila melanogaster*: Genomic organization of Responder sequences. *Genetics* **1996**, *144*, 1365–1371. [CrossRef] [PubMed]
27. Marsano, R.M.; Moschetti, R.; Barsanti, P.; Caggese, C.; Caizzi, R. A survey of the DNA sequences surrounding the Bari1 repeats in the pericentromeric h39 region of *Drosophila melanogaster*. *Gene* **2003**, *307*, 167–174. [CrossRef]
28. Zhao, W.; Bidwai, A.P.; Glover, C.V.C. Interaction of casein kinase II with ribosomal protein L22 of *Drosophila melanogaster*. *Biochem. Biophys. Res. Commun.* **2002**, *298*, 60–66. [CrossRef]
29. Bayona-Feliu, A.; Casas-Lamesa, A.; Reina, O.; Bernués, J.; Azorín, F. Linker histone H1 prevents R-loop accumulation and genome instability in heterochromatin. *Nat. Commun.* **2017**, *8*, 283. [CrossRef]
30. Koyama, Y.; Katagiri, S.; Hanai, S.; Uchida, K.; Miwa, M. Poly(ADP-ribose) polymerase interacts with novel *Drosophila* ribosomal proteins, L22 and L23a, with unique histone-like amino-terminal extensions. *Gene* **1999**, *226*, 339–345. [CrossRef]
31. Kearse, M.G.; Ireland, J.A.; Prem, S.M.; Chen, A.S.; Ware, V.C. RpL22e, but not RpL22e-like-PA, is SUMOylated and localizes to the nucleoplasm of *Drosophila* meiotic spermatocytes. *Nucleus* **2013**, *4*, 241–258. [CrossRef] [PubMed]
32. Mageeney, C.M.; Ware, V.C. Specialized eRpL22 paralogue-specific ribosomes regulate specific mRNA translation in spermatogenesis in *Drosophila melanogaster*. *Mol. Biol. Cell* **2019**, *30*, 2240–2253. [CrossRef]
33. Gietz, R.D. Yeast transformation by the LiAc/SS carrier DNA/PEG method. *Methods Mol. Biol.* **2014**, *1205*, 1–12.
34. Colloms, S.D.; van Luenen, H.G.; Plasterk, R.H. DNA binding activities of the Caenorhabditis elegans Tc3 transposase. *Nucleic Acids Res.* **1994**, *22*, 5548–5554. [CrossRef]
35. Marsano, R.M.; Moschetti, R.; Caggese, C.; Lanave, C.; Barsanti, P.; Caizzi, R. The complete Tirant transposable element in *Drosophila melanogaster* shows a structural relationship with retrovirus-like retrotransposons. *Gene* **2000**, *247*, 87–95. [CrossRef]
36. James, T.C.; Eissenberg, J.C.; Craig, C.; Dietrich, V.; Hobson, A.; Elgin, S.C. Distribution patterns of HP1, a heterochromatin-associated nonhistone chromosomal protein of *Drosophila*. *Eur. J. Cell Biol.* **1989**, *50*, 170–180.
37. Marck, C. 'DNA Strider': A 'C' program for the fast analysis of DNA and protein sequences on the Apple Macintosh family of computers. *Nucleic Acids Res.* **1988**, *16*, 1829–1836. [CrossRef]
38. Kosugi, S.; Hasebe, M.; Fau-Tomita, M.; Tomita, M.; Fau-Yanagawa, H.; Yanagawa, H. Systematic identification of cell cycle-dependent yeast nucleocytoplasmic shuttling proteins by prediction of composite motifs. *Proc. Natl. Acad. Sci. USA* **2009**, *106*, 10171–10176. [CrossRef]
39. Brameier, M.; Krings A Fau-MacCallum, R.M.; MacCallum, R.M. NucPred—Predicting nuclear localization of proteins. *Bioinformatics* **2007**, *23*, 1159–1160. [CrossRef] [PubMed]
40. Coelho, P.A.; Nurminsky, D.; Hartl, D.; Sunkel, C.E. Identification of Porto-1, a new repeated sequence that localises close to the centromere of chromosome 2 of *Drosophila melanogaster*. *Chromosoma* **1996**, *105*, 211–222. [CrossRef] [PubMed]
41. Csink, A.K.; Henikoff, S. Something from nothing: The evolution and utility of satellite repeats. *Trends Genet.* **1998**, *14*, 200–204. [CrossRef]
42. Zhou, R.; Hotta, I.; Denli, A.M.; Hong, P.; Perrimon, N.; Hannon, G.J. Comparative analysis of argonaute-dependent small RNA pathways in *Drosophila*. *Mol. Cell* **2008**, *32*, 592–599. [CrossRef]
43. Anger, A.M.; Armache, J.-P.; Berninghausen, O.; Habeck, M.; Subklewe, M.; Wilson, D.N.; Beckmann, R. Structures of the human and *Drosophila* 80S ribosome. *Nature* **2013**, *497*, 80–85. [CrossRef] [PubMed]
44. Yang, Y.; Xu, S.; Xia, L.; Wang, J.; Wen, S.; Jin, P.; Chen, D. The Bantam microRNA Is Associated with *Drosophila* Fragile X Mental Retardation Protein and Regulates the Fate of Germline Stem Cells. *PLoS Genet.* **2009**, *5*, e1000444. [CrossRef]
45. Celniker, S.E.; Dillon, L.A.L.; Gerstein, M.B.; Gunsalus, K.C.; Henikoff, S.; Karpen, G.H.; Kellis, M.; Lai, E.C.; Lieb, J.D.; MacAlpine, D.M.; et al. Unlocking the secrets of the genome. *Nature* **2009**, *459*, 927–930. [CrossRef]

46. Caizzi, R.; Moschetti, R.; Piacentini, L.; Fanti, L.; Marsano, R.M.; Dimitri, P. Comparative Genomic Analyses Provide New Insights into the Evolutionary Dynamics of Heterochromatin in *Drosophila*. *PLoS Genet.* **2016**, *12*, e1006212. [CrossRef]
47. Tzika, E.; Dreker, T.; Imhof, A. Epigenetics and Metabolism in Health and Disease. *Front. Genet.* **2018**, *9*, 361. [CrossRef]
48. Signor, S. Transposable elements in individual genotypes of *Drosophila* simulans. *Ecol. Evol.* **2020**, *10*, 3402–3412. [CrossRef]
49. Spana, C.; Harrison, D.A.; Corces, V.G. The *Drosophila melanogaster* suppressor of Hairy-wing protein binds to specific sequences of the gypsy retrotransposon. *Genes Dev.* **1988**, *2*, 1414–1423. [CrossRef] [PubMed]
50. Georgiev, P.G.; Corces, V.G. The su(Hw) protein bound to gypsy sequences in one chromosome can repress enhancer-promoter interactions in the paired gene located in the other homolog. *Proc. Natl. Acad. Sci. USA* **1995**, *92*, 5184–5188. [CrossRef]
51. Minervini, C.F.; Marsano, R.M.; Casieri, P.; Fanti, L.; Caizzi, R.; Pimpinelli, S.; Rocchi, M.; Viggiano, L. Heterochromatin protein 1 interacts with 5′UTR of transposable element ZAM in a sequence-specific fashion. *Gene* **2007**, *393*, 1–10. [CrossRef] [PubMed]
52. Bhavsar, R.B.; Makley, L.N.; Tsonis, P.A. The other lives of ribosomal proteins. *Hum. Genom.* **2010**, *4*, 327. [CrossRef] [PubMed]
53. Brogna, S.; Sato, T.-A.; Rosbash, M. Ribosome Components Are Associated with Sites of Transcription. *Mol. Cell* **2002**, *10*, 93–104. [CrossRef]
54. Schroder, P.A.; Moore, M.J. Association of ribosomal proteins with nascent transcripts in S. cerevisiae. *RNA* **2005**, *11*, 1521–1529. [CrossRef]
55. Wool, I.G. Extraribosomal functions of ribosomal proteins. *Trends Biochem. Sci.* **1996**, *21*, 164–165. [CrossRef]
56. Warner, J.R.; McIntosh, K.B. How Common Are Extraribosomal Functions of Ribosomal Proteins? *Mol. Cell* **2009**, *34*, 3–11. [CrossRef]
57. De, S.; Varsally, W.; Falciani, F.; Brogna, S. Ribosomal proteins' association with transcription sites peaks at tRNA genes in Schizosaccharomyces pombe. *RNA* **2011**, *17*, 1713–1726. [CrossRef]
58. Dai, M.-S.; Arnold, H.; Sun, X.-X.; Sears, R.; Lu, H. Inhibition of c-Myc activity by ribosomal protein L11. *Embo J.* **2007**, *26*, 3332–3345. [CrossRef]
59. Dai, M.-S.; Sun, X.-X.; Lu, H. Ribosomal protein L11 associates with c-Myc at 5 S rRNA and tRNA genes and regulates their expression. *J. Biol. Chem.* **2010**, *285*, 12587–12594. [CrossRef]
60. Wan, F.; Anderson, D.E.; Barnitz, R.A.; Snow, A.; Bidere, N.; Zheng, L.; Hegde, V.; Lam, L.T.; Staudt, L.M.; Levens, D.; et al. Ribosomal Protein S3: A KH Domain Subunit in NF-κB Complexes that Mediates Selective Gene Regulation. *Cell* **2007**, *131*, 927–939. [CrossRef] [PubMed]
61. Wan, F.; Weaver, A.; Gao, X.; Bern, M.; Hardwidge, P.R.; Lenardo, M.J. IKKβ phosphorylation regulates RPS3 nuclear translocation and NF-κB function during infection with Escherichia coli strain O157:H7. *Nat. Immunol.* **2011**, *12*, 335–343. [CrossRef] [PubMed]
62. Ni, J.-Q.; Liu, L.-P.; Hess, D.; Rietdorf, J.; Sun, F.-L. *Drosophila* ribosomal proteins are associated with linker histone H1 and suppress gene transcription. *Genes Dev.* **2006**, *20*, 1959–1973. [CrossRef]
63. Kharchenko, P.V.; Alekseyenko, A.A.; Schwartz, Y.B.; Minoda, A.; Riddle, N.C.; Ernst, J.; Sabo, P.J.; Larschan, E.; Gorchakov, A.A.; Gu, T.; et al. Comprehensive analysis of the chromatin landscape in *Drosophila melanogaster*. *Nature* **2011**, *471*, 480–485. [CrossRef]
64. Bonnet, J.; Lindeboom, R.G.H.; Pokrovsky, D.; Stricker, G.; Çelik, M.H.; Rupp, R.A.W.; Gagneur, J.; Vermeulen, M.; Imhof, A.; Müller, J. Quantification of Proteins and Histone Marks in *Drosophila* Embryos Reveals Stoichiometric Relationships Impacting Chromatin Regulation. *Dev. Cell* **2019**, *51*, 632–644.e6. [CrossRef] [PubMed]
65. Sinzelle, L.; Izsvak, Z.; Ivics, Z. Molecular domestication of transposable elements: From detrimental parasites to useful host genes. *Cell Mol. Life Sci.* **2009**, *66*, 1073–1093. [CrossRef]
66. Cosby, R.L.; Chang, N.C.; Feschotte, C. Host-transposon interactions: Conflict, cooperation, and cooption. *Genes Dev.* **2019**, *33*, 1098–1116. [CrossRef] [PubMed]
67. Pane, A.; Jiang, P.; Zhao, D.Y.; Singh, M.; Schupbach, T. The Cutoff protein regulates piRNA cluster expression and piRNA production in the *Drosophila* germline. *EMBO J.* **2011**, *30*, 4601–4615. [CrossRef]
68. Zhao, K.; Cheng, S.; Miao, N.; Xu, P.; Lu, X.; Zhang, Y.; Wang, M.; Ouyang, X.; Yuan, X.; Liu, W.; et al. A Pandas complex adapted for piRNA-guided transcriptional silencing and heterochromatin formation. *Nat. Cell Biol.* **2019**, *21*, 1261–1272. [CrossRef]
69. Cherbas, L.; Willingham, A.; Zhang, D.; Yang, L.; Zou, Y.; Eads, B.D.; Carlson, J.W.; Landolin, J.M.; Kapranov, P.; Dumais, J.; et al. The transcriptional diversity of 25 *Drosophila* cell lines. *Genome Res.* **2011**, *21*, 301–314. [CrossRef]
70. Lee, H.; McManus, C.J.; Cho, D.-Y.; Eaton, M.; Renda, F.; Somma, M.P.; Cherbas, L.; May, G.; Powell, S.; Zhang, D.; et al. DNA copy number evolution in *Drosophila* cell lines. *Genome Biol.* **2014**, *15*, R70.
71. Hammond Mp Fau-Laird, C.D.; Laird, C.D. Control of DNA replication and spatial distribution of defined DNA sequences in salivary gland cells of *Drosophila melanogaster*. *Chromosoma* **1985**, *91*, 279–286. [CrossRef] [PubMed]
72. Spradling, A.; Orr-Weaver, T. Regulation of DNA replication during *Drosophila* development. *Annu. Rev. Genet.* **1987**, *21*, 373–403. [CrossRef]
73. Goshima, G.; Wollman, R.; Goodwin, S.S.; Zhang, N.; Scholey, J.M.; Vale, R.D.; Stuurman, N. Genes required for mitotic spindle assembly in *Drosophila* S2 cells. *Science* **2007**, *316*, 417–421. [CrossRef] [PubMed]
74. Lindström, M.S. Emerging functions of ribosomal proteins in gene-specific transcription and translation. *Biochem. Biophys. Res. Commun.* **2009**, *379*, 167–170. [CrossRef] [PubMed]
75. Marsano, R.M.; Caizzi, R.; Moschetti, R.; Junakovic, N. Evidence for a functional interaction between the Bari1 transposable element and the cytochrome P450 cyp12a4 gene in *Drosophila melanogaster*. *Gene* **2005**, *357*, 122–128. [CrossRef]

76. Palazzo, A.; Caizzi, R.; Viggiano, L.; Marsano, R.M. Does the Promoter Constitute a Barrier in the Horizontal Transposon Transfer Process? Insight from Bari Transposons. *Genome Biol. Evol.* **2017**, *9*, 1637–1645. [PubMed]
77. Palazzo, A.; Lorusso, P.; Miskey, C.; Walisko, O.; Gerbino, A.; Marobbio, C.M.T.; Ivics, Z.; Marsano, R.M. Transcriptionally promiscuous "blurry" promoters in Tc1/mariner transposons allow transcription in distantly related genomes. *Mob. DNA* **2019**, *10*, 13. [CrossRef] [PubMed]
78. Suresh, S.; Ahn, H.W.; Joshi, K.; Dakshinamurthy, A.; Kananganat, A.; Garfinkel, D.J.; Farabaugh, P.J. Ribosomal protein and biogenesis factors affect multiple steps during movement of the Saccharomyces cerevisiae Ty1 retrotransposon. *Mob. DNA* **2015**, *6*, 22. [CrossRef]

genes

MDPI

Article

Epigenetic Silencing of P-Element Reporter Genes Induced by Transcriptionally Active Domains of Constitutive Heterochromatin in *Drosophila melanogaster*

Giovanni Messina [1], Emanuele Celauro [1,†], Renè Massimiliano Marsano [2,†], Yuri Prozzillo [1,†] and Patrizio Dimitri [1,*]

[1] Dipartimento di Biologia e Biotecnologie "Charles Darwin", Sapienza Università di Roma, 00185 Roma, Italy
[2] Dipartimento di Bioscienze, Biotecnologie e Ambiente, Università di Bari, 70125 Bari, Italy
* Correspondence: patrizio.dimitri@uniroma1.it
† These authors contributed equally to the work.

Abstract: Reporter genes inserted via P-element integration into different locations of the *Drosophila melanogaster* genome have been routinely used to monitor the functional state of chromatin domains. It is commonly thought that P-element-derived reporter genes are subjected to position effect variegation (PEV) when transposed into constitutive heterochromatin because they acquire heterochromatin-like epigenetic modifications that promote silencing. However, sequencing and annotation of the *D. melanogaster* genome have shown that constitutive heterochromatin is a genetically and molecularly heterogeneous compartment. In fact, in addition to repetitive DNAs, it harbors hundreds of functional genes, together accounting for a significant fraction of its entire genomic territory. Notably, most of these genes are actively transcribed in different developmental stages and tissues, irrespective of their location in heterochromatin. An open question in the genetic and molecular studies on PEV in *D. melanogaster* is whether functional heterochromatin domains, i.e., heterochromatin harboring active genes, are able to silence reporter genes therein transposed or, on the contrary, can drive their expression. In this work, we provide experimental evidence showing that strong silencing of the Pw^{+} reporters is induced even when they are integrated within or near actively transcribed loci in the pericentric regions of chromosome 2. Interestingly, some Pw^{+} reporters were found insensitive to the action of a known PEV suppressor. Two of them are inserted within *Yeti*, a gene expressed in the deep heterochromatin of chromosome 2 which carries active chromatin marks. The difference sensitivity to suppressors-exhibited Pw^{+} reporters supports the view that different epigenetic regulators or mechanisms control different regions of heterochromatin. Together, our results suggest that there may be more complexity regarding the molecular mechanisms underlying PEV.

Keywords: transposable elements; P-elements; constitutive heterochromatin; position effect variegation (PEV); *Drosophila melanogaster*

Citation: Messina, G.; Celauro, E.; Marsano, R.M.; Prozzillo, Y.; Dimitri, P. Epigenetic Silencing of P-Element Reporter Genes Induced by Transcriptionally Active Domains of Constitutive Heterochromatin in *Drosophila melanogaster*. *Genes* **2023**, *14*, 12. https://doi.org/10.3390/genes14010012

Academic Editor: Michael J. Palladino

Received: 15 November 2022
Revised: 12 December 2022
Accepted: 16 December 2022
Published: 21 December 2022

1. Introduction

Transposable elements (TEs) represent a conspicuous fraction of eukaryotic genomes, varying from 3% in yeast, 15% in *D. melanogaster*, 45% in humans, over 50% in maize and up to 90% in other plants [1–5].

Based on their structure and transposition mechanism, eukaryotic TEs are classified into two main classes. Class I contains retrotransposons that are mobilized via RNA transposition intermediates, whereas class II elements transpose either using a "cut and paste" mechanism (Subclass I), rolling-circle replication [6], or single strand excision followed by extrachromosomal replication [7].

TEs have shaped eukaryotic genomes [8] including regions of constitutive heterochromatin where a build-up of TEs has been documented in many evolutionarily distant organisms [9,10].

Constitutive heterochromatin has been originally defined based on cytological features, i.e., the compact state during all phases of the cell cycle and C-banding staining. In addition, low gene density, low rate of meiotic recombination, high content of repetitive DNAs and specific epigenetic signatures are hallmarks of the constitutive heterochromatin [11]. However, the lack of functional genes is no more considered a general feature of constitutive heterochromatin of *D. melanogaster* and other species [11]. Indeed, genome sequencing and annotation have shown that this peculiar genomic compartment contains hundreds of transcriptionally active genes, both unique and repetitive [12–14], which together account for a large fraction of the genomic territory of *D. melanogaster* constitutive heterochromatin [11].

Among the *Drosophila* TEs that have been extensively characterized both at the genetic and molecular levels, P and I elements are naturally occurring transposons which have been found to insert themselves into the constitutive heterochromatin of *D. melanogaster* and eventually target the genes therein located [15–22].

Engineered P-elements have been routinely used as DNA integration vectors or to generate collections of single-element insertion lines useful for both forward and reverse genetic studies of gene disruption, enhancer trapping and protein trapping [23–32].

Beside the above applications, engineered P-elements represent useful tools to monitor the functional state of diverse chromatin domains of *D. melanogaster* genome. When inserted into regions of constitutive heterochromatin, P-associated reporters undergo strong silencing or expressed following a mosaic pattern, in which some cells fully or partially display the reporter phenotype (known as variegated phenotype), while other cells do not [33–35]. The variegated phenotypes produced by transgene insertions into constitutive heterochromatin are very similar to those usually ascribed to position effect variegation (PEV) caused by chromosome rearrangements that relocate a euchromatic gene nearby to constitutive heterochromatin [36]. It is generally believed that the PEV phenotypes displayed by a variety of P-element-reporter genes reflect the silenced state of a given heterochromatin region where the elements are inserted. The variegated phenotype of P-reporter genes inserted in constitutive heterochromatin can be suppressed by known genetic modifiers of PEV [21] and by the Y chromosome, a known suppressor of PEV [37].

P-element-based vectors containing the *mini-white* gene driven by an hsp70 promoter were also used to characterize the structure of chromatin domains of the *D. melanogaster* fourth chromosome [38,39]. The results of these works have shown that this chromosome contains heterochromatic domains showing a variegating phenotype alternating with euchromatic domains showing a fully pigmented eye phenotype.

The resistance to exogenous nuclease digestion [40] and to methylation [41] exhibited by P-element reporters inserted within constitutive heterochromatin and their relatively ordered nucleosome arrays [42] together suggested that heterochromatic silencing is due to the loss of sites for binding of transcription factors and/or RNA polymerase.

In conclusion, based on the aforementioned results, P-element reporter genes have been considered a useful tool to investigate the genetic and molecular bases underlying the PEV phenomenon.

Histone modifications have been shown to play a role in the occurrence of epigenetic silencing and thus contribute to PEV [36]. Notably, actively transcribed genes in constitutive heterochromatin must be accessible to RNA pol II and cis-acting transcriptional regulators, although they have regulatory requirements different from those of euchromatic genes [43]. In particular, combinations of negative and active histone modification marks, together with the contribution of key epigenetic regulators, such as the HP1 and SU(VAR)3-9 proteins, are likely to contribute to establishing gene expression in constitutive heterochromatin [11,44–52].

Here, we asked whether a given P-element reporter gene inserted into constitutive heterochromatin, within or nearby a transcribed gene, would "sense" the active chromatin state of the surrounding domain, with its expression being driven by the regulatory elements of that region or, on the contrary, would it undergo PEV. To answer this question, we examined 12 lines each containing a *mini-white* reporter gene carried by a P-element (from here indicated as Pw^+ reporters) cytologically and molecularly mapped to the pericentric heterochromatin of chromosome 2 (Table 1 [21,35,53]).

Our results provide new hints on the mechanisms of epigenetic silencing induced by constitutive heterochromatic in *D. melanogaster*.

Table 1. Features of the insertion lines described in this work.

Line	P-Element	P-Element Position	Affected Gene(s)	Gene Position	Insert Position Relative to Nearby Genes	Sensitivity to Suppressors	Chromatin State (Gene)	Chromatin State (Insertion Site)	Natural TE Insertions (+/− 5 kb)
KV00462	*P{SUPor-P}*	2L:22839699 [−]	*lnRNA:CR43266* *CG17715*	2L:22840517..22843668 [−] 2L:22844220..22854541 [+]	818 bp downstream 4521 bp upstream	Y	1 1,2,7	7	1360
KV00524	*P{SUPor-P}*	2R:1260501 [+]	*CG40191*	2R:1264140..1267747 [+]	3639 bp upstream	Y/N	1,7	7	1360
19.74.3	*P{Fab7-w#un1}*	2R:1341511 [+]	*Yeti (CG40218)*	2R:1343403..1345119 [−]	1892 bp upstream	Y	1	NR	
LP1	*P{Fab7-w#un1}*	2R:1345084 [+]	*Yeti (CG40218)*	2R:1343403..1345119 [−]	Start codon	N	1	1	ND
KV00363	*P{SUPor-P}*	2R:1345097 [+]	*Yeti (CG40218)*	2R:1343403..1345119 [−]	5′UTR	N	1	1	
KV00158	*P{SUPor-P}*	2R:1358001 [−]	*l(2)41Ab (CG41265)*	2R:1349392..1443935 [+]	1st intron	Y/N	NR	NR	
KV00299	*P{SUPor-P}*	2R:1366708 [+]	*l(2)41Ab (CG41265)*	2R:1349392..1443935 [+]	1st intron	N	NR	NR	invader1 1360 (3 copies)
KV00249	*P{SUPor-P}*	2R:1370175 [+]	*l(2)41Ab (CG41265)* *lnRNA:CR44043*	2R:1349392..1443935 [+] 2R:1368921..1371468 [−]	2nd intron Gene body	N	NR ND	NR	
KV00376	*P{SUPor-P}*	2R:1412310 [+]	*l(2)41Ab (CG41265)*	2R:1349392..1443935 [+]	2nd intron	Y/N	NR	NR	
KV00171	*P{SUPor-P}*	2R:3147095 [−]	*dpr21 (CG42596)*	2R:3066499..3191011 [+]	Intron 5	Y/N	ND	ND	Gypsy4 1360
KV00384	*P{SUPor-P}*	2R: 4014047 [+]	*Haspin (CG40080)*	2R:4014111..4049342 [+]	64 bp upstream	Y	1	NR	F 297
KV00369	*P{SUPor-P}*	2R:4593333 [−]	*p120ctn (CG17484)*	2R:4595288..4609492 [+]	1955 bp upstream	Y	1,2,7	7	Doc3 (2 copies) Doc 1360 (5 copies) INE-1 (5 copies) Diver (2 copies) Cr1a Burdock Gypsy12

Genomic positions are relative to the dm6 genome assembly. [+] and [−] indicate the orientation of genes and inserts relative to the genome assembly. The chromatin state detected in the gene body and at the P-element insertion site are described in Kharchenko, et al. [54] (chromatin state 1: Active promoter/transcription start site region; chromatin state 2: Actively transcribed exon; chromatin state 7: Heterochromatin). 2L: Heterochromatin of the left arm of chromosome 2; 2R: Heterochromatin of the right arm of chromosome 2. NR: Not Reported. ND: None Detected. Y: Suppressed; N: Not Suppressed.

2. Materials and Methods

2.1. *Drosophila Strains*

Fly cultures and crosses were carried out at 25 °C in standard cornmeal yeast medium. The KV lines, each carrying a single *SUPor-P* element [35], were gifts of Robert Levis. The *Su(var)205*^{2} and *aubergne*^{QC42} mutants were gifts of Sarah Elgin, while the *Su(var)30*^{71} and *Su(var)3-9*^{1} were gifts of Gunter Reuter. The *LP1* mutant was generated by Messina et al. [53].

2.2. *Mapping of the P White+ (Pw*$^{+}$*) Reporters*

The flanking sequences of the KV lines were obtained by inverse PCR (done by Robert Levis) and their locations are reported in FlyBase. The mapping of *19.74.3* and *LP1* lines was reported by Messina et al. [53].

2.3. *Complementation Test with Df(2Rh)MS41-10*

Pw^{+}*/SM1, Cy* females were crossed *to Df(2Rh)MS41-10/SM1, Cy* males, and the progeny was scored for the presence of *Cy*$^{+}$ and *Cy* flies. The expected ratio between *Cy* and *Cy*$^{+}$ among the offspring of these crosses was 2:1 because *Cy* is homozygous lethal. The absence of the *Cy*$^{+}$ progeny indicated that a given *Pw*$^{+}$ insert was lethal over the deficiency. *KV00363* and *LP1* were found to be lethal over *Df(2Rh)MS41-10* (all progeny showing the *Cy* phenotype), while for the other inserts the ratio between *Cy* and *Cy*$^{+}$ phenotypes was close to 2:1. For each complementation test, over 100 adults were counted.

2.4. *Genetic Crosses with PEV Suppressors*

We crossed w/w; *Su(var)205*^{2}*/SM6, CyS; +/+* or *w/w; +/+; Su(var)30*^{71}*/TM3, Sb* females to males *w/Y; Pw*$^{+}$*/SM1,Cy* from 10 selected lines carrying a *Pw*$^{+}$ reporter mapped to 2R heterochromatin balanced over the SM1,Cy chromosome (Figure 1). For the crosses in Figure 1A, we isolated the F1 male progeny carrying the suppressor (*w/Y; Pw*$^{+}$*/Suvar*) and the control progeny carrying only the variegated *Pw*$^{+}$ reporter (*w/Y; Pw*$^{+}$*/SM1, Cy*). Similarly, for the crosses in Figure 1B, we isolated the F1 male progeny carrying the suppressor (*w/Y; Pw*$^{+}$*/+; Suvar/+*), and the control progeny with only the variegated *Pw+* reporter (*w/Y; Pw*$^{+}$*/+; TM3, Sb/+*). All the aforementioned male progeny were isolated for eye pigment quantification.

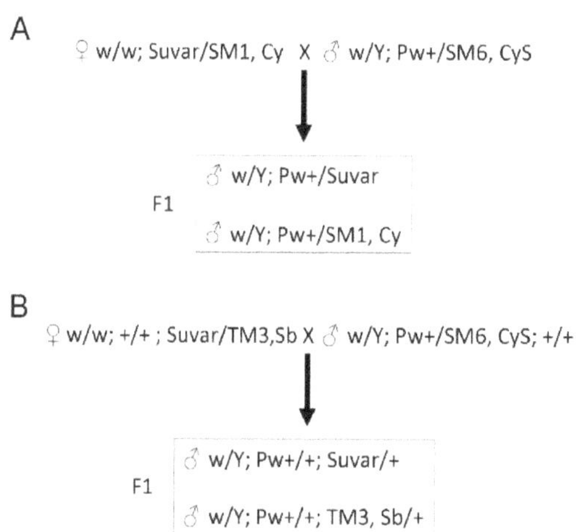

Figure 1. Scheme of genetic crosses. (**A**) Crosses with PEV suppressor mapping to chromosome 2 (*Su(var)205*^{2} and *aubergne*^{QC4}); (**B**) Crosses with PEV suppressor mapping to chromosome 3 (*Su(var)30*^{71} and *Su(var)3-9*^{1}). For a detailed explanation of the crosses see Material and Methods.

2.5. Eye Pigment Assays

The extraction of red eye pigment was performed according to Ephrussi and Herold [55], as described in Prozzillo et al. [56]. For each genotype, three replicate samples of 10 heads were performed. Absorbance at 480 nm was then measured using a 96-well plate in a VICTOR Multilabel Plate Reader spectrophotometer (PerkinElmer, Waltham, MA, USA). Photographs of representative adult fly eyes were taken using a Nikon SMZ745T stereoscopic microscope (Minato, Tokyo, Japan) equipped with a digital C-mount camera.

2.6. Distribution of Epigenetic Marks

The distribution of the epigenetic marks in the regions of constitutive heterochromatin is available in Genome Browser of FlyBase and was also retrieved from the ModEncode data.

3. Results

3.1. Analysis of P-Element Insertions

First, we asked whether Pw^+ reporters inserted into actively transcribed domains of constitutive heterochromatin are subjected to PEV. To answer this question, among a collection of KV lines [21] showing strong *mini-white* silencing, we selected 10 lines with elements located within the pericentric portions of chromosome 2 (Figure 2). These regions have been extensively characterized at both genetic and molecular levels [11,14,20,43,53,57–61]. Each KV line carries a single *SUPor-P* element which has two reporter genes, the *mini-white* gene and the intron-less *yellow* gene, as well as two "suppressor of hairy wing" [*Su(Hw)*] binding sites flanking the *mini-white* gene [62]. This design allows a robust expression of the *yellow* reporter and, together with the presence of the Y chromosome, facilitates the recovery of *SUPor-P* pericentric insertions that would be otherwise missed due to the strong silencing of the *mini-white* reporter [21,63].

The analysis of the insertion sites showed that the *SUPor-P* element are located within or close to genes that are expressed (Table 1). In particular, as shown in Figure 3, according to FlyBase and FlyAtlas, the genes linked to the inserted elements are expressed during different developmental stages and in the eyes of both male and female. We also examined the *LP1* line carrying a Pw^+ reporter into the *Yeti* gene and its progenitor insert of line *19.74.3*, which maps about 2 kb downstream *Yeti* ([53], Table 1). In addition to the *mini-white* reporter, these elements contain Fab-7 insulator sequences characteristic of the *bithorax* complex [64].

All the P-insertions mapping to the right arm of chromosome 2 heterochromatin (2Rh) were tested in complementation analyses with *Df(2Rh)MS41-10*. This deletion lacks almost the entire mitotic heterochromatin of 2Rh, spanning from the deep region h40 to the very distal 46 [58,59,65], together with all the known genes, including *rolled, CG40191, CG41265, Yeti, dpr21, Haspin* and *p120 ctn* (See Figure 1, from Marsano et al. [11]). These results showed that, with the exception of *LP1* and *KV00363* (lethal alleles of *Yeti*), the analyzed P-inserts in *2Rh* were fully viable over *Df(2Rh)MS41-10*. The insert in line *KV00462* which maps to *2Lh* was homozygous viable. Thus, it is conceivable that, while the expression of *Yeti* was clearly affected, the expression of the genes linked to the other insertions was not significantly perturbed. Indeed, in the *LP1* homozygotes no *Yeti* transcripts were detected [53].

Together, from these analyses it emerged that, despite their location in expressed domains of constitutive heterochromatin, all the examined lines exhibited a strong silencing of the Pw^+ reporters. The males of some lines (*KV00524, KV00158, KV00384, KV00376* and *19.74.3*) exhibited a slightly stronger red-eye pigmentation compared to that shown by the females of the same line (Figure 2), an effect that could be due to the PEV suppression exerted by the Y chromosome [36,37].

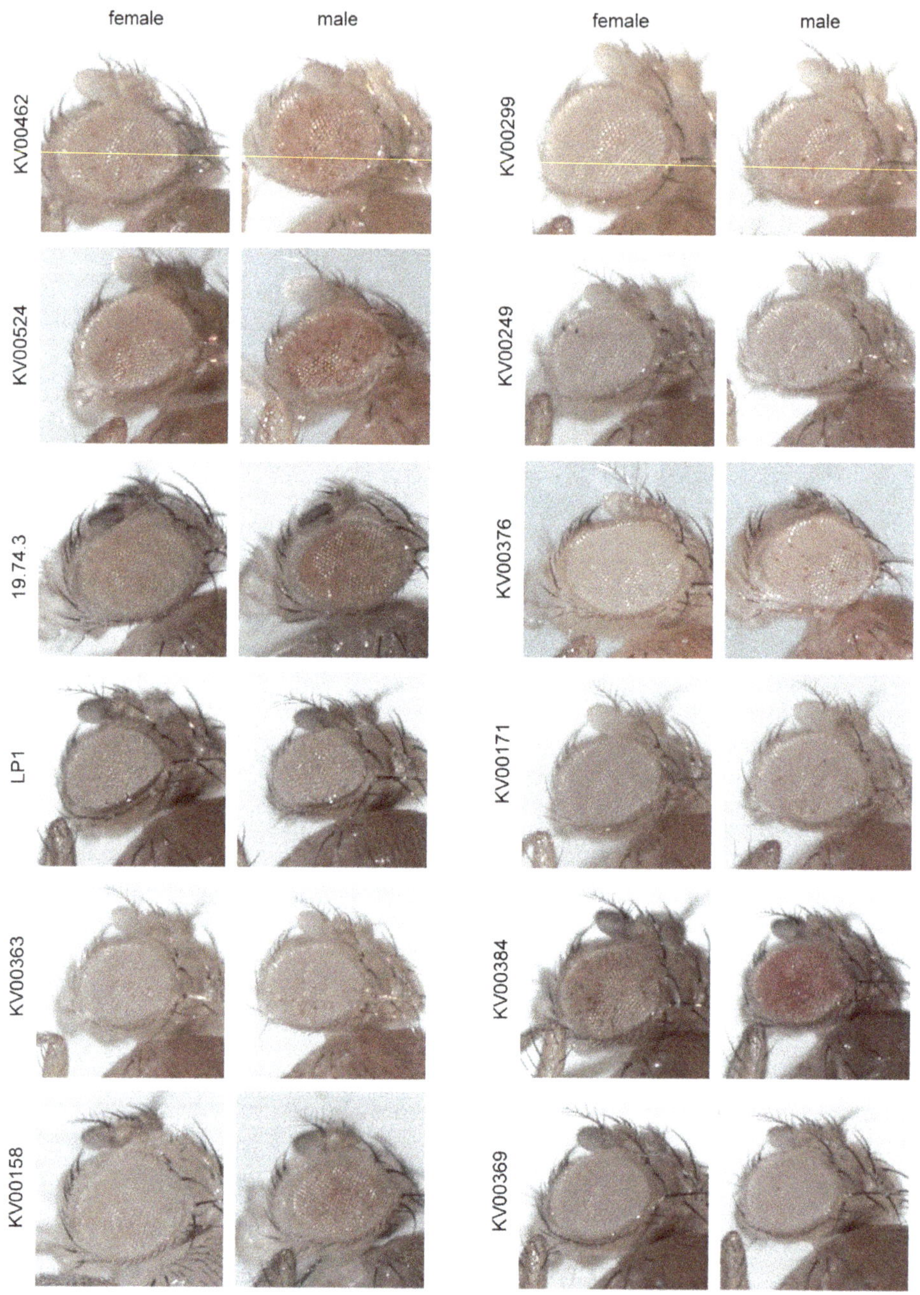

Figure 2. Strong silencing of Pw^+ reporters. Phenotypic analysis of the 12 lines carrying a single Pw^+ reporter inserted in constitutive heterochromatin of chromosome 2. All the Pw^+ reporters are balanced over the *SM1, Cy* chromosome and exhibited a strong silencing giving rises to extremely variegated eyes which in some cases are almost white. The males of lines *KV00462, KV00524, 19.74.3 KV00158* and *KV00384* exhibited a stronger red-eye pigmentation compared to that of females, a difference that could be due to the presence of Y chromosome.

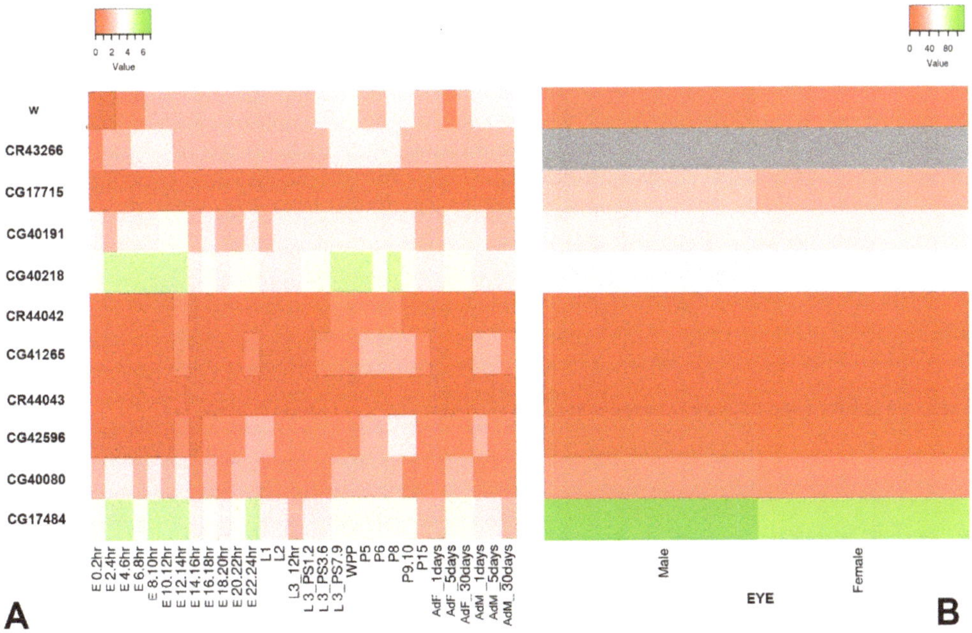

Figure 3. Heatmaps showing expression profile of the heterochromatic genes linked to the examined Pw^+ reporters compared to that of *white* gene. Expression across developmental stages (**A**) and in the eyes of males and females (**B**). Shades of color from red to green indicate the expression bin classification from 1 (no/extremely low expression) to 7 (very high expression). The gray color means that no data about *CR43266* expression are available. Developmental stage and eye expression data were obtained from FlyBase and FlyAtlas, respectively.

3.2. Experiments with Genetic Modifiers of PEV

Loss-of-function mutations resulting in dominant suppression of PEV were shown to identify crucial epigenetic regulators, such as the HP1 protein [36]. Thus, first we investigated whether the variegation of Pw^+ reporters showed by the lines under investigation can be suppressed by *Su(var)205* and *Su(var)307*, two dominant PEV suppressors [36]. To do this, we crossed *w/w; Su(var)205*^{2}*/SM6, Cy,S; +/+* or *w/w; +/+; Su(var)3-7*^{1}*/TM3, Sb* females to males of 10 selected Pw^+ reporters (Figure 2) balanced over the *SM1, Cy* chromosome (Figure 1; Section 2).

The results of this analysis are summarized in Figure 4. *Su(var)205*^{2} clearly suppressed the variegation of *Pw+* reporters in lines *KV00158, KV00376, KV00369* and *KV00171*. However, this was not the case for the remaining five Pw^+ reporters in lines *KV00249, KV00299, KV00363, KV00524* and *LP1*, which were quite insensitive to the suppression. *Su(var)307*^{1} suppressed most Pw^+ reporters, although with a variable efficiency. However, the reporters of *LP1* and *KV00299* lines showed only a slight suppression effect, while the one of *KV00363* was clearly insensitive.

Moreover, mutations in *piwi, aubergine* or *homeless* genes, encoding components of the piRNA pathway [66,67], also suppress the silencing of Pw^+ inserted in heterochromatin as a result of reduction of H3 Lys9 methylation and delocalization of HP1. Thus, we also tested the effect of the *aubergine*^{QC42} mutant allele on the variegation of the Pw^+ reporters under investigation (see Section 2 for the scheme of crosses). As shown in Figure 4, the trend of the effect was very similar to that found with *Su(var)205*^{2} in that the Pw^+ reporters of *KV00249, KV00299, KV00363* and *LP1* lines were insensitive to suppression, with the exception of *KV00524*.

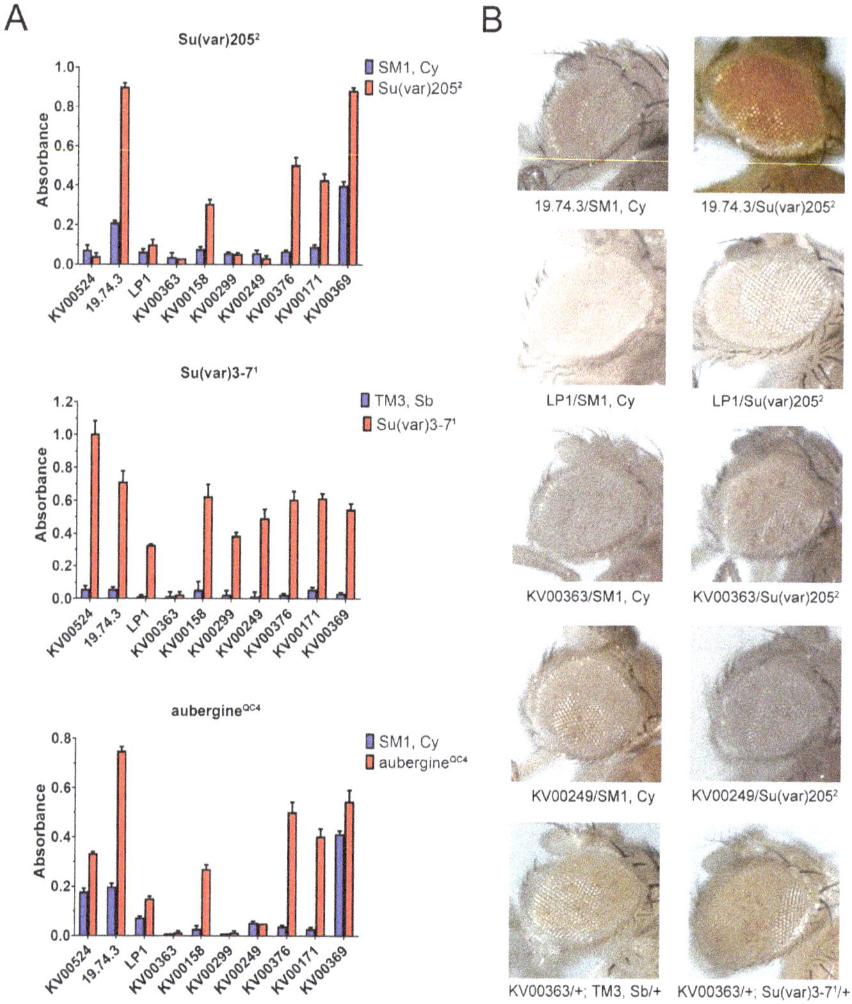

Figure 4. Effects of dominance of *Su(var)205*^{2}, *Su(var)3-7*^{1} and *aubergine*^{QC4} on the silencing of Pw^+ reporters. (**A**) Histograms summarizing the quantification of eye pigment in presence and absence of PEV suppressors; (**B**) Examples of eye phenotypes of Pw^+ reporters in flies with or without suppressors. The silencing of Pw^+ reporter in *19.74.3* line was efficiently suppressed by *Su(var)205*^{2}, *Su(var)3-7*^{1} and *aubergine*^{QC4}; silencing in *KV00363* and *LP1* lines was unaffected by *Su(var)205*^{2} and *aubergine*^{QC4}, while silencing in *LP1* was only partially suppressed by *Su(var)3-7*^{1}.

The product of the wild-type *Su(var)205* gene, the heterochromatin-associated HP1 protein, interacts with the histone methyltransferase (HMTase) encoded by the wild-type *Su(var)3-9* gene. The SU(VAR)3-9 methyltransferase selectively methylates histone H3 at lysine 9 (H3-K9), and it is generally accepted that H3K9me stabilizes the binding of HP1a to chromatin [68–70]. The *Su(var)3-9* mutant alleles are known to be dominant modifiers of PEV [36]. We then tested the effect of the *Su(var)3-9*^{1} mutation on the variegation of Pw^+ reporters from *KV00249*, *KV00299*, *KV00363*, *LP1* and *19.74.3* lines. As shown in Figure 5, the variegation of Pw^+ reporters in *KV00249*, *KV00299*, *KV00363* and *LP1* lines was insensitive to suppression by *Su(var)3-9*^{1}, whereas the Pw^+ variegation in the *19.74.3* was efficiently suppressed. In conclusion, using different PEV suppressors (*Su(var)205*^{2}, *Su(var)3-9*^{1} and *aubergine*^{QC4}), we obtained comparable results with the same *Pw*$^{+}$ reporters (Figures 4 and 5).

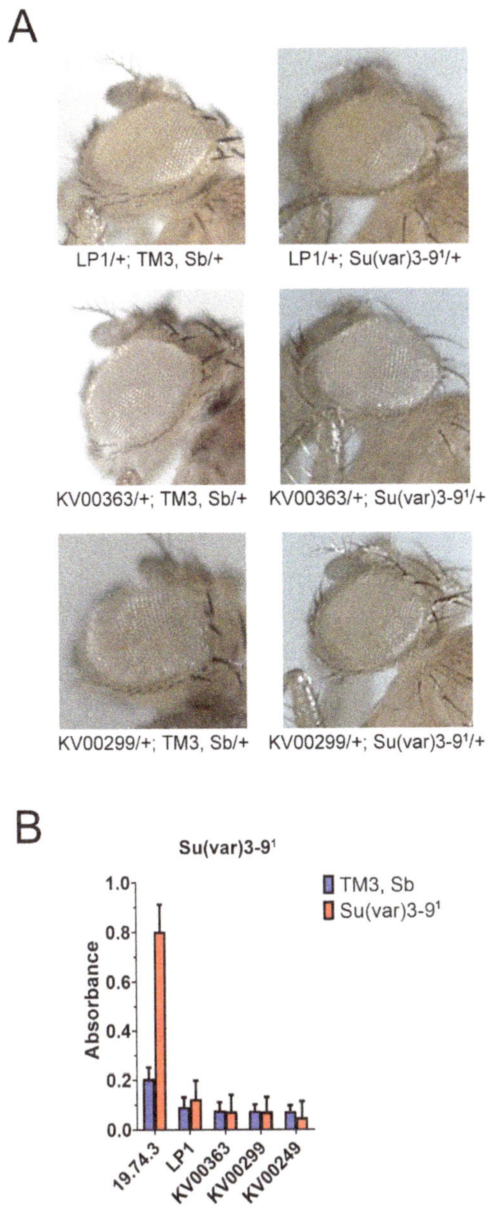

Figure 5. Effects of *Su(var)3-9*^{1} on silencing of *Pw*$^{+}$ reporters. (**A**) Examples of eye phenotypes of *Pw*$^{+}$ reporters in flies with or without *Su(var)3-9*^{1}. (**B**) Histograms summarizing the quantification of eye pigment in presence and absence of *Su(var)3-9*^{1}. The silencing of the *Pw*$^{+}$ reporter in *19.74.3* was efficiently suppressed, while that of *KV00363, LP1, KV00299* and *KV00249* was unaffected. For the scheme of crosses see Figure 1 and Materials and Methods.

4. Discussion

It is generally believed that PEV is induced by transcriptionally inactive constitutive heterochromatin. The aim of this work was to further investigate this aspect by studying the heterochromatin-induced silencing on a number of Pw^{+} reporters inserted into different positions of the pericentromeric regions of chromosome 2, within or nearby a transcribed gene (Table 1; Figures 3 and 6). We found that all the examined reporters subjected to strong PEV (Figure 2) map within or close to genes/domains that support an appreciable level of transcription during development and in the eye (Figure 3). Moreover, some reporters were suppressed by classical dominant PEV suppressors encoding epigenetic factors/regulators required for heterochromatin establishment, while other reporters in lines *KV00249*, *KV00299*, *KV00363* and *LP1* were less sensitive or even insensitive to the action of the suppressors (Figures 4–6). In general, there appears to be no significant correlation between the *mini-white* expression levels for each Pw^{+} reporter line and the transcription levels of the heterochromatic gene(s) linked to it. For example, CG40191 is on average more expressed than *dpr21*, and yet the respective inserts have the same behavior towards suppression.

Although the heterogeneity of the genetic background might partially contribute to explaining the different sensitivity to epigenetic regulators tested, these results suggest that the chromatin state characteristic of transcriptionally active heterochromatin may not be sufficient to guarantee the proper expression of a given reporter gene inserted within or nearby. In other words, the regulatory requirements for the expression of the autochthonous genes in heterochromatin seem to be different, or even antipodal, from those needed for adventitious DNA sequences, such as the Pw^{+} reporters. However, the observation that the silencing of some reporters is abolished by the tested suppressors also suggests that the product of wild-type alleles of some suppressors, on one hand, can induce the silencing of the reporters, but, on the other hand, may be required for, or is compatible with, the expression of the genes located in pericentric heterochromatin. The effect of the product of wild-type alleles of suppressors could be performed directly or indirectly.

Indeed, the expression of *light* and *rolled* heterochromatic genes of chromosome 2 was found to be affected when moved away from their native location to euchromatin by chromosome rearrangements [70,71]. Moreover, genetic and molecular studies have shown that the proper expression of *light, rolled,* and other heterochromatic genes such as *Rpl15* and *Dbp80* depends on the HP1a protein [72–75]. In accordance, large-scale mapping experiments carried out in *Drosophila* Kc cells have shown that HP1a and SU(VAR)3-9 are associated with pericentric genes, such as *light* and *rolled* genes [44,46,47].

Together, the available findings support the idea that histone marks, together with HP1a and other epigenetic regulators, are involved to ensure the expression of genes located in constitutive heterochromatin. However, some apparently contradictory results [44,51] indicated that differences in the epigenetic regulation of heterochromatic genes may exist between in vivo and in vitro systems.

The insensitivity to suppressors showed by Pw^{+} reporters in *KV363* and *LP1* lines is intriguing for several reasons. Firstly, *Yeti* is an efficiently expressed gene in the heterochromatin of chromosome 2 (Figure 2 [11,53,76–79]) and carries active chromatin marks (Table 1). Secondly, the P-insertion of the *19.74.3* line, the progenitor of *LP1*, maps roughly 2 kb downstream the 5′ of *Yeti* (Figure 6), but its silencing is efficiently suppressed by $Su(var)205^{2}$, $Su(var)307^{1}$, $Su(var)309^{2}$ and $aubergne^{QC42}$ (Figures 4 and 5). Thirdly, *Su(var)205*, *Su(var)309* and *Su(var)307* are among the most general and effective PEV suppressors and were already found to efficiently erase the silencing of many pericentric reporters [36,80]. Finally, the distribution of HP1 and SU(VAR)309 proteins along the pericentric regions was found to be enriched in the heterochromatin of 2R where *Yeti* is located [44,46,47].

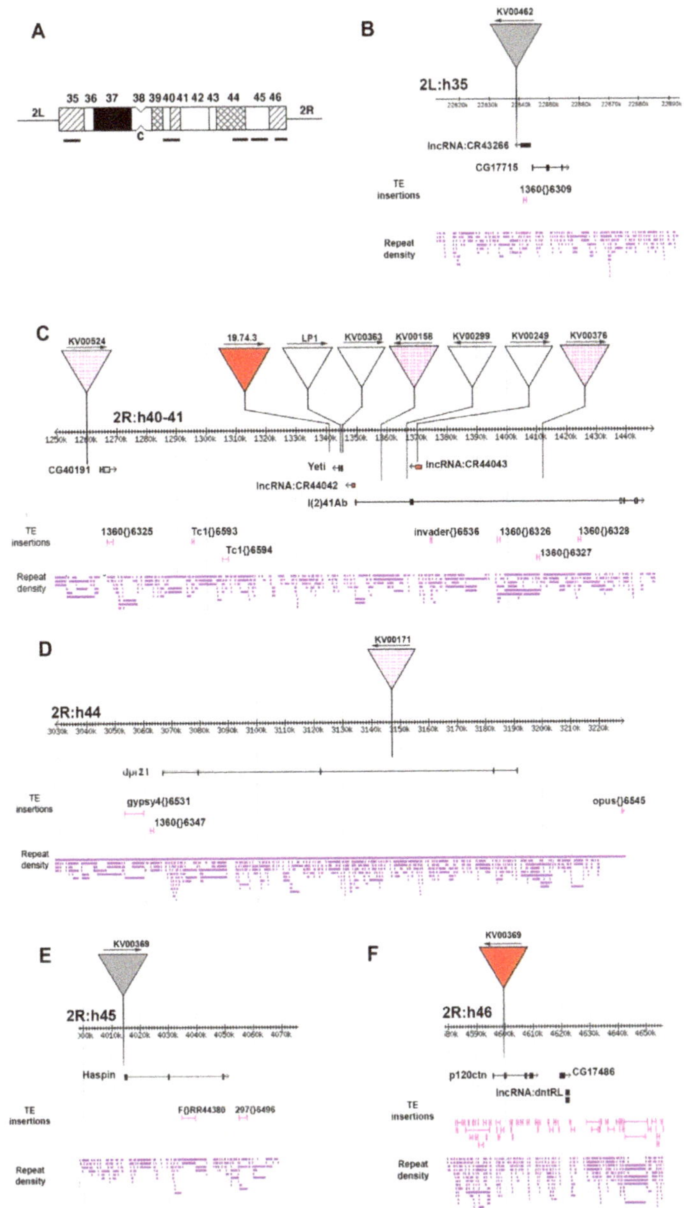

Figure 6. Genomic localization of the Pw^{+} reporters under investigation. (**A**) Cytogenetic map of the heterochromatin of chromosome 2 showing the location of the genomic regions analyzed in this work. From left to right: 2L:h35, 2R:h40-41, 2R: h44, 2R:h45-46, 2R:h46. The genomic regions in (**B–F**) show the localization of the Pw^{+} reporter insertions relative to genes and the distribution of TEs and repetitive sequences. Red triangle = fully suppressed Pw^{+} reporter; dashed triangle = Pw^{+} reporter suppressed only in some cases; white triangle = Pw^{+} reporter insensitive to suppression; grey triangle = Pw^{+} reporter not tested for suppression.

The molecular bases underpinning insensitivity to suppressors exhibited by *Pw*+ reporters in *KV363* and *LP1* lines are currently elusive. A trivial explanation for the apparently strong silencing could be that the *Pw*+ element in these lines is structurally defective so that the expression of the *mini-white* is compromised. However, we found that in both lines the element can still transpose in the presence of the *Delta-2-3* element, a known source of the P-transposase, giving rise to de novo fully pigmented red-eye insertions (not show). This result excludes the occurrence of significant DNA sequence alterations in the *Pw+* element of both *KV363* and *LP1* lines.

It is possible that the chromatin state of the *Yeti* gene is regulated by epigenetic regulators/mechanisms different from those controlled by the wild-type alleles of the tested PEV suppressors.

We did not find recurrent repeated sequences flanking the reporters, suggesting the absence of specific relationships between location and sensitivity to suppression. However, two non-coding RNAs, *CR44042* and *CR44043*, were associated with *Yeti* and *l(2)41Ab* genes (Figure 6), respectively, with the *Pw*+ reporter of *KV00299* being inserted within *CR44043*. It could be possible that the transcription of *CR44042* and *CR44043* may interfere somehow with that of the reporters.

A consistent body of experimental evidence suggested that nascent RNA may play a direct role in transcription and chromatin regulation, in that it may function as regulator of its own expression, giving rise to positive-feedback loops in which active gene expression states can be maintained [81]. This model could explain the resistance to all PEV suppressor exhibited by the *Pw*+ reporters in *LP1* and *KV00363* lines. We hypothesize that in wild-type conditions the *Yeti*-encoded RNA interacts with its own gene at the level of DNA/chromatin, acting as an epigenetic insulator or as a positive effector, giving rise to an active chromatin state which allows a stable expression of *Yeti*. However, following a P-insertional mutation, a consequent lack of functional *Yeti* transcripts (or a low level of transcription) would shut-down the expression of the gene disrupting its active chromatin state with the acquisition of a closed chromatin structure. As a consequence, the *Pw*+ reporters inserted within *Yeti* would become silenced (Figure 7). Such positive RNA feedback mechanism, if existing, should be insensitive to the action of classical PEV suppressors.

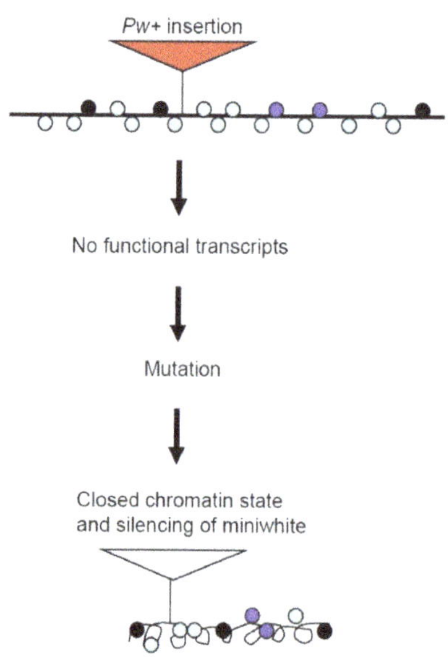

Figure 7. *Pw*+ silencing by perturbation of a positive RNA feedback mechanism. The insertion of a *Pw+* reporter within the *Yeti* gene affects its expression. According to a positive RNA feedback model,

the consequent lack of functional *Yeti* transcripts would induce the formation of a closed chromatin structure of the gene itself. As a result, the *mini-white* gene is silenced. Colored circles represent different chromatin marks.

5. Conclusions

In conclusion, we have shown that transcriptionally active domains of constitutive heterochromatin can induce epigenetic silencing of Pw^{+} reporters. Whatever the genetic bases of this phenomenon are, our results represent a remarkable paradox suggesting that there may be more complexity regarding the molecular mechanisms underlying PEV.

Author Contributions: Conceptualization P.D., E.C., R.M.M., G.M. and Y.P.; Methodology P.D., E.C., R.M.M., G.M. and Y.P.; Validation P.D., E.C., R.M.M., G.M. and Y.P.; Experiments E.C., P.D., R.M.M., Y.P. and G.M.; Writing P.D., G.M. and R.M.M. All authors have read and agreed to the published version of the manuscript.

Funding: This work was supported by grants from Sapienza University of Rome, Progetti di Ricerca di Ateneo #RM120172B851A176 (P.D.) and the University of Bari, Progetti di Ricerca di Ateneo #00869718Ricat (R.M.M.).

Data Availability Statement: Data are contained within the article.

Acknowledgments: We thank Robert Levis for gift of KV lines; we also thank Sarah Elgin and Gunter Reuter for gift of PEV suppressors. Finally, we are grateful to the two reviewers for their help in improving the manuscript.

Conflicts of Interest: The authors declare no conflict of interest.

References

1. Flavell, R.B. Repetitive DNA and chromosome evolution in plants. *Philos. Trans. R. Soc. B Biol. Sci.* **1986**, *312*, 227–242. [CrossRef]
2. SanMiguel, P.; Tikhonov, A.; Jin, Y.-K.; Motchoulskaia, N.; Zakharov, D.; Melake-Berhan, A.; Springer, P.S.; Edwards, K.J.; Lee, M.; Avramova, Z.; et al. Nested Retrotransposons in the Intergenic Regions of the Maize Genome. *Science* **1996**, *274*, 765–768. [CrossRef] [PubMed]
3. Kidwell, M.G.; Lisch, D. Transposable elements as sources of variation in animals and plants. *Proc. Natl. Acad. Sci.USA* **1997**, *94*, 7704–7711. [CrossRef] [PubMed]
4. Kaminker, J.S.; Bergman, C.; Kronmiller, B.; Carlson, J.; Svirskas, R.; Patel, S.; Frise, E.; A Wheeler, D.; E Lewis, S.; Rubin, G.; et al. The transposable elements of the *Drosophila melanogaster* euchromatin: A genomics perspective. *Genome Biol.* **2002**, *3*, research0084.1. [CrossRef] [PubMed]
5. Wells, J.N.; Feschotte, C. A Field Guide to Eukaryotic Transposable Elements. *Annu. Rev. Genet.* **2020**, *54*, 539–561. [CrossRef]
6. Kapitonov, V.V.; Jurka, J. Rolling-circle transposons in eukaryotes. *Proc. Natl. Acad. Sci. USA* **2001**, *98*, 8714–8719. [CrossRef]
7. Kapitonov, V.V.; Jurka, J. Self-synthesizing DNA transposons in eukaryotes. *Proc. Natl. Acad. Sci. USA* **2006**, *103*, 4540–4545. [CrossRef]
8. Romano, N.C.; Fanti, L. Transposable Elements: Major Players in Shaping Genomic and Evolutionary Patterns. *Cells* **2022**, *11*, 1048. [CrossRef]
9. Dimitri, P.; Corradini, N.; Rossi, F.; Mei, E.; Zhimulev, I.; Vernì, F. Transposable elements as artisans of the heterochromatic genome in *Drosophila melanogaster. Cytogenet. Genome Res.* **2005**, *110*, 165–172. [CrossRef]
10. Marsano, R.M.; Dimitri, P. Constitutive Heterochromatin in Eukaryotic Genomes: A Mine of Transposable Elements. *Cells* **2022**, *11*, 761. [CrossRef]
11. Marsano, R.M.; Giordano, E.; Messina, G.; Dimitri, P. A New Portrait of Constitutive Heterochromatin: Lessons from *Drosophila melanogaster. Trends Genet.* **2019**, *35*, 615–631. [CrossRef]
12. A Hoskins, R.; Smith, C.D.; Carlson, J.W.; Carvalho, A.B.; Halpern, A.; Kaminker, J.S.; Kennedy, C.; Mungall, C.J.; A Sullivan, B.; Sutton, G.G.; et al. Heterochromatic sequences in a Drosophila whole-genome shotgun assembly. *Genome Biol.* **2002**, *3*, Rresearch0085.1. [CrossRef]
13. Hoskins, R.A.; Carlson, J.W.; Kennedy, C.; Acevedo, D.; Evans-Holm, M.; Frise, E.; Wan, K.H.; Park, S.; Mendez-Lago, M.; Rossi, F.; et al. Sequence Finishing and Mapping of *Drosophila melanogaster* Heterochromatin. *Science* **2007**, *316*, 1625–1628. [CrossRef]
14. Hoskins, R.A.; Carlson, J.W.; Wan, K.H.; Park, S.; Mendez, I.; Galle, S.E.; Booth, B.W.; Pfeiffer, B.D.; George, R.A.; Svirskas, R.; et al. The Release 6 reference sequence of the *Drosophila melanogaster* genome. *Genome Res.* **2015**, *25*, 445–458. [CrossRef]
15. Kidwell, M.G.; Kidwell, J.F.; Sved, J.A. Hybrid Dysgenesis in *Drosophila melanogaster*: A Syndrome of Aberrant Traits Including Mutation, Sterility and Male Recombination. *Genetics* **1977**, *86*, 813–833. [CrossRef]
16. Rubin, G.; Kidwell, M.G.; Bingham, P.M. The molecular basis of P-M hybrid dysgenesis: The nature of induced mutations. *Cell* **1982**, *29*, 987–994. [CrossRef]

17. Sang, H.; Pélisson, A.; Bucheton, A.; Finnegan, D. Molecular lesions associated with white gene mutations induced by I-R hybrid dysgenesis in *Drosophila melanogaster*. *EMBO J.* **1984**, *3*, 3079–3085. [CrossRef]
18. Bucheton, A. I transposable elements and I-R hybrid dysgenesis in Drosophila. *Trends Genet.* **1990**, *6*, 16–21. [CrossRef]
19. Zhang, P.; Spradling, A.C. Insertional mutagenesis of Drosophila heterochromatin with single P elements. *Proc. Natl. Acad. Sci. USA* **1994**, *91*, 3539–3543. [CrossRef]
20. Dimitri, P.; Arcà, B.; Berghella, L.; Mei, E. High genetic instability of heterochromatin after transposition of the LINE-like *I* factor in *Drosophila melanogaster*. *Proc. Natl. Acad. Sci. USA* **1997**, *94*, 8052–8057. [CrossRef]
21. Konev, A.Y.; Yan, C.M.; Acevedo, D.; Kennedy, C.; Ward, E.; Lim, A.; Tickoo, S.; Karpen, G.H. Genetics of P-element transposition into *Drosophila melanogaster* centric heterochromatin. *Genetics* **2003**, *165*, 2039–2053. [CrossRef] [PubMed]
22. Dimitri, P.; Bucheton, A. I element distribution in mitotic heterochromatin of *Drosophila melanogaster* reactive strains: Identification of a specific site which is correlated with the reactivity levels. *Cytogenet. Genome Res.* **2005**, *110*, 160–164. [CrossRef] [PubMed]
23. Rubin, G.M.; Spradling, A.C. Genetic transformation of Drosophila with transposable element vectors. *Science* **1982**, *218*, 348–353. [CrossRef] [PubMed]
24. Robertson, H.M.; Preston, C.R.; Phillis, R.W.; Johnson-Schlitz, D.M.; Benz, W.K.; Engels, W.R. A stable genomic source of P element transposase in *Drosophila melanogaster*. *Genetics* **1988**, *118*, 461–470. [CrossRef] [PubMed]
25. Cooley, L.; Kelley, R.; Spradling, A. Insertional Mutagenesis of the *Drosophila* Genome with Single P Elements. *Science* **1988**, *239*, 1121–1128. [CrossRef]
26. Cooley, L.; Berg, C.; Kelley, R.; McKearin, D.; Spradling, A. Identifying and cloning Drosophila genes by single P element insertional mutagenesis. *Prog. Nucleic Acid Res. Mol. Biol.* **1989**, *36*, 99–109. [CrossRef]
27. Spradling, A.C.; Stern, D.; Beaton, A.; Rhem, E.J.; Laverty, T.; Mozden, N.; Misra, S.; Rubin, G. The Berkeley Drosophila Genome Project Gene Disruption Project: Single P-Element Insertions Mutating 25% of Vital Drosophila Genes. *Genetics* **1999**, *153*, 135–177. [CrossRef]
28. Zhai, R.G.; Hiesinger, P.R.; Koh, T.-W.; Verstreken, P.; Schulze, K.L.; Cao, Y.; Jafar-Nejad, H.; Norga, K.K.; Pan, H.; Bayat, V.; et al. Mapping *Drosophila* mutations with molecularly defined *P* element insertions. *Proc. Natl. Acad. Sci. USA* **2003**, *100*, 10860–10865. [CrossRef]
29. O'Kane, C.J.; Gehring, W.J. Detection in situ of genomic regulatory elements in Drosophila. *Proc. Natl. Acad. Sci. USA* **1987**, *84*, 9123–9127. [CrossRef]
30. Wilson, C.; Pearson, R.K.; Bellen, H.J.; O'Kane, C.J.; Grossniklaus, U.; Gehring, W.J. P-element-mediated enhancer detection: An efficient method for isolating and characterizing developmentally regulated genes in Drosophila. *Genes Dev.* **1989**, *3*, 1301–1313. [CrossRef]
31. Hazelrigg, T.; Levis, R.; Rubin, G.M. Transformation of white locus DNA in Drosophila: Dosage compensation, zeste interaction, and position effects. *Cell* **1984**, *36*, 469–481. [CrossRef]
32. Bellen, H.; Levis, R.; Liao, G.; He, Y.; Carlson, J.W.; Tsang, G.; Evans-Holm, M.; Hiesinger, P.R.; Schulze, K.L.; Rubin, G.; et al. The BDGP Gene Disruption Project. *Genetics* **2004**, *167*, 761–781. [CrossRef]
33. Levis, R.; Hazelrigg, T.; Rubin, G.M. Effects of Genomic Position on the Expression of Transduced Copies of the *white* Gene of *Drosophila*. *Science* **1985**, *229*, 558–561. [CrossRef]
34. Karpen, G.H.; Spradling, A.C. Analysis of subtelomeric heterochromatin in the Drosophila minichromosome Dp1187 by single P element insertional mutagenesis. *Genetics* **1992**, *132*, 737–753. [CrossRef]
35. Roseman, R.; Pirrotta, V.; Geyer, P. The su(Hw) protein insulates expression of the *Drosophila melanogaster* white gene from chromosomal position-effects. *EMBO J.* **1993**, *12*, 435–442. [CrossRef]
36. Elgin, S.C.; Reuter, G. Position-Effect Variegation, Heterochromatin Formation, and Gene Silencing in Drosophila. *Cold Spring Harb. Perspect. Biol.* **2013**, *5*, a017780. [CrossRef]
37. Dimitri, P.; Pisano, C. Position effect variegation in *Drosophila melanogaster*: Relationship between suppression effect and the amount of Y chromosome. *Genetics* **1989**, *122*, 793–800. [CrossRef]
38. Sun, F.-L.; Haynes, K.; Simpson, C.L.; Lee, S.D.; Collins, L.; Wuller, J.; Eissenberg, J.C.; Elgin, S.C.R. *cis* -Acting Determinants of Heterochromatin Formation on *Drosophila melanogaster* Chromosome Four. *Mol. Cell. Biol.* **2004**, *24*, 8210–8220. [CrossRef]
39. Riddle, N.C.; Leung, W.; Haynes, K.A.; Granok, H.; Wuller, J.; Elgin, S.C.R. An Investigation of Heterochromatin Domains on the Fourth Chromosome of *Drosophila melanogaster*. *Genetics* **2008**, *178*, 1177–1191. [CrossRef]
40. Wallrath, L.L.; Elgin, S.C. Position effect variegation in Drosophila is associated with an altered chromatin structure. *Genes Dev.* **1995**, *9*, 1263–1277. [CrossRef]
41. Boivin, A.; Dura, J.-M. In Vivo Chromatin Accessibility Correlates with Gene Silencing in Drosophila. *Genetics* **1998**, *150*, 1539–1549. [CrossRef] [PubMed]
42. Sun, F.-L.; Cuaycong, M.H.; Elgin, S.C.R. Long-Range Nucleosome Ordering Is Associated with Gene Silencing in *Drosophila melanogaster* Pericentric Heterochromatin. *Mol. Cell. Biol.* **2001**, *21*, 2867–2879. [CrossRef] [PubMed]
43. Howe, M.; Dimitri, P.; Berloco, M.; Wakimoto, B.T. Cis-effects of heterochromatin on heterochromatic and euchromatic gene activity in *Drosophila melanogaster*. *Genetics* **1995**, *140*, 1033–1045. [CrossRef] [PubMed]
44. Greil, F.; van der Kraan, I.; Delrow, J.; Smothers, J.F.; de Wit, E.; Bussemaker, H.J.; van Driel, R.; Henikoff, S.; van Steensel, B. Distinct HP1 and Su(var)3-9 complexes bind to sets of developmentally coexpressed genes depending on chromosomal location. *Genes Dev.* **2003**, *17*, 2825–2838. [CrossRef] [PubMed]

45. Yasuhara, J.C.; Wakimoto, B.T. Molecular landscape of modified histones in Drosophila heterochromatic genes and euchromatin-heterochromatin transition zones. *PLoS Genet.* **2008**, *4*, e16. [CrossRef]
46. de Wit, E.; Greil, F.; van Steensel, B. Genome-wide HP1 binding in *Drosophila*: Developmental plasticity and genomic targeting signals. *Genome Res.* **2005**, *15*, 1265–1273. [CrossRef]
47. De Wit, E.; Greil, F.; Van Steensel, B. High-Resolution Mapping Reveals Links of HP1 with Active and Inactive Chromatin Components. *PLoS Genet.* **2007**, *3*, e38. [CrossRef]
48. Fanti, L.; Pimpinelli, S. HP1: A functionally multifaceted protein. *Curr. Opin. Genet. Dev.* **2008**, *18*, 169–174. [CrossRef]
49. Riddle, N.C.; Minoda, A.; Kharchenko, P.V.; Alekseyenko, A.A.; Schwartz, Y.B.; Tolstorukov, M.Y.; Gorchakov, A.A.; Jaffe, J.D.; Kennedy, C.; Linder-Basso, D.; et al. Plasticity in patterns of histone modifications and chromosomal proteins in Drosophila heterochromatin. *Genome Res.* **2011**, *21*, 147–163. [CrossRef]
50. Saha, P.; Sowpati, D.T.; Mishra, R.K. Epigenomic and genomic landscape of *Drosophila melanogaster* heterochromatic genes. *Genomics* **2019**, *111*, 177–185. [CrossRef]
51. Saha, P.; Sowpati, D.T.; Soujanya, M.; Srivastava, I.; Mishra, R.K. Interplay of pericentromeric genome organization and chromatin landscape regulates the expression of *Drosophila melanogaster* heterochromatic genes. *Epigenetics Chromatin* **2020**, *13*, 41. [CrossRef]
52. Wallrath, L.L.; Rodriguez-Tirado, F.; Geyer, P.K. Shining Light on the Dark Side of the Genome. *Cells* **2022**, *11*, 330. [CrossRef]
53. Messina, G.; Damia, E.; Fanti, L.; Atterrato, M.T.; Celauro, E.; Mariotti, F.R.; Accardo, M.C.; Walther, M.; Verni, F.; Picchioni, D.; et al. Yeti, an essential *Drosophila melanogaster* gene, encodes a protein required for chromatin organization. *J. Cell Sci.* **2014**, *127*, 2577–2588. [CrossRef]
54. Kharchenko, P.V.; Alekseyenko, A.A.; Schwartz, Y.B.; Minoda, A.; Riddle, N.C.; Ernst, J.; Sabo, P.J.; Larschan, E.; Gorchakov, A.A.; Gu, T.; et al. Comprehensive analysis of the chromatin landscape in *Drosophila melanogaster*. *Nature* **2011**, *471*, 480–485. [CrossRef]
55. Ephrussi, B.; Herold, J.L. Studies of Eye Pigments of Drosophila. I. Methods of Extraction and Quantitative Estimation of the Pigment Components. *Genetics* **1944**, *29*, 148–175. [CrossRef]
56. Prozzillo, Y.; Cuticone, S.; Ferreri, D.; Fattorini, G.; Messina, G.; Dimitri, P. In Vivo Silencing of Genes Coding for dTip60 Chromatin Remodeling Complex Subunits Affects Polytene Chromosome Organization and Proper Development in *Drosophila melanogaster*. *Int. J. Mol. Sci.* **2021**, *22*, 4525. [CrossRef]
57. Hilliker, A.J. Genetic analysis of the centromeric heterochromatin of chromosome 2 of *Drosophila melanogaster*: Deficiency mapping of EMS-induced lethal complementation groups. *Genetics* **1976**, *83*, 765–782. [CrossRef]
58. Dimitri, P. Cytogenetic analysis of the second chromosome heterochromatin of *Drosophila melanogaster*. *Genetics* **1991**, *127*, 553–564. [CrossRef]
59. Corradini, N.; Rossi, F.; Verni, F.; Dimitri, P. FISH analysis of *Drosophila melanogaster* heterochromatin using BACs and P elements. *Chromosoma* **2003**, *112*, 26–37. [CrossRef]
60. Rossi, F.; Moschetti, R.; Caizzi, R.; Corradini, N.; Dimitri, P. Cytogenetic and molecular characterization of heterochromatin gene models in *Drosophila melanogaster*. *Genetics* **2007**, *175*, 595–607. [CrossRef]
61. Coulthard, A.B.; Alm, C.; Cealiac, I.; Sinclair, D.A.R.S.A.; Honda, B.M.H.M.; Rossi, F.; Dimitri, P.; Hilliker, A.J. Essential Loci in Centromeric Heterochromatin of *Drosophila melanogaster*. I: The Right Arm of Chromosome 2. *Genetics* **2010**, *185*, 479–495. [CrossRef] [PubMed]
62. Roseman, R.R.; Johnson, E.A.; Rodesch, C.K.; Bjerke, M.; Nagoshi, R.N.; Geyer, P.K. A P element containing suppressor of hairy-wing binding regions has novel properties for mutagenesis in *Drosophila melanogaster*. *Genetics* **1995**, *141*, 1061–1074. [CrossRef]
63. Yan, C.M.; Dobie, K.W.; Le, H.D.; Konev, A.Y.; Karpen, G.H. Efficient Recovery of Centric Heterochromatin *P*-Element Insertions in *Drosophila melanogaster*. *Genetics* **2002**, *161*, 217–229. [CrossRef] [PubMed]
64. Hagstrom, K.; Muller, M.; Schedl, P. Fab-7 functions as a chromatin domain boundary to ensure proper segment specification by the Drosophila bithorax complex. *Genes Dev.* **1996**, *10*, 3202–3215. [CrossRef] [PubMed]
65. Andreyeva, E.N.; Kolesnikova, T.D.; Demakova, O.V.; Mendez-Lago, M.; Pokholkova, G.V.; Belyaeva, E.S.; Rossi, F.; Dimitri, P.; Villasante, A.; Zhimulev, I.F. High-resolution analysis of Drosophila heterochromatin organization using SuUR Su(var)3-9 double mutants. *Proc. Natl. Acad. Sci. USA* **2007**, *104*, 12819–12824. [CrossRef] [PubMed]
66. Pal-Bhadra, M.; Leibovitch, B.A.; Gandhi, S.G.; Rao, M.; Bhadra, U.; Birchler, J.A.; Elgin, S.C.R. Heterochromatic Silencing and HP1 Localization in *Drosophila* Are Dependent on the RNAi Machinery. *Science* **2004**, *303*, 669–672. [CrossRef]
67. Nishida, K.M.; Saito, K.; Mori, T.; Kawamura, Y.; Nagami-Okada, T.; Inagaki, S.; Siomi, H.; Siomi, M.C. Gene silencing mechanisms mediated by Aubergine–piRNA complexes in *Drosophila* male gonad. *RNA* **2007**, *13*, 1911–1922. [CrossRef]
68. Bannister, A.J.; Zegerman, P.; Partridge, J.F.; Miska, E.A.; Thomas, J.O.; Allshire, R.C.; Kouzarides, T. Selective recognition of methylated lysine 9 on histone H3 by the HP1 chromo domain. *Nature* **2001**, *410*, 120–124. [CrossRef]
69. Lachner, M.; O'Carroll, D.; Rea, S.; Mechtler, K.; Jenuwein, T. Methylation of histone H3 lysine 9 creates a binding site for HP1 proteins. *Nature* **2001**, *410*, 116–120. [CrossRef]
70. Nakayama, J.-I.; Rice, J.C.; Strahl, B.D.; Allis, C.D.; Grewal, S.I.S. Role of Histone H3 Lysine 9 Methylation in Epigenetic Control of Heterochromatin Assembly. *Science* **2001**, *292*, 110–113. [CrossRef]
71. Eberl, D.F.; Duyf, B.J.; Hilliker, A.J. The role of heterochromatin in the expression of a heterochromatic gene, the rolled locus of *Drosophila melanogaster*. *Genetics* **1993**, *134*, 277–292. [CrossRef]

72. Hearn, M.G.; Hedrick, A.; Grigliatti, T.A.; Wakimoto, B.T. The effect of modifiers of position-effect variegation on the variegation of heterochromatic genes of *Drosophila melanogaster*. *Genetics* **1991**, *128*, 785–797. [CrossRef]
73. Clegg, N.; Honda, B.; Whitehead, I.; Grigliatti, T.; Wakimoto, B.; Brock, H.; Lloyd, V.; Sinclair, D. Suppressors of position-effect variegation in *Drosophila melanogaster* affect expression of the heterochromatic gene light in the absence of a chromosome rearrangement. *Genome* **1998**, *41*, 495–503. [CrossRef]
74. Lu, B.Y.; Emtage, P.C.R.; Duyf, B.J.; Hilliker, A.J.; Eissenberg, J.C. Heterochromatin Protein 1 Is Required for the Normal Expression of Two Heterochromatin Genes in *Drosophila*. *Genetics* **2000**, *155*, 699–708. [CrossRef]
75. Schulze, S.R.; Sinclair, D.A.R.; Fitzpatrick, K.A.; Honda, B.M. A Genetic and Molecular Characterization of Two Proximal Heterochromatic Genes on Chromosome 3 of *Drosophila melanogaster*. *Genetics* **2005**, *169*, 2165–2177. [CrossRef]
76. Prozzillo, Y.; Monache, F.D.; Ferreri, D.; Cuticone, S.; Dimitri, P.; Messina, G. The True Story of Yeti, the "Abominable" Heterochromatic Gene of *Drosophila melanogaster*. *Front. Physiol.* **2019**, *10*, 1093. [CrossRef]
77. Messina, G.; Atterrato, M.T.; Fanti, L.; Giordano, E.; Dimitri, P. Expression of human Cfdp1 gene in Drosophila reveals new insights into the function of the evolutionarily conserved BCNT protein family. *Sci. Rep.* **2016**, *6*, 25511. [CrossRef]
78. Messina, G.; Celauro, E.; Atterrato, M.T.; Giordano, E.; Iwashita, S.; Dimitri, P. The Bucentaur (BCNT) protein family: A long-neglected class of essential proteins required for chromatin/chromosome organization and function. *Chromosoma* **2014**, *124*, 153–162. [CrossRef]
79. Messina, G.; Prozzillo, Y.; Bizzochi, G.; Marsano, R.M.; Dimitri, P. The Green Valley of *Drosophila melanogaster* Constitutive Heterochromatin: Protein-Coding Genes Involved in Cell Division Control. *Cells* **2022**, *11*, 3058. [CrossRef]
80. Haynes, K.A.; Gracheva, E.; Elgin, S.C.R. A Distinct Type of Heterochromatin Within *Drosophila melanogaster* Chromosome 4. *Genetics* **2007**, *175*, 1539–1542. [CrossRef]
81. Skalska, L.; Beltran-Nebot, M.; Ule, J.; Jenner, R. Regulatory feedback from nascent RNA to chromatin and transcription. *Nat. Rev. Mol. Cell Biol.* **2017**, *18*, 331–337. [CrossRef] [PubMed]

genes

Review

Leukocyte Telomere Length as a Molecular Biomarker of Coronary Heart Disease

Olga V. Zimnitskaya [1,*], Marina M. Petrova [1], Natalia V. Lareva [2], Marina S. Cherniaeva [3], Mustafa Al-Zamil [4], Anastasia E. Ivanova [5] and Natalia A. Shnayder [1,5,*]

1 Department of Outpatient Therapy and General Practice with Course of Postgraduate Education, V.F. Voino-Yasenetsky Krasnoyarsk State Medical University, 660022 Krasnoyarsk, Russia; stk99@yandex.ru
2 Department of Therapy, Faculty of Postgraduate Education, Chita State Medical Academy, 672000 Chita, Russia; larevanv@mail.ru
3 Department of Internal and Preventive Medicine, Central State Medical Academy of the Presidential Administration, 121359 Moscow, Russia; doctor@cherniaeva.ru
4 Department of Physiotherapy, Faculty of Continuing Medical Education, Peoples' Friendship University of Russia, 117198 Moscow, Russia; alzamil@mail.ru
5 V.M. Bekhterev National Medical Research Center for Psychiatry and Neurology, Institute of Personalized Psychiatry and Neurology, 192019 St. Petersburg, Russia; anastasiae.ivanova@bekhterev.ru
* Correspondence: zvezda_5786@mail.ru (O.V.Z.); naschnaider@yandex.ru (N.A.S.); Tel.: +7-(391)-228-0628 (O.V.Z.); +7-(812)-670-02-20 (N.A.S.)

Abstract: Background. This work is a review of preclinical and clinical studies of the role of telomeres and telomerase in the development and progression of coronary heart disease (CHD). Materials and methods. A search for full-text publications (articles, reviews, meta-analyses, Cochrane reviews, and clinical cases) in English and Russian was carried out in the databases PubMed, Oxford University Press, Scopus, Web of Science, Springer, and E-library electronic library using keywords and their combinations. The search depth is 11 years (2010–2021). Results. The review suggests that the relative leukocyte telomere length (LTL) is associated with the development of socially significant and widespread cardiovascular diseases such as CHD and essential hypertension. At the same time, the interests of researchers are mainly focused on the study of the relative LTL in CHD. Conclusions. Despite the scientific and clinical significance of the analyzed studies of the relative length of human LTL as a biological marker of cardiovascular diseases, their implementation in real clinical practice is difficult due to differences in the design and methodology of the analyzed studies, as well as differences in the samples by gender, age, race, and ethnicity. The authors believe that clinical studies of the role of the relative length of leukocyte telomeres in adult patients with coronary heart disease are the most promising and require large multicenter studies with a unified design and methodology.

Keywords: telomere length; stable coronary heart disease; people; adults; coronary atherosclerosis; acute coronary syndrome; acute myocardial infarction; early vascular aging; molecular predictors

Citation: Zimnitskaya, O.V.; Petrova, M.M.; Lareva, N.V.; Cherniaeva, M.S.; Al-Zamil, M.; Ivanova, A.E.; Shnayder, N.A. Leukocyte Telomere Length as a Molecular Biomarker of Coronary Heart Disease. *Genes* **2022**, *13*, 1234. https://doi.org/10.3390/genes13071234

Academic Editors: Luigi Viggiano and Renè Massimiliano Marsano

Received: 7 June 2022
Accepted: 9 July 2022
Published: 12 July 2022

1. Introduction

Cardiovascular diseases (CVDs) are prevalent worldwide. In Europe, there were 19.9 million new cases of cardiovascular disease in 2017. There were 2.5 million new cases of CVD in Germany, 1.115 million in France, and 1.209 million new cases of CVD in the United Kingdom [1]. CVDs are the leading cause of death in most European countries [1] and Russia [2]. In European countries, CVDs cause about 2.2 million female and 1.9 million male deaths per year [1], and in Russia, the mortality rate from CVDs is about 938,500 per year [3]. Coronary artery disease (CAD) or coronary heart disease (CHD) are responsible for over 50% of CVD deaths in Russia [2,3] and cause about 40% of CVD deaths in Europe [1].

Genetic factors have a significant impact on the risk of CVD. A history of CVD increases their future risk from 40% to 75% depending on the degree of the relationship [4]. Aging is a major risk factor for CVDs and cerebrovascular diseases but it has been established that people age at different rates. Therefore, aging is characterized by chronological and biological aging. Chronological aging refers to the time elapsed since a person was born, and biological aging refers to the decline in the function of a person's tissues and organs. In people who age normally, chronological age is equated with biological age [5]. In 2008, Nisson et al. [6] formulated the concept of early vascular aging (EVA), according to which the biological age of a person depends on the age of his or her blood vessels, and persons with EVA syndrome due to early vascular aging have an increased risk of developing CVDs and their complications. There is no clear list of EVA criteria, but some authors refer to physiological biomarkers of EVA as increased arterial vascular wall stiffness (arterial stiffness) assessed by pulse wave velocity or by calculation of the cardiovascular ankle index; a thickening of the intima-media complex; endothelial dysfunction [7–9]; the presence of atherosclerotic plaques in arteries; and the deposition of calcium phosphate crystals in arterial intima [10,11].

Deoxyribonucleic acid (DNA) methylation and telomere shortening are considered potential molecular biomarkers of EVA [5]. In a progressively aging world population, early diagnosis based on the development and implementation into real clinical practice of such molecular biomarkers is very important, as it may allow effective identification of people with a high risk of EVA in different climatic–geographical regions and racial and ethnic groups. Such a personalized strategy is expected to reduce the socio-economic burden of age-associated diseases including CVDs in general but in particular CAD.

In recent years, blood leukocyte telomere length (LTL) has been considered a "mitotic clock" fixing human biological age [12] and as a potential molecular biomarker of EVA, but research in this area is ongoing and no unequivocal decision on this issue has been made yet [13].

Also, as other early prognostic biochemical biomarkers of EVA, CVDs (atherosclerosis and CAD) are being actively studied: neutrophil gelatinase-associated lipocalin (NGAL) [14]; tissue inhibitor of metalloproteinase 2 (TIMP-2) [14]; fibroblast growth factor 23 (FGF-23) [15,16]; syndecan-1 [16]; interleukin 6 (IL-6) [17]; and galectin-3 [18].

In addition, biomarkers of the adverse outcomes of atherosclerosis and CAD (total mortality and mortality from CVDs) are being developed, among which LTL is of undoubted scientific and clinical interest [19,20]. This is due to the fact that the mechanisms of LTL shortening in adults, which lead to stable CAD, acute coronary syndrome (ACS), and acute myocardial infarction (AMI), have not yet been sufficiently studied [21]. LTL is also actively studied in other CVDs (essential arterial hypertension [22,23], atrial fibrillation [24–27], cardiomyopathy [28], cerebrovascular diseases (stroke [29–33], vascular cognitive disorders [34,35], and vascular dementia [36]).

Telomeres (from the Greek telos "end" and meros "part") are nucleoprotein structures located at the ends of chromosomes (Figure 1), consisting of a noncoding repetitive DNA sequence (-TTAGGG-), a single-stranded region called the protruding part of the G-chain [37,38], and proteins that compose the Shelterin complex. Due to the Shelterin proteins, telomeric DNA is folded into a complex three-dimensional structure [39,40]. The telomere length of an adult human is approximately 10–15 thousand base pairs (bp). The protruding part of the G-chain, including 150–200 bp, can bend and form a loop structure (T-loop). Also, a D-loop can be formed [39]. The T-loop protects the 3′OH ends of the chromosomes from recognition of 3′OH as a double break in the DNA chain [40,41].

Telomeres are bound to telomere-specific proteins that are part of the Shelterin complex. The Shelterin complex provides greater telomere stability and consists of six telomere-specific proteins: Telomeric Repeat Binding Factor 1 (TRF1); Telomeric Repeat Binding Factor 2 (TRF2); Telomeric Interacting Nuclear Factor 2 (TINF2 or TIN2), a complex consisting of TERF1 and Nuclear Factor 2; Protection of Telomeres 1 (POT1), a protein that provides telomere protection; Shelterin complex subunit and telomerase recruitment factor (TPP1), a subunit of Shelterin complex and telomerase recruitment factor; and TERF2 interacting protein (TERF2IP or RAP1), a protein that interacts with TERF2. All six proteins regulate telomere length [42].

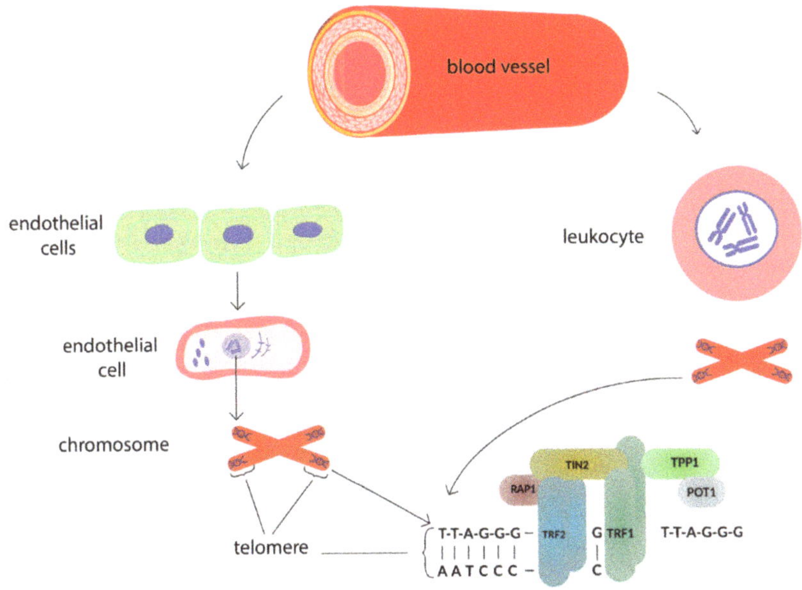

Figure 1. The relationship between the telomere length of endotheliocytes and leukocytes.

The interaction of the Shelterin protein complex with the telomere DNA sequence ensures the stabilization of the telomere structure and regulates the access of proteins involved in DNA elongation and repair. The TRF1 complex is involved in telomere length control by regulating telomerase access to the telomere sequence [42]. The TRF2 complex protects the G-chain protrusion from degradation and prevents telomere fusion [43].

Telomeres were first identified by Hermann Müller in 1938 and he and McClintock determined the protective role of telomeres in 1941 [44]. The first human telomeres were isolated by Moyzic et al. [45] in 1988. Since then, telomere biology has been extensively studied.

Telomeres protect chromosome ends, maintaining genome integrity and stability [38]; telomeres prevent loss of coding DNA during DNA replication [39]. Chromosome telomere fusion can lead to gene amplification, chromosome imbalance, non-reciprocal translocations, and changes in gene expression [40,45]. Telomere length is shortened due to end replication problems and nucleolytic DNA degradation. Each cell division results in the loss of 50–200 bp of telomere sequence [40,46,47]. It is believed that telomere shortening is the reason for the limited number of divisions in most human cells. This phenomenon was first described by Hayflick [48] on diploid human cells. Also, he found that each cloned cell in the population is endowed with the same doubling potential—50 ± 10 cell divisions. This phenomenon was later called the "Hayflick Limit," meaning that a cell can divide a limited number of times after which cell division stops [49].

In contrast to most somatic cells, hematopoietic stem cells, germ cells, keratinocytes in the basal layer of the epidermis, uterine endometrial cells, and cells from various tumors avoid telomere shortening by activating telomerase [40]. Telomerase, also called terminal transferase, is a ribonucleoprotein that adds a species-dependent telomere repeat sequence to the 3′ end of telomeres [50]. Although telomerase activity has been vigorously investigated over the last few decades, many questions remain open regarding the mechanisms of physiological regulation in normal cells [51]. The complex regulation at the levels of transcription, splicing, and post-transcriptional activation certainly contributes to that.

Moreover, mutational analysis and knockdown experiments showed that telomerase deficiency led to telomere loss and uncapping, causing progressive atrophy of renewal tissues, a gradual depletion of stem cells, and the eventual failure of organ systems. Above all, telomerase may play a critical role in cellular and organismal aging and could be a potential target for anti-aging therapies [52].

Telomere shortening is associated with non-genetic (physiological) and genetic mechanisms of aging (inflammation, oxidative stress, chronic diseases, cellular aging, mortality), as well as with social factors of aging (gender, race, ethnicity, low socioeconomic status, stress, smoking) [47,53]. EVA syndrome, as well as age-associated diseases (CVDs, type 2 diabetes mellitus, cancer, or chronic obstructive pulmonary disease), are associated with telomere shortening and/or dysfunction [46]. For example, people with different degrees of atherosclerosis and CAD have significantly different LTL [52].

The purpose of this systematic review is to find, analyze, and systematize studies on the relationship between LTL and CAD.

2. Materials and Methods

Full-text publications were searched in the following databases: PubMed, Web of Science, Springer, Google Scholar, Oxford Press, Clinical Cases, Cochrane, and e-Library. We analyzed articles published between 10 January 2010 and 10 December 2021. Key words and their combinations were used to search for: "telomere length," "stable coronary heart disease," "humans," "adults," "coronary atherosclerosis," "acute coronary syndrome," "acute myocardial infarction," "early vascular aging," and "molecular predictors."

Publications were searched and selected using the Preferred Reporting Items for Systematic Reviews and Meta-Analyses (PRISMA) 2020 guidelines. A total of 278 articles were found using the keywords (Figure 2). We analyzed original studies, systematic reviews and meta-analyses, clinical cases, and Cochrane reviews. After reading the relevant titles and abstracts, we excluded publications with irrelevant topics as well as duplicate publications, open access preprints, and conference posters. We excluded original studies that did not provide primary data, including demographic data, and original articles with retrospective studies. We also excluded reviews, editorials, animal experiments, and publications with questionable or insufficiently proven results of the authors' investigated molecular biomarkers of CAD and the relationship between CAD and LTL. In addition, we analyzed earlier publications of historical interest.

Finally, we selected 14 suitable publications for our systematic review in which the authors used comparable methodological approaches to measure LTL using a T/S ratio and/or bp. This approach was important to be able to systematize the results of the analyzed studies. We analyzed but excluded from further processing publications with alternative methods of LTL calculation.

We selected articles in which the relative LTL was measured by the method of Cawthon R. M. [54]. This technique uses real-time quantitative polymerase chain reaction (PCR) to measure LTL. The advantages of this technique are speed of performance and a small amount of DNA. Relative LTL is defined as the ratio of telomeric repeats to a single copy of a standard gene (T/S) and is measured in conventional units (CU). The T/S ratio reflects the average LTL in all human leukocytes [55].

Statistical analysis of the obtained data was performed using the SPSS software package, version 23 (Stat Soft, Tulsa, AK, USA,). Since the size of compared samples was small ($n \leq 30$), nonparametric statistics were used. Median (Me) and percentiles [25; 75] were calculated for each of the three groups (healthy adults without CVDs, adults with stable CAD, and adults with AMI). The groups were compared using the Mann–Whitney test. Significance of differences was considered significant at a p-value < 0.05.

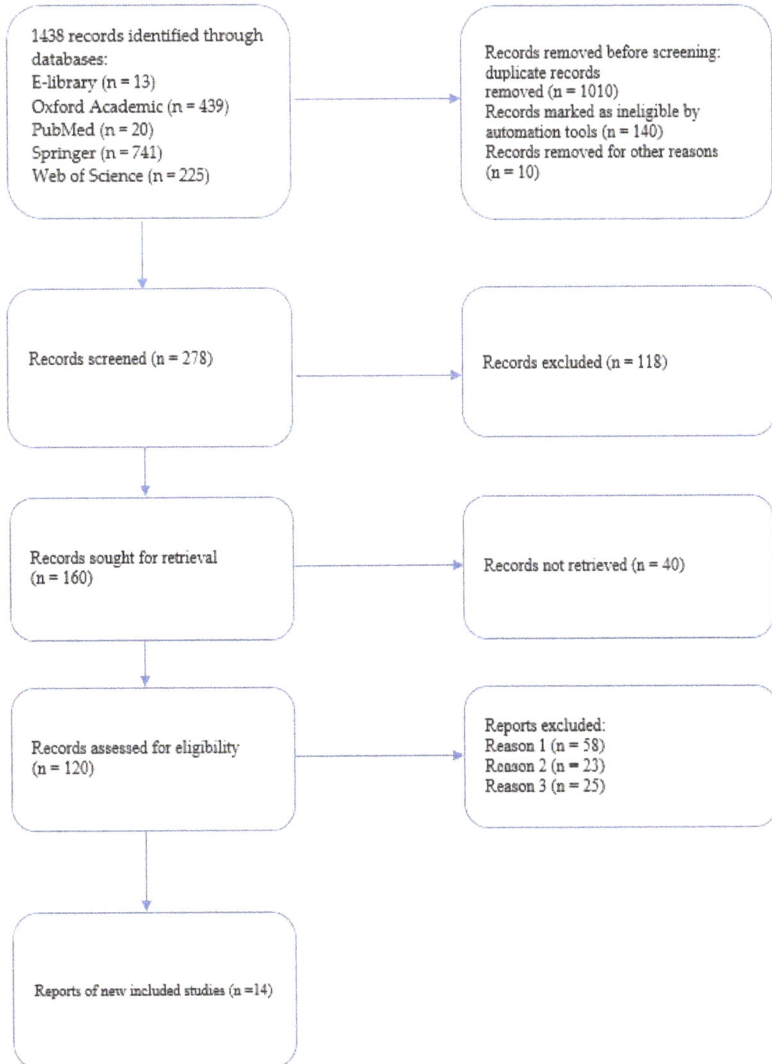

Figure 2. Flow chart diagram visualizing the database searches and the number of publications identified, screened, and the final full texts included in the present systematic review. Reason 1—there is no primary data on relative LTL in the article. Reason 2—the method of measuring relative LTL was carried out using an alternative method to the Ca. thon R. M. method. Reason 3—review and meta-analysis.

3. Results

3.1. Leukocyte Telomere Length in Patients with Stable Coronary Heart Disease

Our analysis of the studies demonstrates that the issue of LTL changing in patients with CAD has been extensively studied worldwide. The number of LTL studies on this disease is increasing in different age groups of patients and in different racial and ethnic groups compared with healthy controls (Table 1). Thus, in healthy adults without CVD, the relative LTL ranged from 0.69 [56] to 1.52 CU [55] and the median relative LTL was 0.93 [0.70; 1.10] CU.

Table 1. Mean Relative Telomere Length of Leukocytes in Adults without Cardiovascular Disease.

Authors [References]	Study Characteristics	Group Characteristics	Participants (n)	Age, in Years (M ± SE or Me [P25; P75])	Sex (Male/Female, %)	Method	Telomere Length	
							Absolute, bp	Relative (T/S Ratio), CU
Williet et al., 2010 [55]	Prospective, population-based study	Austrians without CAD	712	61.8 ± 10.8	47.6/52.4	Real-time quantitative PCR	N/A	1.52 ± 0.81
Dlouha et al., 2016 [57]	Observational, cross-sectional case-control study	Czechs without CAD	642	50 ± 2.7	0/100	Real-time quantitative PCR	N/A	0.93 ± 0.38
Tian et al., 2017 [58]	Observational, cross-sectional study	Chinese without CAD	128	48.5 ± 7.33	57.8/42.2	Real-time quantitative PCR	N/A	1.1 ± 0.57
Pejenaute et al., 2020 [59]	Observational, cross-sectional study	Spaniards without CAD	389	54 ± 1	80/20	Real-time quantitative PCR	8591 ± 84	N/A
Gupta et al., 2020 [60]	Observational, cross-sectional study	Indians without CVDs	77	34.38 ± 5.86	75/25	Real-time quantitative PCR	N/A	0.792
Starnino et al., 2021 [61]	Observational, cross-sectional study	Canadians without CVDs	25	55.68 ± 0.19	56/44	Real-time quantitative PCR	N/A	0.94 ± 0.15
Mazidi et al., 2021 * [62]	Mendelian randomized trial	British without CAD	20	22.3 ± 1.8	100/0	Real-time quantitative PCR	12 420 ± 80	N/A
Mazidi et al., 2021 * [62]	Mendelian randomized trial	British without CAD	20	62.75 ± 2.1	100/0	Real-time quantitative PCR	6 380 ± 60	N/A
Hassler et al., 2021 ** [56]	Observational, cross-sectional study	Austrians without CVDs	90	40.77 ± 11.62	100/0	Real-time quantitative PCR	N/A	0.7 ± 0.28
Hassler et al., 2021 ** [56]	Observational, cross-sectional study	Austrians without CVDs	90	44.71 ± 10.96	0/100	Real-time quantitative PCR	N/A	0.69 ± 0.31

Notes: bp—base pairs; CU—conventional units; N/A—no data; PCR—polymerase chain reaction. * There were two groups in the Mazidi study. One group of patients is young (mean age 22 years old), and the other group is elderly (mean age 62.75 years old). For each of the groups, this author calculated the relative LTL. ** There were two groups in the Hassler study: men and women without CVDs. Separately for each group, the author measured the relative LTL. Therefore, these results are listed in two rows of Table 1.

Hassler et al. [56] studied relative LTL in healthy adults (mean age of men 40.77 ± 11.62 years, mean age of women 44.71 ± 10.96 years). The authors found no significant difference in relative LTL in men and women of the study age (0.70 ± 0.28 vs. 0.69 ± 0.31 CU, p-value = 0.75).

In patients with stable CAD (Table 2), the relative LTL ranged from 0.82 [63] to 1.13 CU [55]; the median relative LTL was 0.86 [0.82; 1.07] CU. The differences in median LTL in patients with stable CAD compared to healthy controls were not statistically significant (p-value = 0.850).

Table 2. Mean Relative Telomere Length of Leukocytes in Adults with Stable Coronary Artery Disease.

Authors [References]	Study Characteristics	Group Characteristics	Participants (n)	Age, in Years (M ± SE or Me [P25; P75])	Sex (Male/Female, %)	Method	Telomere Length	
							Absolute, bp	Relative (T/S Ratio), CU
Williet et al., 2010 [55]	Prospective, population-based study	Austrians with a stable CAD	88	70 ± 10.5	63.6/26.4	Real-time quantitative PCR	N/A	1.13 ± 0.52
Yakhontov et al., 2017 * [64]	Observational, cross-sectional study	Russians with stable CAD I-III FC	59	52 [46.5; 55]	100/0	Real-time quantitative PCR	N/A	0.84 [0.2; 1.9]
Yakhontov et al., 2017 * [64]	Observational, cross-sectional study	Russians with stable CAD I-III FC	47	64 [62; 67]	100/0	Real-time quantitative PCR	N/A	0.3 [0.09; 1.2]
Hammadah et al., 2017 [63]	Observational, cross-sectional study	Canadians with stable CAD	566	63 ± 9,0	63.6/26.4	Real-time quantitative PCR	N/A	0.82 ± 0.14
Tian et al., 2017 [58]	Observational, cross-sectional study	Chinese with premature CAD	128	48.6 ± 7.26	57.8/42.2	Real-time quantitative PCR	N/A	0.88 ± 0.86
Yakhontov et al., 2018 [65]	Observational, cross-sectional study	Russians with essential hypertension and stable CAD I-III FC	43	52 [46.5; 55.0]	100/0	Real-time quantitative PCR	N/A	0.7 [0.12; 0.92]
Pejenaute et al., 2020 [59]	Observational, cross-sectional study	Spaniards with coronary atherosclerosis	116	61 ± 1	88/12	Real-time quantitative PCR	8315 ± 98	N/A
Starnino et al., 2021 [61]	Observational, cross-sectional study	Canadians with stable CAD	598	66.13 ± 6.25	80.6/19.4	Real-time quantitative PCR	N/A	0.83 ± 0.18

Notes: bp—base pairs; CAD—coronary artery disease; CU—conventional units; N/A—no data; PCR—polymerase chain reaction; FC—functional class. * There were two groups in Yakhontov's study: middle-aged patients with stable CAD (mean age—52 years) and elderly patients with stable CAD (mean age—64 years). LTL was determined for each group. Therefore, both groups studied by Yakhontov are included in Table 2.

Mazidi et al. [62] found that older men without CAD had a shorter relative LTL than younger men without CAD (6380 ± 80 bp vs. 12,420 ± 60 bp, p-value < 0.05).

Willeit et al. [55] in a prospective population-based PCR study, evaluated relative LTL in individuals aged 45 to 84 years without CVD and in patients with stable CAD. The authors found that LTL in patients with stable CAD was significantly shorter than in healthy individuals of the same age (1.13 ± 0.52 CU in patients with CAD versus 1.52 ± 0.81 CU in healthy individuals, p-value < 0.001). An interesting finding was that LTL was shorter in men than in women (1.41 [1.33–1.49] CU versus 1.55 [1.47–1.62] CU; p-value = 0.02). The authors attributed this phenomenon to the higher estrogen levels in women. In addition, the study found that LTL was inversely correlated with age (r = −0.22, p-value < 0.001).

The possibility of using LTL as a molecular biomarker of human biological aging as it reflects the telomere length of endothelial cells has been confirmed by several studies. The study of Hammadah et al. [63] established the relationship between LTL shortening and low level of CD34+ expression on human endothelial progenitor cells. After adjustment for age, sex, race, body mass index, smoking, and previous myocardial infarction, a shorter LTL was associated with lower CD34+ cell levels; for every 10% shorter relative LTL, CD34+ levels were 5.2% lower (p-value < 0.001). This is indirect evidence of the decreased regenerative capacity of bone marrow cells and the decreased repair of blood vessel endothelium. A study by Wilson et al. [66] revealed a strong correlation between endotheliocyte telomere length and relative LTL in the blood (r = 0.62, p-value < 0.001). Thus, LTL reflects vascular endotheliocyte telomere length, which allows us to use LTL assessment as a biomarker of vascular age, EVA, and human biological aging in various CVDs in adults.

Starnino et al. [61] found that patients with stable CAD had a shorter LTL compared with healthy volunteers and people without CVD (p-value < 0.001).

Tian et al. [58] analyzed cases of stable CAD in men younger than 55 years and women younger than 65 years. The authors measured relative LTL in Chinese patients with premature CAD compared with a control group (people of comparable age without CAD). The effects of oxidative stress on LTL shortening were assessed. Patients with premature CAD had a shorter relative LTL compared with those without CAD (0.88 ± 0.86 CU versus 1.1 ± 0.57 CU, p-value = 0.015). Thus, there was an association between LTL shortening and decreased plasma antioxidant capacity in patients with CAD.

Huang et al. [66] studied the relationship between LTL and all-cause mortality, cardiovascular mortality, and cerebrovascular mortality among adults in the USA. The study included 7827 participants (48.18% men). The researchers conventionally divided all participants by LTL into three groups: short LTL (0.77 ± 0.09 CU); medium relative LTL (1.00 ± 0.06 CU); and large LTL (1.32 ± 0.26 CU). After 158.26 months of follow-up, there were an average of 1876 (23.97%), 87 (1.11%), and 243 (3.10%) all-cause, cerebrovascular, and cardiovascular deaths. The authors showed that LTL was nonlinearly correlated with all-cause mortality (OR—95% CI: 0.03 to 0.09; p-value < 0.0001) but not with mortality from cerebrovascular disease and CVDs (p-value > *0.05*).

Yakhontov et al. [64] studied the relationship between LTL in men with stable angina I-III functional classes according to the Canadian Cardiovascular Society classifications [67] in different age groups: the middle-aged group (median age 52 [46.5; 55] years) and the elderly group (median age 64 [62; 67] years). The authors found no statistically significant differences in LTL in patients with stable angina pectoris as a function of mean and old age (p-value = 0.058) [63]. Another study by these authors [64] examined LTL in young (median age 52 [46.5; 55] years) and middle-aged (median age 64 [62; 67] years) men with arterial hypertension and with stable angina I-III functional classes with and without EVA. The criteria for the inclusion of patients in the subgroup with EVA were a young age of arterial hypertension debut (before 45 years), a young age of CAD debut (before 45 years), and increased vascular wall stiffness according to the cardio-ankle vascular index (according to sonography). The authors showed that in men with arterial hypertension, CAD, and EVA, the relative LTL was statistically significantly shorter than in men with arterial hypertension and CAD but without EVA (p-value = 0.026) [65].

Thus, in recent years, researchers and clinicians have been very interested in studying the relationship between LTL and stable CAD. The number of ongoing studies is increasing. However, the results obtained vary over a wide range, which may be due to differences in patient age, ethnicity and race, and region of residence. Nevertheless, there is no doubt that LTL is reduced in middle-aged and elderly adults developing stable CAD compared with healthy adult controls without CVD including CAD [68].

3.2. Leukocyte Telomere Length in Patients with Acute Myocardial Infarction

The number of LTL studies in patients with AMI is still significantly lower compared to LTL studies in patients with stable CAD. We found and analyzed five studies (Table 3). In adults with AMI, the relative LTL ranged from 0.115 CU [60] to 0.86 CU [57]; the median relative LTL was 0.62 [0.20; 0.84] CU. Differences in mean LTL in patients with AMI compared with healthy controls were not statistically significant (p-value = 0.089).

Table 3. Mean Relative Telomere Length of Leukocytes in Adults with Acute Myocardial Infarction.

Authors [References]	Study Characteristics	Group Characteristics	Participants (n)	Age, in Years (M ± SE or Me [P25; P75])	Sex (Male/Female, %)	Method	Telomere Length	
							Absolute, bp	Relative (T/S Ratio), CU
Russo A. et al., 2012 [69]	Observational, open, cross-sectional, longitudinal study.	Italians with AMI	199	40.1 ± 5	89.4/10.6	Real-time quantitative PCR	N/A	0.77 ± 0.2
Dlouha, D. et al., 2016 [57]	Observational, cross-sectional case-control study	Czechs with AMI	505	61 ± 9.7	0/100	Real-time quantitative PCR	N/A	0.86 ± 0.32
Margaritis, M. et al., 2017 [70]	Observational, open, cross-sectional, longitudinal study	British with AMI	290	63 ± 12.7	85.2/14.8	Real-time quantitative PCR	N/A	1.08 [0.41—2.66] *
Gupta M.D. et al., 2020 [60]	Observatio-nal, open, cross-sectional study	Indians with AMI	77	35.33 ± 6.22	84.4/15.6	Real-time quantitative PCR	N/A	0.115
Chan D. et al., 2020 [71]	Prospective, observation, cohort, longitudinal study.	British with AMI	135	81 ± 4	64/36	Real-time quantitative PCR	N/A	0.47 ± 0.25

Notes: bp—base pairs; CU—conventional units; N/A—no data; PCR—polymerase chain reaction; *—median [P10—P90].

Russo et al. [69] found no statistically significant association between relative LTL and the risk of AMI in an Italian cohort of younger patients ($\leq$ 48 years).

However, Gupta et al. [60] demonstrated that the relative LTL adjusted for sex, age, and body mass index was statistically significantly greater in the control group (0.792 CU) compared with AMI patients (0.115 CU, p -value < 0.001).

Chan et al. [71] investigated LTL in 135 patients with ACS without ST elevations who underwent percutaneous coronary intervention. The mean age of the patients was 81 ± 4 years and 64% of them were men. The mean LTL was found to be 0.47 ± 0.25 CU. Then, patients were divided into 3 groups according to the relative LTL to assess the risk of adverse clinical outcomes (death, recurrent AMI, unplanned revascularization, stroke, significant bleeding) recorded 1 year from the time of the ASC diagnosis. Long LTL was taken as 0.74 ± 0.27 CU, medium LTL as 0.42 ± 0.05 CU, and short LTL as 0.25 ± 0.27 CU. The authors found no statistically significant association between relative LTL and adverse ACS outcomes in older Chinese people.

Dlouha et al. [57] found that the mean relative LTL in 505 elderly (mean age 61 ± 9.7 years) Czech women with AMI was statistically significantly lower than those in the control group (0.86 ± 0.32 CU vs. 0.93 ± 0.38 CU; p-value < 0.001). However, after adjusting for age, smoking status, and type 2 diabetes mellitus, the differences between the groups were no longer significant (p-value = 0.25). Thus, the authors concluded to the contrary that AMI was not associated with relative LTL in Czech women.

A study by Margaritis et al. [70] determined relative LTL in AMI patients in the United Kingdom. The results were presented as median and percentiles. Short LTL was considered to be less than 0.96 CU and long LTL $\geq$ 0.96 CU. The authors showed that LTL is a molecular biomarker of cardiovascular outcomes after AMI regardless of patients' ages. Also, they demonstrated that short LTL (T/S < 0.916 CU) in patients with AMI is a predictor of the high risk of all-cause mortality (p-value = 0.008) and mortality from CVDs within the first year after AMI (p-value = 0.005).

Thus, in patients with AMI, the relative LTL ranged from 0.115 CU [60] to 0.86 CU [57].

4. Discussion

We found and analyzed a total of 17 publications including 5 studies conducted in Russia. However, three [72–74] of the five Russian studies were excluded from the subsequent systematic analysis due to an alternative methodological approach to LTL measurement. The results of the 14 publications were systematized and ranked into 3 groups (adults without CVDs, adults with stable CAD, and adults with AMI), as demonstrated in Tables 1–3.

We demonstrated that in studies of the mean relative LTL in young, middle-aged, and elderly adults, the value of this molecular biomarker decreases not only in relation to the physiological biological aging of the human body but also in relation to the premature development of stable CAD [64] and the early development of AMI [60]. At the same time, the most convincing results were obtained in studies of relative LTL in middle-aged and elderly patients with stable CAD [55,58,59,61,63–65]. It seems important from a practical point of view because patients with stable CAD and with shortened relative LTL can have unfavorable prognoses concerning general mortality [75]. However, the prognosis of cardiovascular and cerebrovascular mortality in patients with stable CAD and with shortened LTL need to be clarified in the future.

Interestingly, the relative LTL in patients with AMI (Table 3) is shorter compared with the relative LTL in patients with CAD (Table 2) and with healthy controls (Table 1). However, we must admit that the small number of studies of relative LTL in patients with AMI does not allow us to draw any definitive conclusions.

Studies of relative LTL in patients with AMI (predominantly) and with stable CAD (to a lesser extent) are in their infancy. Such studies are still in the minority, the sample sizes are small, and the results obtained in some studies are contradictory [60,69–71]. The ethnic and racial heterogeneity of the samples in the publications we analyzed draws attention, which does not allow us to assess the additional influences of ethnicity and region of residence of adults with and without the studied CVDs on the absolute and relative LTL.

However, our systematic review of available publications demonstrated a trend toward shorter relative LTL in patients with AMI (0.62 [0.20; 0.84] CU) compared with patients with stable CAD (0.86 [0.82; 1.07] CU) and a healthy control group (0.93 [0.70; 1.10] CU) (Figure 3). However, these results did not reach statistical significance (p-value > 0.05). Nevertheless, this trend could be clarified by new, large, and multicenter studies with a similar design in the future.

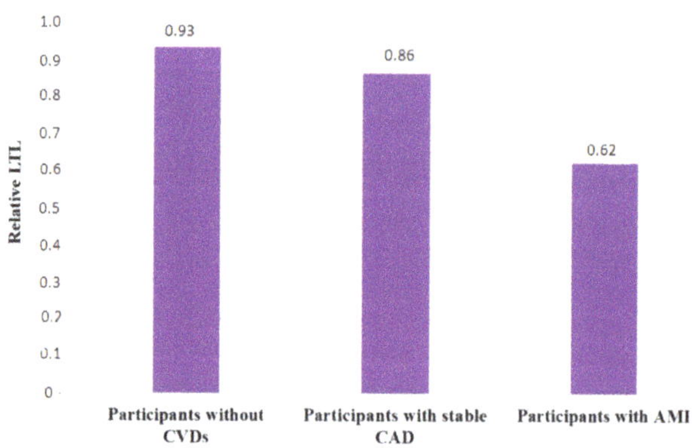

Figure 3. Median Relative Leukocyte Telomere Length in Patients with Coronary Heart Disease and Acute Myocardial Infarction Compared to Healthy Controls: AMI—acute myocardial infarction; CVDs—cardiovascular disease; CAD—coronary artery disease; LTL—leukocyte telomere length (CU—conventional units).

Thus, CAD and arterial hypertension [76–81], CAD and atrial fibrillation [82–91] and CAD and vascular cognitive disorders [92–95] are common comorbid conditions with overlap syndrome. Therefore, the study of LTL seems important not only in isolated CAD and AMI but also in these syndromes of mutual aggravation.

5. Limitations

The limitations of this systematic review include the analysis of publications in English and Russian only. It is possible that we missed some studies published in other languages that were not represented in the databases we analyzed.

Another limitation of the review of published and available studies of relative LTL in adults with CVDs is the different methodological approaches of the investigators with regard to the study design, inclusion/exclusion criteria (e.g., age and sex of patients, ethnic group, etc.), and study duration (most studies were cross-sectional, not longitudinal). In addition, some authors used % rather than CU to estimate relative LTL.

Due to methodological problems (differences in how LTL is calculated), three Russian-language publications were excluded from the review. Thus, an interesting study is by Maximov et al. [72], in which the association of LTL with various risk factors of age-associated diseases in the Russian population was studied. The authors identified a group of patients with stable CAD in whom LTL was determined, but no recalculation of mean age and percentage by gender was performed for this group of patients, which did not allow us to include this study in our systematic review. Strajesko et al. [73] studied the relationship between risk factors for CVDs and LTL, but the authors used an alternative approach to LTL measurement, unlike the methods in the English-language publications we analyzed. Thus, the authors took 9.75 units as a short LTL and more than or equal to 9.75 units as a long LTL. Thus, the approach to determining the relative LTL in the English-language and some Russian-language publications differed significantly. In addition, the authors used three models based on multivariate linear regression analysis [73] to assess the relationship between LTL and CVD risk factors, which is of undoubted scientific interest. However, this method of statistical analysis was not used in other publications we analyzed. The authors showed that relative LTL was associated with the mean and old age of the individuals included in the study, but this publication lacked baseline data on LTL in patients with stable CAD and healthy adults. This limitation prevented us from including the authors' findings in our systematic review. Doroshchuk et al. [74] used LTL in patients with stable CAD compared to a group of age-matched healthy volunteers as the study index. The authors demonstrated that LTL statistically significantly decreased with increasing cardiovascular mortality risk according to the Systematic Coronary Risk Evaluation (SCORE) scale, p-value < 0.005. However, the authors used an alternative method of measuring LTL—not in CU, but as a percentage compared to healthy controls. Due to the alternative LTL calculation methodology, the results of this study were also not included in our systematic review.

The other main concern is that relative measured LTL cannot be compared between studies, but relative measured LTLs are comparable between groups inside one study. This is due to the fact that relative LTL measured as the T/S ratio measures the relative amount of telomeric DNA (T) to a single copy gene (S), calibrated to a plate reference genomic DNA sample. Therefore, and as this reference sample is unique for each study, the results on relative quantified LTLs are not directly comparable between studies.

The association between LTL and EVA has had a limited focus in the present review and we therefore plan to include it in a future review of the relationship between EVA and LTL in young, middle-aged, and elderly patients.

6. Conclusions

Despite the scientific and clinical significance of the analyzed studies on relative LTL as a molecular biomarker of CVDs, their translation into real clinical practice is difficult due to disparities in the design and methodology of the analyzed studies, including the studies on cell cultures and humans, as well as the differences in samples by gender, age, race, and ethnicity. The authors believe that clinical studies of the role of relative LTL in adult patients with CAD are the most promising and require large multicenter studies with a unified design and methodology.

Author Contributions: Conceptualization, N.A.S.; methodology, O.V.Z. and N.A.S.; software, O.V.Z.; validation, O.V.Z. and M.A.-Z.; formal analysis, O.V.Z. and N.A.S.; investigation, O.V.Z. and M.S.C.; resources, O.V.Z. and N.V.L.; data curation, O.V.Z., M.M.P. and M.A.-Z.; writing—original draft preparation, O.V.Z., N.A.S. and A.E.I.; writing—review and editing, N.A.S. & M.M.P.; visualization, O.V.Z. and N.A.S.; supervision, N.A.S.; project administration, M.M.P. All authors have read and agreed to the published version of the manuscript.

Funding: This research received no external funding.

Institutional Review Board Statement: Not applicable.

Informed Consent Statement: Not applicable.

Data Availability Statement: Not applicable.

Acknowledgments: The authors thank Ekaterina V. Kachura (katy.kachura@mail.ru) for help in preparing the figures.

Conflicts of Interest: The authors declare no conflict of interest.

References

1. Timmis, A.; Townsend, N.; Gale, C.P.; Torbica, A.; Lettino, M.; Petersen, S.E.; Mossialos, E.A.; Maggioni, A.P.; Kazakiewicz, D.; May, H.T.; et al. European Society of Cardiology: Cardiovascular Disease Statistics 2019. *Eur. Heart J.* **2020**, *41*, 12–85. [CrossRef] [PubMed]
2. Shalnova, S.A.; Drapkina, O.M. The trends of cardiovascular and cancer mortality in Russian men and women from 2000 to 2016 years. *Ration. Pharmacother. Cardiol.* **2019**, *15*, 77–83. [CrossRef]
3. Ageeva, L.I.; Alexandrova, G.A.; Golubev, N.A.; Kirillova, G.N.; Ogryzko, E.V.; Oskov, Y.I.; Nam, P.D.; Kharkov, T.L.; Chumarina, V.Z. Healthcare in Russia. 2021: Stat.sat./Rosstat.—M., 2021, 171 p. Available online: https://ghdx.healthdata.org/organizations/federal-state-statistics-service-russia (accessed on 1 June 2022).
4. Kolber, M.R.; Scrimshaw, C. Family history of cardiovascular disease. *Can. Fam Physician* **2014**, *60*, 1016. [PubMed]
5. Hamczyk, M.R.; Nevado, R.M.; Barettino, A.; Fuster, V.; Andres, V. Biological versus chronological aging: JACC focus seminar. *J. Am. Coll. Cardiol.* **2020**, *75*, 919–930. [CrossRef]
6. Nilsson, P.M. Early vascular aging (EVA): Consequences and prevention. *Vasc. Health Risk Manag.* **2008**, *4*, 547–552. [CrossRef]
7. Thijssen, D.H.J.; Bruno, R.M.; Mil, A.C.C.M.; Holder, S.M.; Faita, F.; Greyling, A.; Zock, P.L.; Taddei, S.; Deanfield, J.E.; Luscher, T.; et al. Expert consensus and evidence-based recommendations for the assessment of flow mediated dilation in humans. *Eur. Heart J.* **2019**, *40*, 2534–2547. [CrossRef]
8. Bauer, M.; Caviezel, S.; Teynor, A.; Erbel, R.; Mahabadi, A.A.; Schmidt-Trucksass, A. Carotid intima-media thickness as a bio-marker of subclinical atherosclerosis. *Swiss Med. Wkly.* **2012**, *142*, w13705. [CrossRef]
9. Grillo, A.; Lonati, A.M.; Guida, V.; Parati, G. Cardio-ankle vascular stiffness index (CAVI) and 24-h blood pressure profiles. *Eur. Heart J. Suppl.* **2017**, *19* (Suppl. B), 17–23. [CrossRef]
10. Townsend, R.R.; Wilkinson, I.B.; Schiffrin, E.L.; Townsend, R.R.; Wilkinson, I.B.; Schiffrin, E.L.; Avolio, A.P.; Chirinos, J.A.; Cockcroft, J.R.; Heffernan, K.S.; et al. Recommendations for improving and standardizing vascular research on arterial stiffness: A scientific statement from the American Heart Association. *Hypertension* **2015**, *66*, 698–722. [CrossRef]
11. Lanzer, P.; Boehm, M.; Sorribas, V.; Thiriet, M.; Janzen, J.; Zeller, T.; Hilaire, C.S.; Shanahan, C. Medial vascular calcification revisited: Review and perspectives. *Eur. Heart J.* **2014**, *35*, 1515–1525. [CrossRef]
12. De Meyer, T.; Nawrot, T.; Bekaert, S.; De Buyzere, M.L.; Rietzschel, E.R.; Andres, V. Telomere length as cardiovascular aging biomarker: JACC review topic of the week. *J. Am. Coll. Cardiol.* **2018**, *72*, 805–813. [CrossRef] [PubMed]
13. Burko, N.V.; Avdeeva, I.V.; Oleynikov, V.E.; Boytsov, S.A. The concept of early vascular aging. *Ration. Pharmacother. Cardiol.* **2019**, *15*, 742–749. [CrossRef]
14. Marco, L.D.; Bellasi, A.; Raggi, P. Cardiovascular biomarkers in chronic kidney disease: State of current research and clinical applicability. *Dis. Markers* **2015**, 586569. [CrossRef]
15. Gregoli, K.D.; George, S.J.; Jackson, C.L.; Newby, A.C.; Johnson, J.L. Differential effects of tissue inhibitor of metalloproteinase (TIMP)-1 and TIMP-2 on atherosclerosis and monocyte/macrophage invasion. *Cardiovasc. Res.* **2016**, *109*, 318–330. [CrossRef]
16. Freitas, I.A.; Lima, N.A.; Silva, G.B.; Castro, R.L.; Patel, P.; Vasconcelos Lima, C.C.; Costa Lino, D.O. Novel biomarkers in the prognosis of patients with atherosclerotic coronary artery disease. *Port. J. Cardiol.* **2020**, *39*, 667–672. [CrossRef]

17. Wainstein, M.V.; Mossmann, M.; Araujo, G.N.; Gonçalves, S.C.; Gravina, G.L.; Sangalli, M.; Veadrigo, F.; Matte, R.; Reich, R.; Costa, F.G.; et al. Elevated serum interleukin-6 is predictive of coronary artery disease in intermediate risk over-weight patients referred for coronary angiography. *Diabetol. Metab. Syndr.* **2017**, *9*, 67. [CrossRef]
18. Velde, A.R.; Lexis, C.P.H.; Meijers, W.C.; Horst, I.C.; Lipsic, E.; Dokter, M.M.; Veldhuisen, D.J.; Harst, P.; Boer, R.A. Galectin-3 and sST2 in prediction of left ventricular ejection fraction after myocardial infarction. *Clin. Chim. Acta* **2016**, *452*, 50–57. [CrossRef] [PubMed]
19. Sun, Y.; Zhao, J.Q.; Jiao, Y.R.; Ren, J.; Zhou, Y.H.; Li, L.; Yao, H.C. Predictive value of leukocyte telomere length for the severity of coronary artery disease. *Pers. Med.* **2020**, *17*, 175–183. [CrossRef]
20. Sun, Y.; Wang, W.; Jiao, Y.R.; Ren, J.; Gao, L.; Li, Y.; Hu, P.; Ren, T.Y.; Han, Q.F.; Chen, C.; et al. Leukocyte telomere length: A potential biomarker for the prognosis of coronary artery disease. *Biomark. Med.* **2020**, *14*, 933–941. [CrossRef]
21. Xu, X.; Hu, H.; Lin, Y.; Huang, F.; Ji, H.; Li, Y.; Lin, S.; Chen, X.; Duan, S. Differences in leukocyte telomere length between coronary heart disease and normal population: A Multipopulation Meta-Analysis. *BioMed Res. Int.* **2019**, 5046867. [CrossRef]
22. Ma, L.N.; Li, Y.; Wang, J.Y. Telomeres, and essential hypertension. *Clin. Biochem.* **2015**, *48*, 1195–1199. [CrossRef] [PubMed]
23. Cheng, G.; Wang, L.; Dai, M.; Wei, F.; Xu, D. Shorter leukocyte telomere length coupled with lower expression of telomerase genes in patients with essential hypertension. *Int. J. Med. Sci.* **2020**, *17*, 2180–2186. [CrossRef]
24. Allende, M.; Molina, E.; González-Porras, J.R.; Toledo, E.; Lecumberri, R.; Hermida, J. Short leukocyte telomere length is associated with cardioembolic stroke risk in patients with atrial fibrillation. *Stroke* **2016**, *47*, 863–865. [CrossRef] [PubMed]
25. Wang, S.; Gao, Y.; Zhao, L.; Hu, R.; Yang, X.; Liu, Y. Shortened leukocyte telomere length as a potential biomarker for predicting the progression of atrial fibrillation from paroxysm to persistence in the short-term. *Medicine* **2021**, *100*, e26020. [CrossRef] [PubMed]
26. Hayashi, T. Vascular senescense and endothelial function—Can we apply it to atrial fibrillation? *Circ. J.* **2019**, *83*, 1439–1440. [CrossRef] [PubMed]
27. Nikulina, S.I.; Shishkova, K.I.; Shulman, V.A.; Chernova, A.A.; Maksimov, V.N. Peripheral blood leukocyte telomere length as a possible prognostic marker for the development of atrial fibrillation. *CardioSomatics* **2020**, *11*, 50–54. [CrossRef]
28. Chatterjee, S.; Gonzalo-Calvo, D.; Derda, A.A.; Schimmel, K.; Sonnenschein, K.; Bavendiek, U.; Bauersachs, J.; Bar, C.; Thum, T. Leukocyte telomere length correlates with hypertrophic cardiomyopathy severity. *Sci. Rep.* **2018**, *8*, 11227. [CrossRef]
29. Wang, Y.; Jiao, F.; Zheng, H.; Kong, Q.; Li, R.; Zhang, X.; Yan, L.; Hao, Y.; Wu, Y. Gender difference in associations between telomere length and risk factors in patients with stroke. *Front. Aging Neurosci.* **2021**, *13*, 719538. [CrossRef]
30. Yetim, E.; Topcuoglu, M.A.; Kutlay, N.Y.; Tukun, A.; Oguz, K.K.; Arsava, E.M. The association between telomere length and ischemic stroke risk and phenotype. *Sci. Rep.* **2021**, *11*, 10967. [CrossRef]
31. Cao, W.; Zheng, D.; Zhang, J.; Wang, A.; Liu, D.; Zhang, J.; Singh, M.; Maranga, I.E.; Cao, M.; Wu, L.; et al. Association between telomere length in peripheral blood leukocytes and risk of ischemic stroke in a Han Chinese population: A linear and non-linear Mendelian randomization analysis. *J. Transl. Med.* **2020**, *18*, 385. [CrossRef]
32. Tian, Y.J.; Wang, S.; Jiao, F.J.; Kong, Q.; Liu, C.; Wu, Y. Telomere length: A potential biomarker for the risk and prognosis of stroke. *Front. Neurol.* **2019**, *10*, 624. [CrossRef] [PubMed]
33. Li, J.; Feng, C.; Li, L.; Yang, S.; Chen, Y.; Hui, R.; Zhang, M.; Zhang, W. The association of telomere attrition with first-onset stroke in Southern Chinese: A case-control study and meta-analysis. *Sci. Rep.* **2018**, *8*, 2290. [CrossRef] [PubMed]
34. Yu, J.; Kanchi, M.M.; Rawtaer, I.; Feng, L.; Kumar, A.P.; Kua, E.H.; Mahendran, R. The functional and structural connectomes of telomere length and their association with cognition in mild cognitive impairment. *Cortex* **2020**, *132*, 29–40. [CrossRef] [PubMed]
35. Yang, T.; Wang, H.; Xiong, Y.; Chen, C.; Duan, K.; Jia, J.; Ma, F. Vitamin D supplementation improves cognitive function through reducing oxidative stress regulated by telomere length in older adults with mild cognitive impairment: A 12-month randomized controlled trial. *J. Alzheimers Dis.* **2020**, *78*, 1509–1518. [CrossRef]
36. Hinterberger, M.; Fischer, P.; Huber, K.; Krugluger, W.; Zehetmayer, S. Leukocyte telomere length is linked to vascular risk factors not to Alzheimer's disease in the VITA study. *J. Neural Transm.* **2017**, *124*, 809–819. [CrossRef]
37. Blackburn, E.H.; Appel, E.S.; Link, J. Human telomere biology: A contributory and interactive factor in aging, disease risks, and protection. *Science* **2015**, *350*, 1193–1198. [CrossRef]
38. Yeh, J.K.; Wang, C.Y. Telomeres and telomerase in cardiovascular diseases. *Genes* **2016**, *7*, 58. [CrossRef]
39. Herrmann, M.; Pusceddu, I.; Marz, W.; Herrmann, W. Telomere biology and age-related diseases. *Clin. Chem. Labor-Atory Med.* **2018**, *56*, 1210–1222. [CrossRef]
40. Pusceddu, I.; Farrell, C.J.L.; Di Pierro, A.M.; Jani, E.; Herrmann, W.; Herrmann, M. The role of telomeres and vitamin D in cellular aging and age-related diseases. *Clin. Chem. Lab. Med.* **2015**, *53*, 1661–1678. [CrossRef]
41. Dorajoo, R.; Chang, X.; Gurung, R.L.; Li, Z.; Wang, L.; Wang, R.; Beckman, K.B.; Adams-Haduch, J.; M, Y.; Liu, S.; et al. Loci for human leukocyte telomere length in the Singaporean Chinese population and trans-ethnic genetic studies. *Nat. Commun.* **2019**, *10*, 2491. [CrossRef]
42. Turner, K.J.; Vasu, V.; Darren, K.; Griffin, D.K. Telomere biology and human phenotype. *Cells* **2019**, *8*, 73. [CrossRef] [PubMed]
43. Salakhov, R.R.; Ponasenko, A.V. Telomere length and cardiovascular diseases. *Complex. Issues Cardiovasc. Dis.* **2018**, *7*, 101–107. [CrossRef]
44. McClintock, B. The stability of broken ends of chromosomes in zea mays. *Genetics* **1941**, *26*, 234–282. [CrossRef] [PubMed]

45. Moyzis, R.K.; Buckingham, J.M.; Cram, L.S.; Dani, M.; Deaven, L.L.; Jones, M.D.; Meyne, J.; Ratliff, R.L.; Wu, J.R. A highly con-served repetitive DNA sequence, (TTAGGG)n, present at the telomeres of human chromosomes. *Proc. Natl. Acad. Sci. USA* **1988**, *85*, 6622–6666. [CrossRef]
46. Armanios, M. Telomeres and age-related disease: How telomere biology informs clinical paradigms. *J. Clin. Investig.* **2013**, *123*, 996–1002. [CrossRef]
47. Brown, L.L.; Zhang, Y.S.; Mitchell, C.; Ailshire, J. Does telomere length indicate biological, physical, and cognitive health among older adults? Evidence from the Health and Retirement Study. *J. Gerontol.* **2018**, *73*, 1626–1632. [CrossRef]
48. Hayflick, L. The limited in vitro lifetime of human diploid cell strains. *Exp. Cell Res.* **1965**, *37*, 614–636. [CrossRef]
49. Samani, N.J.; Boultby, R.; Butler, R.; Thompson, J.R.; Goodall, A.H. Telomere shortening in atherosclerosis. *Lancet* **2001**, *358*, 472–473. [CrossRef]
50. Rubtsova, M.; Dontsova, O. Human telomerase RNA: Telomerase component or more? *Biomolecules* **2020**, *10*, 873. [CrossRef]
51. Arai, Y.; Martin-Ruiz, C.M.; Takayama, M.; Abe, Y.; Takebayashi, T.; Koyasu, S.; Suematsu, M.; Hirose, N.; von Zglinicki, T. Inflammation, but not telomere length, predicts successful ageing at extreme old age: A longitudinal study of semi-supercentenarians. *EBioMedicine* **2015**, *2*, 1549–1558. [CrossRef]
52. Allsopp, R.C.; Morin, G.B.; DePinho, R.; Harley, C.B.; Weissman, I.L. Telomerase is required to slow telomere shortening and extend replicative lifespan of HSCs during serial transplantation. *Blood* **2003**, *102*, 517–520. [CrossRef]
53. Bhattacharyya, J.; Mihara, K.; Bhattacharjee, D.; Mukherjee, M. Telomere length as a potential biomarker of coronary artery disease. *Indian J. Med. Res.* **2017**, *145*, 730–737. [CrossRef] [PubMed]
54. Cawthon, R.M. Telomere measurement by quantitative PCR. *Nucleic Acids Res.* **2002**, *30*, e47. [CrossRef] [PubMed]
55. Willeit, P.; Willeit, J.; Brandstätter, A.; Ehrlenbach, S.; Mayr, A.; Gasperi, A.; Weger, S.; Oberhollenzer, F.; Reindl, M.; Kronenberg, F.; et al. Cellular aging reflected by leukocyte telomere length predicts advanced atherosclerosis and cardiovascular disease risk. *Arterioscler. Thromb. Vasc. Biol.* **2010**, *30*, 1649–1656. [CrossRef] [PubMed]
56. Hassler, E.; Almer, G.; Reishofer, G.; Marsche, G.; Mangge, H.; Deutschmann, H.; Herrmann, M.; Leber, S.; Gunzer, F.; Renner, W. Sex-specific association of serum antioxidative capacity and leukocyte telomere length. *Antioxidants* **2021**, *10*, 1908. [CrossRef]
57. Dlouha, D.; Pitha, J.; Mesanyova, J.; Mrazkova, J.; Fellnerova, A.; Stanek, V.; Lanska, V.; Hubacek, J.A. Genetic variants within telomere-associated genes, leukocyte telomere length and the risk of acute coronary syndrome in Czech women. *Clin. Chim. Acta* **2016**, *454*, 62–65. [CrossRef]
58. Tian, R.; Zhang, L.N.; Zhang, T.T.; Pang, H.Y.; Chen, L.F.; Shen, Z.J.; Liu, Z.; Fang, Q.; Zhang, S.Y. Association between oxidative stress and peripheral leukocyte telomere length in patients with premature coronary artery disease. *Med. Sci. Monit.* **2017**, *23*, 4382–4390. [CrossRef]
59. Pejenaute, A.; Cortes, A.; Marques, J.; Montero, L.; Beloqui, O.; Fortuno, A.; Martí, A.; Orbe, J.; Zalba, G. NADPH oxidase overactivity underlies telomere shortening in human atherosclerosis. *Int. J. Mol. Sci.* **2020**, *21*, 1434. [CrossRef]
60. Gupta, M.D.; Miglani, M.; Bansal, A.; Jain, V.; Arora, S.; Kumar, S.; Virani, S.S.; Kalra, A.; Yadav, R.; Pasha, Q.; et al. Telomere length in young patients with acute myocardial infarction without conventional risk factors: A pilot study from a South Asian population. *Indian Heart J.* **2020**, *72*, 619–622. [CrossRef]
61. Starnino, L.; Dupuis, G.; Busque, L.; Bourgoin, V.; Dube, M.P.; Busseuil, D.; D'Antono, B. The associations of hostility and defensiveness with telomere length are influenced by sex and health status. *Biol. Sex. Differ.* **2021**, *12*, 2. [CrossRef]
62. Mazidi, M.; Shekoohi, N.; Katsiki, N.; Rakowski, M.; Mikhailidis, D.P.; Banach, M. Serum anti-inflammatory and inflammatory markers have no causal impact on telomere length: A Mendelian randomization study. *Arch. Med. Sci.* **2021**, *17*, 739–751. [CrossRef] [PubMed]
63. Hammadah, M.; Mheid, I.A.; Wilmot, K.; Ramadan, R.; Abdelhadi, N.; Alkhoder, A.; Obideen, M.; Pimple, P.M.; Levantsevych, O.; Kelli, H.M.; et al. Telomere shortening, regenerative capacity, and cardiovascular outcomes. *Circ. Res.* **2017**, *120*, 1130. [CrossRef] [PubMed]
64. Yakhontov, D.A.; Ostanina, J.O.; Pakharukova, M.Y.; Mordvinov, V.A. Clinical sings and symptoms of polyvascular disease in coronary artery disease patients of different age groups. *Complex. Issues Cardiovasc. Dis.* **2017**, *6*, 36–43. [CrossRef]
65. Yakhontov, D.A.; Ostanina, J.O. Early vascular aging syndrome in young and middle age patients with hypertension and coronary artery disease. *Med. Alph.* **2018**, *1*, 33–36.
66. Wilson, R.W.; Herbert, K.E.; Mistry, Y.; Stevens, S.E.; Patel, H.R.; Hastings, R.A.; Thompson, M.M.; Williams, B. Blood leukocyte telomere DNA content predicts vascular telomere DNA content in humans with and without vascular disease. *Eur. Heart J.* **2008**, *29*, 2689–2694. [CrossRef] [PubMed]
67. Huang, Y.Q.; Lo, K.; Feng, Y.Q.; Zhang, B. The association of mean telomere length with all-cause, cerebrovascular and cardiovascular mortality. *Biosci. Rep.* **2019**, *39*, BSR20192306. [CrossRef]
68. Clinical Recommendations. Stable Coronary Heart Disease. Russian Society of Cardiology 2020.—Text: Electronic. Available online: https://scardio.ru/content/Guidelines/2020/Clinic_rekom_IBS.pdf (accessed on 20 November 2021).
69. Russo, A.; Palumbo, L.; Fornengo, C. Telomere length variation in juvenile acute myocardial infarction. *PLoS ONE* **2012**, *7*, e49206. [CrossRef] [PubMed]
70. Margaritis, M.; Sanna, F.; Lazaros, G.; Akoumianakis, I.; Patel, S.; Antonopoulos, A.S.; Duke, C.; Herdman, L.; Psarros, C.; Oikonomou, E.K.; et al. Predictive value of telomere length on outcome following acute myocardial infarction: Evidence for contrasting effects of vascular vs. blood oxidative stress. *Eur. Heart J.* **2017**, *38*, 3094–3104. [CrossRef] [PubMed]

71. Chan, D.; Martin-Ruiz, C.; Saretzki, G.; Neely, D.; Qiu, W.; Kunadian, V. The association of telomere length and telomerase activity with adverse outcomes in older patients with non-ST-elevation acute coronary syndrome. *PLoS ONE* **2020**, *15*, e0227616. [CrossRef] [PubMed]
72. Maximov, V.N.; Malyutina, S.K.; Orlov, P.S.; Ivanoschuk, D.E.; Voropaeva, E.N.; Bobak, M.; Voevoda, M.I. Length telomere leukocytes as aging markers and risk factors for age-related disease in humans. *Adv. Gerontol.* **2016**, *29*, 702–708. [CrossRef]
73. Strajesko, I.D.; Tkacheva, I.N.; Akasheva, D.U.; Dudinskaya, E.V.; Agaltsov, M.V.; Kruglikova, A.S.; Brailova, N.V.; Pykhtina, V.S.; Plokhova, E.V.; Ozerova, I.N.; et al. Relation of cardiovascular risk factors and leukocyte telomere length. *Cardiovasc. Ther. Prev.* **2016**, *15*, 52–57. [CrossRef]
74. Doroshchuk, N.A.; Tikhase, A.K.; Lankin, V.Z.; Konovalova, G.G.; Mednikova, T.K.; Postnov, A.Y.; Kukharchuk, V.V. The influence of oxidative stress on the length of telomeric repeats in chromosomes o white blood cells in patients with coronary artery disease. *Cardiol. Bull.* **2017**, *12*, 32–37.
75. Wang, H.; Naghavi, M.; Allen, C.; Barber, R.M.; Bhutta, Z.B.; Carter, A.; Casey, D.C.; Charlson, F.J.; Chen, A.Z.; Coates, M.M.; et al. Global, regional, and national life expectancy, all-cause mortality, and cause-specific mortality for 249 causes of death, 1980–2015: A systematic analysis for the Global Burden of Disease Study 2015. *The Lancet.* **2016**, *388*, 1459–1544. [CrossRef]
76. Oganov, R.G.; Simanenkov, V.I.; Bakulin, I.G.; Bakulina, N.V.; Barbarash, O.L.; Boytsov, S.A.; Boldueva, S.A.; Garganeeva, N.P.; Doshchitsin, V.L.; Karateev, A.E.; et al. Comorbidities in clinical practice. Algorithms for diagnostics and treatment. *Cardiovasc. Ther. Prevention.* **2019**, *18*, 5–66. [CrossRef]
77. Paluch, W.; Semczuk, K.; Rys, A.; Szymanski, F.M.; Filipiak, K.J. Anti-hypertensive treatment efficacy in patients with arterial hypertension and coronary artery disease or coronary equivalent. *Arter. Hypertens.* **2017**, *21*, 93–98. [CrossRef]
78. Zhang, W.Y.; Zhang, J.; Jin, F.; Zhou, H. Efficacy of felodipine and enalapril in the treatment of essential hypertension with coronary artery disease and the effect on levels of salusin-β, apelin, and *PON1* gene expression in patients. *Cell. Mol. Biol.* **2021**, *67*, 174–180. [CrossRef]
79. Sarkar, G.; Gaikwad, V.B.; Sharma, A.; Halder, S.K.; Kumar, D.A.; Anand, J.; Agrawal, S.; Kumbhar, A.; Kinholkar, B.; Mathur, R.; et al. Fixed-dose combination of metoprolol, telmisartan, and chlorthalidone for essential hypertension in adults with stable coronary artery disease: Phase III Study. *Adv. Ther.* **2022**, *39*, 923–942. [CrossRef]
80. Zheng, Y.; Li, D.; Zeng, N.; Guo, H.; Li, H.; Shen, S. Trends of antihypertensive agents in patients with hypertension and coronary artery disease in a tertiary hospital of China. *Int. J. Clin. Pharmacol. Ther.* **2020**, *42*, 482–488. [CrossRef]
81. Guo, Q.; Lu, X.; Gao, Y.; Zhang, J.; Yan, B.; Su, D.; Song, A.; Zhao, X.; Wang, G. Cluster analysis: A new approach for identification of underlying risk factors for coronary artery disease in essential hypertensive patients. *Sci. Rep.* **2017**, *7*, 43965. [CrossRef]
82. Steensig, K.; Olesen, K.K.W.; Thim, T.; Nielsen, J.C.; Jensen, S.E.; Jensen, L.O.; Kristensen, S.D.; Botker, H.E.; Lip, G.Y.H.; Maeng, M. CAD is an independent risk factor for stroke among patients with atrial fibrillation. *J. Am. Coll. Cardiol.* **2018**, *72*, 2540–2542. [CrossRef]
83. Michniewicz, E.; Mlodawska, E.; Lopatowska, P.; Tomaszuk-Kazberuk, A.; Malyszko, J. Patients with atrial fibrillation and coronary artery disease—Double trouble. *Adv. Med. Sci.* **2018**, *63*, 30–35. [CrossRef] [PubMed]
84. Gladding, P.A.; Legget, M.; Fatkin, D.; Larsen, P.; Doughty, R. Polygenic risk scores in coronary artery disease and atrial fibrillation. *Heart Lung Circ.* **2020**, *29*, 634–640. [CrossRef] [PubMed]
85. Alkindi, F.A.; Rafie, I.M. Anticoagulation in patients with atrial fibrillation and coronary artery disease. *Heart Views* **2020**, *21*, 32–36. [CrossRef] [PubMed]
86. Nortamo, S.; Kentta, T.V.; Ukkola, O.; Huikuri, H.V.; Perkiomaki, J.S. Supraventricular premature beats and risk of new-onset atrial fibrillation in coronary artery disease. *J. Cardiovasc. Electrophysiol.* **2017**, *28*, 1269–1274. [CrossRef]
87. Zheng, Y.J.; He, J.Q. Common differentially expressed genes and pathways correlating both coronary artery disease and atrial fibrillation. *Excli J.* **2021**, *20*, 126–141. [CrossRef] [PubMed]
88. Pastori, D.; Pignatelli, P.; Sciacqua, A.; Perticone, M.; Violi, F.; Lip, G.Y.H. Relationship of peripheral and coronary artery disease to cardiovascular events in patients with atrial fibrillation. *Int. J. Cardiol.* **2018**, *255*, 69–73. [CrossRef]
89. Inohara, T.; Shrader, P.; Pieper, K.; Blanco, R.G.; Allen, L.A.; Fonarow, G.C.; Gersh, B.J.; Go, A.S.; Ezekowitz, M.D.; Kowey, P.R.; et al. Treatment of atrial fibrillation with concomitant coronary or peripheral artery disease: Results from the outcomes registry for better informed treatment of atrial fibrillation II. *Am. Heart J.* **2019**, *213*, 81–90. [CrossRef]
90. Lamblin, N.; Ninni, S.; Tricot, O.; Meurice, T.; Lemesle, G.; Bauters, C. Secondary prevention and outcomes in outpatients with coronary artery disease, atrial fibrillation or heart failure: A focus on disease overlap. *Open Heart* **2020**, *7*, e001165. [CrossRef]
91. Wakili, R.; Riesinger, L.; Fender, A.C.; Dobrev, D. Double Jeopardy: Will the new trials tell us how to manage patients with atrial fibrillation and coronary artery disease? *IJC Heart Vasc.* **2019**, *23*, 100369. [CrossRef]
92. Suridjan, I.; Herrmann, N.; Adibfar, A.; Saleem, M.; Andreazza, A.; Oh, P.I.; Lanctot, K.L. Lipid peroxidation markers in coronary artery disease patients with possible vascular mild cognitive impairment. *J. Alzheimer's Dis.* **2017**, *58*, 885–896. [CrossRef]
93. Xia, C.; Vonder, M.; Sidorenkov, G.; Oudkerk, M.; de Groot, J.C.; Harst, P.; Bock, G.H.; De Deyn, P.P.; Vliegenthart, R. The relationship of coronary artery calcium and clinical coronary artery disease with cognitive function: A systematic review and meta-analysis. *J. Atheroscler. Thromb.* **2020**, *27*, 934–958. [CrossRef] [PubMed]

94. Saleem, M.; Herrmann, N.; Dinoff, A.; Mazereeuw, G.; Oh, P.I.; Goldstein, B.I.; Kiss, A.; Shammi, P.; Lanctot, K.L. Association between endothelial function and cognitive performance in patients with coronary artery disease during cardiac rehabilitation. *Psychosom. Med.* **2019**, *81*, 184–191. [CrossRef] [PubMed]
95. Tarasova, I.V.; Trubnikova, O.A.; Barbarash, O.L. EEG and clinical factors associated with mild cognitive impairment in coronary artery disease patients. *Dement. Geriatr. Cogn. Disord.* **2018**, *46*, 275–284. [CrossRef] [PubMed]

genes

Review

The Adenine/Thymine Deleterious Selection Model for GC Content Evolution at the Third Codon Position of the Histone Genes in *Drosophila*

Yoshinori Matsuo

Division of Science and Technology, Tokushima University, 2-1 Minamijosanjima-cho, Tokushima 770-8506, Japan; matsuo.yoshinori@tokushima-u.ac.jp; Tel.: +81-88-656-7270

Abstract: The evolution of the GC (guanine cytosine) content at the third codon position of the histone genes (*H1, H2A, H2B, H3, H4, H2AvD, H3.3A, H3.3B*, and *H4r*) in 12 or more *Drosophila* species is reviewed. For explaining the evolution of the GC content at the third codon position of the genes, a model assuming selection with a deleterious effect for adenine/thymine and a size effect is presented. The applicability of the model to whole-genome genes is also discussed.

Keywords: histone gene; GC content; *Drosophila*; codon usage; size effect

check for updates

Citation: Matsuo, Y. The Adenine/Thymine Deleterious Selection Model for GC Content Evolution at the Third Codon Position of the Histone Genes in *Drosophila*. *Genes* **2021**, *12*, 721. https://doi.org/10.3390/genes12050721

Academic Editors: Luigi Viggiano and Renè Massimiliano Marsano

Received: 29 March 2021
Accepted: 7 May 2021
Published: 12 May 2021

1. Introduction

Histones are basic proteins that package and arrange DNA into nucleosomes [1–4]. There are two major types of histones: a replication-dependent (canonical) type and a replication-independent (replacement) type [5]. In addition to these, centromeric proteins [6–8] and histone-like proteins [9] also exist.

In *Drosophila*, five replication-dependent (canonical) histones are known [10,11]: H2A, H2B, H3, and H4, which are core histones that organize the nucleosome core by forming an octamer comprising two copies of each protein, and H1, which is a linker protein that binds to each nucleosome core [1–4]. As for replication-independent (replacement) histones, four kinds are currently known in *Drosophila*: H2AvD, H3.3A, H3.3B, and H4r [12–15]. In addition to histone modification [16–22], the replacement of histones by a different histone type causes chromatin remodeling [23–25]. Nucleosome remodeling is involved in many important biological processes, such as cell division, differentiation, gene expression, and replication [26–28]. Therefore, histone modification and replacement are mechanisms that can lead to epigenetic changes [21,29,30]. In *Drosophila*, the histone genes for the canonical type of histones are clustered in a repetitive unit, and in *Drosophila melanogaster*, the unit repeats about 110 times [10,31]. In contrast, the histone genes for the replacement type of histones are found as single genes or with only a few copies per genome, and they contain a few introns [12–15]. For the detailed structure of the histone genes in *Drosophila*, please refer to another review article [32]. The mode of molecular evolution of a multigene family, compared to a single gene, can be studied by analyzing histone genes [33].

The usage of codons in protein-coding genes is not uniform among synonymous codons and is biased in many species [34,35]. The mechanism of codon bias has been discussed for decades, and candidate factors include mutation bias, natural selection, and genetic drift [36–44]. Unequal usage of codons occurs when the rate of mutations due to nucleotide substitutions is biased or when selection pressure is exerted differently between synonymous codons. Fitness differences among synonymous codons may be present due to differences in the efficiency or speed of translation [45,46]. However, the selection pressure on codons, if any, would be comparatively weaker than that on amino acid sequences; therefore, the codon usage can be influenced by population size [32,39,47–53]. Since the largest difference in codon usage is observed in the nucleotide at the third codon position of genes, the guanine–cytosine (GC) content at the third codon position is strongly related to codon usage bias. In *Drosophila*, the higher the GC content at the third codon position is, the stronger the bias of codons [37,40,54]. Moreover, regarding the relationship with the evolutionary rate, the stronger the bias of codons is, the slower the evolutionary rate [55].

In *Drosophila saltans*, the low GC content of the *Xdh* and *Adh* genes was explained by fluctuating mutation bias [56,57]. However, it may also be explained by changes in selection [32,38,50–53]. Although many *Drosophila* species have been analyzed for their histone genes [31,49,58–60], no changes in the rate of mutations were observed among the species in our analysis [49]. Here, the evolution of the GC content at the third codon position of histone genes in *Drosophila* is reviewed, and a model that can best explain the evolution of the GC content at the third codon position in *Drosophila* is presented.

2. Evolution of the GC Content at the Third Codon Position of the Histone Genes in *Drosophila*

The GC content at the third codon position of the histone genes in 12 *Drosophila* species is shown in Figure 1. Parts of histone genes data have been published from our laboratory [31,49,51,52,58–60]. The rest is obtained from FlyBase (http://flybase.org, accessed on 2017–2019) [61]. Several characteristic points on the evolution of the GC content at the third codon position of histone genes in *Drosophila* are summarized below.

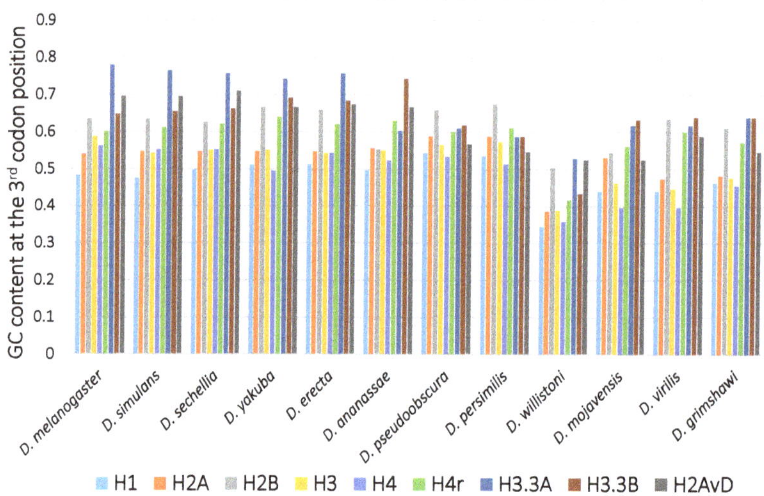

Figure 1. The GC content at the third codon position of the histone genes in *Drosophila*. The data grouped according to the *Drosophila* species.

2.1. Disparity in the GC Content at the Third Codon Position among the Genes

In many *Drosophila* species, the codon usage of the genes was uneven and varied from gene to gene [36,40,50]. Therefore, the GC content at the third codon position differed between the genes. Although the reason remains unclear, codon bias was found to be related to the level of gene expression, which also varied from gene to gene. The positive relationship found between codon bias and the level of gene expression most likely resulted from the difference in translation efficiency [45,46]. Among the canonical histone genes, *H2B* showed the highest GC content at the third codon position, while *H1* showed the lowest GC content at the third codon position [62]. H1, a linker protein, is expressed at approximately half of the level of the other four canonical histones. This is likely the reason why the GC content at the third codon position of *H1* is not as high as those of the core histone genes.

2.2. Disparity in the GC Content at the Third Codon Position between the Genes of the Canonical and Replacement Types of Histones

A comparison of the average GC content at the third codon position of genes in 12 common species revealed a higher GC content at the third codon position in the genes of the replacement type of histones than in those of the canonical type of histones [62–65]. Analysis of codon bias in the histone genes demonstrated that the difference was caused not by an obvious codon bias in a specific amino acid but by a general tendency that was observed for many codons [62]. Differences in functional differentiation or translation efficiency may be the cause of the differences in GC content at the third codon position between the histone types.

2.3. Disparity in GC Content at the Third Codon Position of the Genes among the Different Species

Although variability in the GC content among the genes within a species has been previously noted [36,40,50,51], variability has also been observed between different species [40,51,62]. For example, among 12 *Drosophila* species, the GC content at the third codon position of many genes in *Drosophila willistoni* was relatively lower than in the other 11 species [39,62]. Furthermore, when the GC content at the third codon position of corresponding genes was compared between the *Drosophila* species, nearly parallel differences, similar patterns of ups and downs, were observed for most comparisons (Figure 2). A lower GC content at the third codon position was also observed in the genes of *Drosophila* species other than these 12 species, such as in *Drosophila hydei* and *Drosophila americana* (Figure 1).

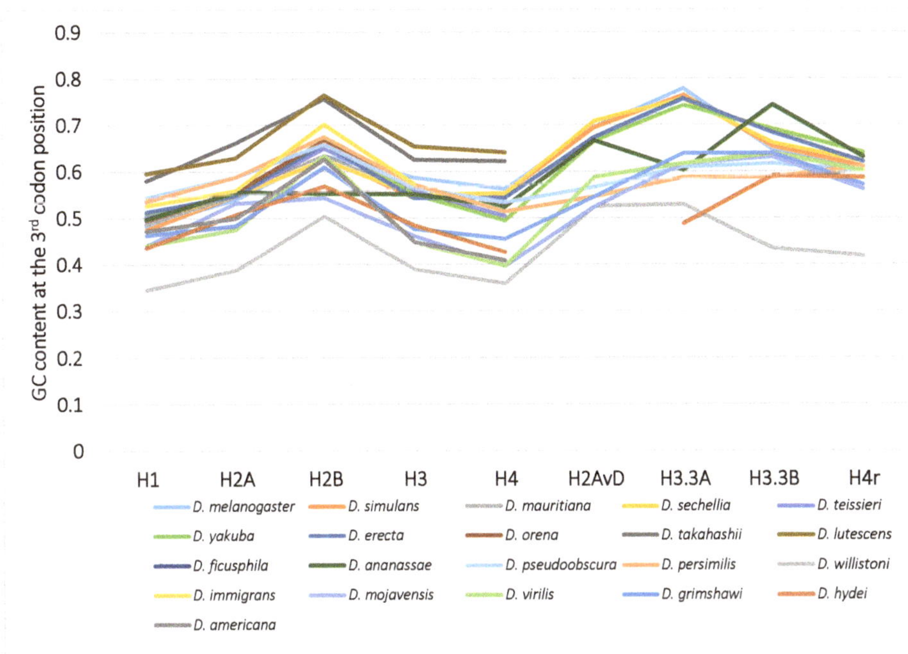

Figure 2. The GC content at the third codon position of the nine histone genes in *Drosophila*. The points from the same species were connected by lines to show the trend for each species.

2.4. Mode of the Evolution of GC Content at the Third Codon Position According to Phylogeny

The differences in GC content at the third codon position according to the *Drosophila* phylogeny were unexpected and lacked consistency with evolution [33,62]. The GC contents at the third codon position of closely related species showed similar values, but those in distantly related species did not always show larger differences. Unlike the case for nucleotide and amino acid substitutions, the relationship between differences in GC content at the third codon position and the evolutionary distances between species is not co-linear. The differences in GC content at the third codon position are independent of phylogenetic distance.

3. Models for the Evolution of GC Content

3.1. *Mutation Bias*

Candidate genetic factors that can explain the evolution of GC content including selection, genetic drift, and mutation [36,41,57]. The patterns and rates of mutations are hard to measure directly by experiments, but they can be estimated by several indirect methods, for example, by analyzing the bases, or GC content, in regions that are free from selection [40,41]; such regions include long introns, intergenic spacers, and regions of transposons without functional constraints. The GC content of such regions is most likely determined by mutation bias alone. Therefore, if the effect of mutations in these regions is stationary, the patterns and rates of mutations can be estimated by the content of each base in these regions. In most cases, these regions are observed to be AT rich, meaning that the mutation is biased for adenine/thymine (A/T) [40,41]. For example, in the *D. melanogaster* histone gene cluster, the GC content of the longest spacer between the *H1–H3* genes was observed to be the lowest at 30% [31,32,62]. Therefore, a mutation in *D. melanogaster* must be biased, at least a little, for A/T.

It is also possible to test whether or not the mutation bias varies among species by comparing the GC content of these regions. If the GC content in a broad range of species is distributed within a narrow range, then the mutation pattern and rate must be stable among those species. For example, the GC content of the *H1–H3* spacer region in *Drosophila* is approximately 30% in a broad range of species [51,62]; thus, the mutation bias in *Drosophila* must be stable, and it is unlikely that a variation in GC content at the third codon position among the species is the result of a fluctuating mutation bias among the species. A low GC content can be easily explained by mutation bias because the mutation was biased for A/T. However, a high GC content cannot be explained merely by mutation bias; to explain a high GC content, other genetic factors, such as selection and drift, need to be considered.

3.2. *Deleterious Selection for A/T*

Since the GC content was lowest in regions free from selection and was most likely determined by mutation bias alone, to explain the observed codon usage and GC content at the third codon position described above, selection resulting in G/C increases or A/T decreases was assumed to have been involved. Here, an evolutionary observation has to be taken into account for considering the assumptions of our model. It is a fact that most new mutations biased for A/T are observed to be either neutral or deleterious [66,67]. This can again be explained by the selection that is deleterious for A/T. Therefore, these observations on the GC content at the third codon position can be most easily explained by assuming that selection for A/T is mostly, but not completely, deleterious. In regions with a high GC content, there is selection acting for the removal of A/T, and in regions with a low GC content, the selection is weak or not acting at all. However, the selection coefficient for A/T nucleotides, which varies from codon to codon in genes, must be much smaller than that of the amino acid sequence. Therefore, the overall selection for codon bias in a gene seems to be a "gene effect." For example, the GC content at the third codon position of the *H1* gene is lower than those of the other histone genes. Another example is the difference in GC content at the third codon position among the different histone types, which must be due to the functional differentiation between the histone types, such as the efficiency and rate of translation [62].

3.3. *Effect of the Population Size*

The selection for A/T at the third codon position was weak enough for it to be affected by the population size. In the nearly neutral model of molecular evolution, the selection is more effective on larger populations but not as effective on smaller populations [47,48]. The deleterious selection for A/T in a large population would thus be expected to be effective, resulting in a high GC content at the third codon position and high codon bias. In contrast, in a small population size, the deleterious selection for A/T would not be expected to be effective, and the GC content at the third codon position would remain low with a low codon bias. Another point to be noted here is that the whole genome is affected by the size effect, that is, all genes in the same genome simultaneously experience the same size effect. Furthermore, in the past, the genomes of species must have experienced repeated changes due to changes in the population size. Therefore, the increases and decreases in GC content are expected to be linked for all genes in the same genome. The accumulation of past changes in the genomic genes must be a "species effect." In *D. willistoni* and several other *Drosophila* species, the GC content of the genomic genes was considerably lower in comparison to that in other *Drosophila* species [39,41,51,62]. In the species with a low GC content, none or a weaker selection must have been acting. Therefore, in those species or their ancestors, a decrease in the population size must have occurred [51].

4. Generality of the Model

Although the evolution of the GC content at the third codon position of the histone genes in *Drosophila* can be most easily explained by the deleterious selection for the A/T model, data for more genes and from more species should be analyzed in detail to confirm whether this model can be applied to other genes in the genome. More than 6000 genomic genes from each of 12 *Drosophila* species were analyzed for codon bias by Vicario et al. [39]. The following facts used for constructing the above model are applicable to histone genes, as well as many other genes in the genome: (1) there is variability in the GC content at the third codon position or codon bias of the genes in a species that is hard to explain by mutation bias alone; (2) the GC content was lowest in the regions with weak or no selection, such as the introns and spacers, in a broad range of species, and the level was similar between the different species [32,39,41,52,60,62]; (3) in *D. willistoni*, the GC content at the third codon position was lower in most genes, and the codon bias was relaxed when compared to the other species [39,41]; (4) the GC content at the third codon position of corresponding genes among *Drosophila* species tended to show a nearly parallel difference for each comparison (Figure 2). Therefore, it is "unlikely" that the deleterious selection for the A/T model is applicable only to the histone genes. Although many more genes and species should be analyzed, the results from the analysis of histone genes appear to be applicable to all genes.

5. Conclusions

The evolution of the GC content at the third codon position of histone genes in *Drosophila* was reviewed. The model that can best explain the observed data is the deleterious selection for the A/T model with the population size effect.

Funding: This research received no external funding.

Institutional Review Board Statement: Not applicable.

Informed Consent Statement: Not applicable.

Data Availability Statement: No report for available data.

Acknowledgments: I would like to thank the researchers in our laboratory who worked on the studies of the histone genes in *Drosophila*.

Conflicts of Interest: The author declares no conflict of interest.

References

1. Kornberg, R.D. Chromatin Structure: A Repeating Unit of Histones and DNA. *Science* **1974**, *184*, 868–871. [CrossRef]
2. Kornberg, R.D. Structure of Chromatin. *Annu. Rev. Biochem.* **1977**, *46*, 931–954. [CrossRef]

3. McGhee, J.D.; Rau, D.C.; Charney, E.; Felsenfeld, G. Orientation of the nucleosome within the higher order structure of chromatin. *Cell* **1980**, *22*, 87–96. [CrossRef]
4. Cutter, A.R.; Hayes, J.J. A brief review of nucleosome structure. *FEBS Lett.* **2015**, *589*, 2914–2922. [CrossRef]
5. Schümperli, D. Cell-cycle regulation of histone gene expression. *Cell* **1986**, *45*, 471–472. [CrossRef]
6. Malik, H.S.; Henikoff, S. Adaptive evolution of Cid, a centromere-specific histone in Drosophila. *Genetics* **2001**, *157*, 1293–1298. [CrossRef]
7. Vermaak, D.; Hayden, H.S.; Henikoff, S. Centromere targeting element within the histone fold domain of Cid. *Mol. Cell. Biol.* **2002**, *22*, 7553–7561. [CrossRef] [PubMed]
8. Dalal, Y.; Furuyama, T.; Vermaak, D.; Henikoff, S. Structure, dynamics, and evolution of centromeric nucleosomes. *Proc. Natl. Acad. Sci. USA* **2007**, *104*, 15974–15981. [CrossRef] [PubMed]
9. Palmer, D.; Snyder, L.A.; Blumenfeld, M. Drosophila nucleosomes contain an unusual histone-like protein. *Proc. Natl. Acad. Sci. USA* **1980**, *77*, 2671–2675. [CrossRef]
10. Lifton, R.P.; Goldberg, M.L.; Karp, R.W.; Hogness, D.S. The Organization of the Histone Genes in *Drosophila melanogaster*: Functional and Evolutionary Implications. *Cold Spring Harb. Symp. Quant. Biol.* **1978**, *42*, 1047–1051. [CrossRef] [PubMed]
11. Pardue, M.L.; Kedes, L.H.; Weinberg, E.S.; Birnstiel, M.L. Localization of sequences coding for histone messenger RNA in the chromosomes of *Drosophila melanogaster*. *Chromosoma* **1977**, *63*, 135–151. [CrossRef]
12. Fretzin, S.; Allan, B.D.; van Daal, A.; Elgin, S.C.R. A *Drosophila melanogaster* H3.3 cDNA encodes a histone variant identical with the vertebrate H3.3. *Gene* **1991**, *107*, 341–342. [CrossRef]
13. Van Daal, A.; White, E.M.; Gorovsky, M.A.; Elgin, C.R. *Drosophila* has a single copy of the gene encoding a highly conserved histone H2A variant of the H2A. F/Z type. *Nucleic Acids Res.* **1988**, *16*, 7487–7497. [CrossRef] [PubMed]
14. Akhmanova, A.S.; Bindels, P.C.T.; Xu, J.; Miedema, K.; Kremer, H.; Hennig, W.; Xu, J.; Hennig, W. Structure and expression of histone H3.3 genes in *Drosophila melanogaster* and *Drosophila hydei*. *Genome* **1995**, *38*, 586–600. [CrossRef]
15. Akhmanova, A.; Miedema, K.; Hennig, W. Identification and characterization of the *Drosophila* histone H4 replacement gene. *FEBS Lett.* **1996**, *388*, 219–222. [CrossRef]
16. Lawrence, M.; Daujat, S.; Schneider, R. Lateral Thinking: How Histone Modifications Regulate Gene Expression. *Trends Genet.* **2016**, *32*, 42–56. [CrossRef]
17. Jenuwein, T.; Allis, C.D. Translating the Histone Code. *Science* **2001**, *293*, 1074–1080. [CrossRef]
18. Musselman, C.A.; Lalonde, M.-E.; Côté, J.; Kutateladze, T.G. Perceiving the epigenetic landscape through histone readers. *Nat. Struct. Mol. Biol.* **2012**, *19*, 1218–1227. [CrossRef] [PubMed]
19. Bannister, A.; Kouzarides, T. Regulation of chromatin by histone modifications. *Cell Res.* **2011**, *21*, 381–395. [CrossRef] [PubMed]
20. Benson, L.J.; Gu, Y.; Yakovleva, T.; Tong, K.; Barrows, C.; Strack, C.L.; Cook, R.G.; Mizzen, C.A.; Annunziato, A.T. Modifications of H3 and H4 during chromatin replication, nucleosome assembly, and histone exchange. *J. Biol. Chem.* **2006**, *281*, 9287–9296. [CrossRef] [PubMed]
21. Zentner, G.E.; Henikoff, S. Regulation of nucleosome dynamics by histone modifications. *Nat. Struct. Mol. Biol.* **2013**, *20*, 259–266. [CrossRef] [PubMed]
22. Cedar, H.; Bergman, Y. Linking DNA methylation and histone modification: Patterns and paradigms. *Nat. Rev. Genet.* **2009**, *10*, 295–304. [CrossRef]
23. Santoro, S.W.; Dulac, C. Histone variants and cellular plasticity. *Trends Genet.* **2015**, *31*, 516–527. [CrossRef] [PubMed]
24. Biterge, B.; Schneider, R. Histone variants: Key players of chromatin. *Cell Tissue Res.* **2014**, *356*, 457–466. [CrossRef] [PubMed]
25. Maze, I.; Noh, K.-M.; Soshnev, A.A.; Allis, C.D. Every amino acid matters: Essential contributions of histone variants to mammalian development and disease. *Nat. Rev. Genet.* **2014**, *15*, 259–271. [CrossRef] [PubMed]
26. Becker, P.B.; Workman, J.L. Nucleosome remodeling and epigenetics. *Cold Spring Harb. Perspect. Biol.* **2013**, *5*, a017905. [CrossRef]
27. Venkatesh, S.; Workman, J.L. Histone exchange, chromatin structure and the regulation of transcription. *Nat. Rev. Mol. Cell Biol.* **2015**, *16*, 178–189. [CrossRef]
28. Grunstein, M. Histone acetylation in chromatin structure and transcription. *Nature* **1997**, *389*, 349–352. [CrossRef]
29. Gaume, X.; Torres-Padilla, M.-E. Regulation of reprogramming and cellular plasticity through histone exchange and histone variant incorporation. *Cold Spring Harb. Symp. Quant. Biol.* **2015**, *80*, 165–175. [CrossRef]
30. Henikoff, S. Epigenetic profiling of histone variants. *Epigenomics* **2009**, 101–118. [CrossRef]
31. Matsuo, Y.; Yamazaki, T. tRNA derived insertion element in histone gene repeating unit of *Drosophila melanogaster*. *Nucleic Acids Res.* **1989**, *17*, 225–238. [CrossRef] [PubMed]
32. Matsuo, Y. Genomic structure and evolution of the histone gene family in *Drosophila*. *Curr. Top. Genet.* **2006**, *2*, 1–14.
33. Matsuo, Y.; Yamazaki, T. Nucleotide variation and divergence in the histone multigene family in *Drosophila melanogaster*. *Genetics* **1989**, *122*, 87–97. [CrossRef] [PubMed]
34. Wells, D.; Bains, W.; Kedes, L. Codon usage in histone gene families of higher eukaryotes reflects functional rather than phylogenetic relationships. *J. Mol. Evol.* **1986**, *23*, 224–241. [CrossRef]
35. Ikemura, T. Codon Usage and tRNA Content in Unicellular and Multicellular Organisms. *Mol. Biol. Evol.* **1985**, *2*, 13–34. [CrossRef]
36. Shields, D.C.; Sharp, P.M.; Higgins, D.G.; Wright, F. "Silent" sites in Drosophila genes are not neutral: Evidence of selection among synonymous codons. *Mol. Biol. Evol.* **1988**, *5*, 704–716. [CrossRef]

37. Poh, Y.-P.; Ting, C.-T.; Fu, H.-W.; Langley, C.H.; Begun, D.J. Population Genomic Analysis of Base Composition Evolution in *Drosophila melanogaster*. *Genome Biol. Evol.* **2012**, *4*, 1245–1255. [CrossRef]
38. Lawrie, D.S.; Messer, P.W.; Hershberg, R.; Petrov, D.A. Strong Purifying Selection at Synonymous Sites in *D. melanogaster*. *PLoS Genet.* **2013**, *9*, e1003527. [CrossRef]
39. Vicario, S.; Moriyama, E.N.; Powell, J.R. Codon usage in twelve species of *Drosophila*. *BMC Evol. Biol.* **2007**, *7*, 226. [CrossRef]
40. Moriyama, E.N.; Hartl, D.L. Codon usage bias and base composition of nuclear genes in *Drosophila*. *Genetics* **1993**, *134*, 847–858. [CrossRef]
41. Powell, J.R.; Moriyama, E.N. Evolution of codon usage bias in Drosophila. *Proc. Natl. Acad. Sci. USA* **1997**, *94*, 7784–7790. [CrossRef] [PubMed]
42. Akashi, H. Inferring weak selection from patterns of polymorphism and divergence at 'silent' sites in Drosophila DNA. *Genetics* **1995**, *139*, 1067–1076. [CrossRef]
43. Li, W.H. Models of nearly neutral mutations with particular implications for nonrandom usage of synonymous codons. *J. Mol. Evol.* **1987**, *24*, 337–345. [CrossRef]
44. Kliman, R.M.; Hey, J. The effects of mutation and natural selection on codon bias in the genes of drosophila. *Genetics* **1994**, *137*, 1049–1056. [CrossRef] [PubMed]
45. Akashi, H. Translational selection and yeast proteome evolution. *Genetics* **2003**, *164*, 1291–1303. [CrossRef] [PubMed]
46. Hartl, D.L.; Moriyama, E.N.; Sawyer, S.A. Selection intensity for codon bias. *Genetics* **1994**, *138*, 227–234. [CrossRef]
47. Ohta, T. Population size and rate of evolution. *J. Mol. Evol.* **1972**, *1*, 305–314. [CrossRef]
48. Ohta, T. The Nearly Neutral Theory of Molecular Evolution. *Annu. Rev. Ecol. Syst.* **2003**, *23*, 263–286. [CrossRef]
49. Matsuo, Y. Molecular evolution of the histone 3 multigene family in the *Drosophila melanogaster* species subgroup. *Mol. Phylogenet. Evol.* **2000**, *16*, 339–343. [CrossRef]
50. Matsuo, Y. Evolutionary change of codon usage for the histone gene family in *Drosophila melanogaster* and *Drosophila hydei*. *Mol. Phylogenet. Evol.* **2000**, *15*, 283–291. [CrossRef]
51. Matsuo, Y. Evolution of the GC content of the histone 3 gene in seven Drosophila species. *Genes Genet. Syst.* **2003**, *78*, 309–318. [CrossRef]
52. Nakashima, Y.; Higashiyama, A.; Ushimaru, A.; Nagoda, N.; Matsuo, Y. Evolution of GC content in the histone gene repeatingunits from *Drosophila lutescens*, *D. takahashii* and *D. pseudoobscura*. *Genes Genet. Syst.* **2016**, *91*, 27–36. [CrossRef]
53. Lanfear, R.; Kokko, H.; Eyre-Walker, A. Population size and the rate of evolution. *Trends Ecol. Evol.* **2014**, *29*, 33–41. [CrossRef]
54. Bernardi, G.; Bernardi, G. Compositional constraints and genome evolution. *J. Mol. Evol.* **1986**, *24*, 1–11. [CrossRef] [PubMed]
55. Fitch, D.H.; Strausbaugh, L.D. Low codon bias and high rates of synonymous substitution in *Drosophila hydei* and *D. melanogaster* histone genes. *Mol. Biol. Evol.* **1993**, *10*, 397–413. [CrossRef] [PubMed]
56. Rodríguez-Trelles, F.; Tarrío, R.; Ayala, F.J. Switch in codon bias and increased rates of amino acid substitution in the *Drosophila saltans* species group. *Genetics* **1999**, *153*, 339–350. [CrossRef]
57. Rodríguez-Trelles, F.; Tarrío, R.; Ayala, F.J. Fluctuating mutation bias and the evolution of base composition in *Drosophila*. *J. Mol. Evol.* **2000**, *50*, 1–10. [CrossRef]
58. Tsunemoto, K.; Matsuo, Y.; Tsunemotoa, K.; Matsuo, Y.; Tsunemoto, K.; Matsuo, Y. Molecular evolutionary analysis of a histone gene repeating unit from *Drosophila simulans*. *Genes Genet. Syst.* **2001**, *76*, 355–361. [CrossRef]
59. Nagoda, N.; Fukuda, A.; Nakashima, Y.; Matsuo, Y. Molecular characterization and evolution of the repeating units of histone genes in *Drosophila americana*: Coexistence of quartet and quintet units in a genome. *Insect Mol. Biol.* **2005**, *14*, 713–717. [CrossRef]
60. Kakita, M.; Shimizu, T.; Emoto, M.; Nagai, M.; Takeguchi, M.; Hosono, Y.; Kume, N.; Ozawa, T.; Ueda, M.; Bhuiyan, M.S.I.; et al. Divergence and heterogeneity of the histone gene repeating units in the *Drosophila melanogaster* species subgroup. *Genes Genet. Syst.* **2003**, *78*, 383–389. [CrossRef] [PubMed]
61. Drosophila 12 Genomes Consortium. Evolution of genes and genomes on the *Drosophila* phylogeny. *Nature* **2007**, *450*, 203–218. [CrossRef] [PubMed]
62. Matsuo, Y. Epigenetics and Codon Usage of the Histone Genes in 12 *Drosophila* Species. *J. Phylogenet. Evol. Biol.* **2017**, *5*, 1–7. [CrossRef]
63. Matsuo, Y.; Kakubayashi, N. Epigenetics Evolution and Replacement Histones: Evolutionary Changes at *Drosophila* H3.3A and H3.3B. *J. Phylogenet. Evol. Biol.* **2016**, *4*, 1–8. [CrossRef]
64. Yamamoto, Y.; Watanabe, T.; Nakamura, M.; Kakubayashi, N.; Saito, Y.; Matsuo, Y. Epigenetics Evolution and Replacement Histones: Evolutionary Changes at Drosophila H4r. *J. Phylogenet. Evol. Biol.* **2016**, *4*, 3. [CrossRef]
65. Matsuo, Y.; Kakubayashi, N. Epigenetics Evolution and Replacement Histones: Evolutionary Changes at *Drosophila* H2AvD. *J. Data Min. Genom. Proteom.* **2017**, *8*, 1. [CrossRef]
66. Mukai, T. The genetic structure of natural populations of *Drosophila melanogaster*. I. Spontaneous mutation rate of polygenes controlling viability. *Genetics* **1964**, *50*, 1–19. [CrossRef]
67. Eyre-Walker, A.; Keightley, P.D. The distribution of fitness effects of new mutations. *Nat. Rev. Genet.* **2007**, *8*, 610–618. [CrossRef]

genes

Review

Saccharomyces cerevisiae as a Tool for Studying Mutations in Nuclear Genes Involved in Diseases Caused by Mitochondrial DNA Instability

Alexandru Ionut Gilea †, Camilla Ceccatelli Berti †, Martina Magistrati, Giulia di Punzio, Paola Goffrini, Enrico Baruffini * and Cristina Dallabona

Department of Chemistry, Life Sciences and Environmental Sustainability, University of Parma, Parco Area delle Scienze 11/A, 43124 Parma, Italy; alexandruionut.gilea@unipr.it (A.I.G.); camilla.ceccatelliberti@unipr.it (C.C.B.); martina.magistrati@unipr.it (M.M.); giulia.dipunzio@unipr.it (G.d.P.); paola.goffrini@unipr.it (P.G.); cristina.dallabona@unipr.it (C.D.)

* Correspondence: enrico.baruffini@unipr.it; Tel.: +39-0521-905679

† These authors contributed equally to this work.

Abstract: Mitochondrial DNA (mtDNA) maintenance is critical for oxidative phosphorylation (OXPHOS) since some subunits of the respiratory chain complexes are mitochondrially encoded. Pathological mutations in nuclear genes involved in the mtDNA metabolism may result in a quantitative decrease in mtDNA levels, referred to as mtDNA depletion, or in qualitative defects in mtDNA, especially in multiple deletions. Since, in the last decade, most of the novel mutations have been identified through whole-exome sequencing, it is crucial to confirm the pathogenicity by functional analysis in the appropriate model systems. Among these, the yeast *Saccharomyces cerevisiae* has proved to be a good model for studying mutations associated with mtDNA instability. This review focuses on the use of yeast for evaluating the pathogenicity of mutations in six genes, *MPV17/SYM1*, *MRM2/MRM2*, *OPA1/MGM1*, *POLG/MIP1*, *RRM2B/RNR2*, and *SLC25A4/AAC2*, all associated with mtDNA depletion or multiple deletions. We highlight the techniques used to construct a specific model and to measure the mtDNA instability as well as the main results obtained. We then report the contribution that yeast has given in understanding the pathogenic mechanisms of the mutant variants, in finding the genetic suppressors of the mitochondrial defects and in the discovery of molecules able to improve the mtDNA stability.

Keywords: mtDNA depletion syndromes; diseases associated with mtDNA deletions; yeast model; *MPV17/SYM1*; *MRM2/MRM2*; *OPA1/MGM1*; *POLG/MIP1*; *RRM2B/RNR2*; *SLC25A4* (*ANT1*)/*AAC2*; drug repurposing

Citation: Gilea, A.I.; Ceccatelli Berti, C.; Magistrati, M.; di Punzio, G.; Goffrini, P.; Baruffini, E.; Dallabona, C. *Saccharomyces cerevisiae* as a Tool for Studying Mutations in Nuclear Genes Involved in Diseases Caused by Mitochondrial DNA Instability. *Genes* **2021**, *12*, 1866. https://doi.org/10.3390/genes12121866

Academic Editor: Luigi Viggiano

Received: 29 October 2021
Accepted: 23 November 2021
Published: 24 November 2021

1. Introduction

Mitochondria are cellular organelles present in most eukaryotic organisms, among which include all fungi, plants, and animals. Their main physiological role, albeit not the only one, is the production of most of the cell energy by means of electron transport through the respiratory chain and the oxidative phosphorylation (OXPHOS). The mitochondrial mass inside a cell depends on many factors, including the species and, in multicellular organisms, the cell type, the cell phase, and the energy requirement of the cell.

Mitochondria are semi-autonomous organelles, since they have their own genome, called mitochondrial DNA (mtDNA), which encodes for genes that are fundamental for oxidative phosphorylation. In mammals, mtDNA, which has been discovered in 1963 [1], is circular, approximately 16.5 Kbp long, and contains 37 genes. Thirteen genes encode for subunits of the respiratory chain complexes and of ATP synthase (ND1, ND2, ND3, ND4, ND4L, ND5, and ND6 of complex I; CYB of complex III, CO1, CO2, and CO3 of complex IV; ATP6 and ATP8 of complex V), 2 genes encode for mitochondrial rRNA, and

22 genes for mitochondrial tRNAs. All the approximately 1500–2000 remaining proteins of the mitochondrial proteome, including all the remaining subunits of the complexes involved in OXPHOS, are encoded by nuclear genes, and are imported into mitochondria (reviewed in [2,3]).

In mammals, all the mtDNA present in cells derived from the mtDNA present in the oocyte, and are thus maternally inherited [4]. In addition, excluding the early divisions during embryonic development, mtDNA is synthesized during the whole cell cycle. Indeed, each cell contains several molecules of mtDNA, whose number depends on the cell type and the energetic requirement. Cells that contain most copies of mtDNA are neurons, muscle fibers, hepatocytes, and oocytes, where the mtDNA copy number is equal to thousands or tens of thousands (reviewed in [5–7]).

MtDNA molecules are not naked but organized in DNA–protein complexes called nucleoids. Several nucleoids are present inside mitochondria, and each of them contains in general a single mtDNA molecule. Nucleoids, which are anchored to the inner mitochondrial membrane (IMM), form replication units that are autonomous in mtDNA replication and segregation [8–10]. According to the most accepted model, mtDNA is replicated in mammals through the strand displacement model [11]. DNA synthesis is continuous on both strands, called H and L; however, the replication is asymmetric, since it starts from two different dedicated replication origins (O_H and O_L) and at different times (reviewed in [12]). Several proteins are involved in the mtDNA replication, and mutations in most of nuclear genes encoding for these proteins are associated with mitochondrial diseases characterized as secondary mutations by mtDNA deletions or depletion.

If mutations occur, mutated mtDNA molecules are produced. If two mtDNA molecules with a different sequence exist inside a tissue, for example, a wild-type one and a mutant one, either homoplasmic or heteroplasmic conditions are possible. In the former case, each cell contains a single type of mtDNA, whereas, in the latter case, each cell contains both mutant and wild-type copies of mtDNA. In most patients affected by mutations in mtDNA, including secondary deletions caused by nuclear mutations, heteroplasmy is present. These deletions can be considered recessive [13], and pathology occurs only when the number of mutant copies is above a threshold level, whose percentage depends on the mutation and on the affected tissues [14].

Mutations in human nuclear genes involved in mtDNA replication, integrity, maintenance, and segregation are associated with a plethora of mitochondrial diseases associated with mtDNA depletion or multiple deletions. In these disorders, the primary cause is one or more mutations in nuclear genes which, in turn, result in secondary qualitative (multiple mtDNA deletions) or quantitative (mtDNA depletion) mtDNA defects (reviewed in [2,3,5,15]). Depending on the gene and on the mutation, the disorder can be inherited by means of either recessive or dominant inheritance. In the latter case, the dominance can be due to a loss of function, i.e., to haploinsufficiency, to a gain-of-function, or to negative dominance. Whereas mutations in some genes are associated primarily with mtDNA depletion and mutations in other genes primarily to multiple mtDNA deletions, mutations in most genes affecting mtDNA maintenance can be associated with either depletion or multiple deletions. Genes associated with such pathologies are reported in Table 1.

Table 1. Genes associated with mitochondrial diseases characterized by mtDNA depletion and/or multiple deletions.

Human Gene	Protein Function	Disease	OMIM Number	Onset	Inheritance	mtDNA Alteration	Main Phenotype	Yeast Gene	Study in Yeast
ABAT	4-aminobutyrate aminotransferase	GABA-transaminase deficiency	613163	Infancy	AR	Multiple deletions	Encephalopathy, myopathy, and elevated GABA	*UGA1*	/
AGK	Acylglycerol kinase	MDDS 10, Sengers syndrome	212350	Neonatal	AR	Depletion	Cardiac and skeletal myopathy and cataract	NP	/
DGUOK	Mitochondrial deoxyguanosine kinase	MDDS 3	251880	Neonatal period, infancy, or childhood	AR	Depletion	Hepatopathy and encephalopathy	NP	/
		PEO, autosomal recessive 4	617070	Early or mid-adulthood	AR	Multiple deletions	Myopathy and ophthalmoplegia		
DNA2	DNA replication helicase/nuclease 2	PEO, autosomal dominant 6	615156	Childhood or early adulthood	AD	Multiple deletions	Myopathy and ophthalmoplegia	*DNA2*	/
FBXL4	F-box and leucine-rich repeat protein 4	MDDS 13	615471	Neonatal period or infancy	AR	Depletion	Encephalopathy and myopathy	NP	/
LIG3	Ligase III	Neurogastrointestinal encephalomyopathy		Infancy to adolescence	AR	Depletion	Gut dysmotility, encephalopathy, and myopathy	NP	/
MFN2	Mitofusin 2	Hereditary motor and sensory neuropathy VIA; DOA	601152	Early childhood	AD	Multiple deletions	Optic atrophy and neuropathy	*FZO1*	/
MGME1	Mitochondrial exonuclease 1	MDDS 11	615084	Childhood or early adulthood	AR	Depletion and multiple deletions	Myopathy	NP	/
MPV17	IMM protein	PEO, autosomal recessive		Adulthood	AR	Multiple deletions	Ophthalmoplegia leukoen-cephalopathy and/or parkinsonism	*SYM1*	[16,17]
		Neuromyopathic MDMD		Adulthood	AR	Multiple deletions	Neuropathy and myopathy		
		MDDS 6	256810	Neonatal period, infancy, or early childhood	AR	Depletion	Neuropathy, hepatopathy and/or encephalopathy		

Table 1. *Cont.*

Human Gene	Protein Function	Disease	OMIM Number	Onset	Inheritance	mtDNA Alteration	Main Phenotype	Yeast Gene	Study in Yeast
MRM2	Mitochondrial ribosomal RNA methyltransferase 2	MDDS 17	618567	Infancy	AR	Depletion	MELAS-like with encephalopathy, lactic acidosis and stroke-like episodes	*MRM2*	[18]
OPA1	Mitochondrial dynamin-like GTPase	MDDS 14	616896	Neonatal or infancy	AR	Depletion	Cardiomyopathy, encephalopathy	*MGM1*	[19–21]
		DOA	165500	Childhood or early adulthood	AD	(Multiple deletions)	Optic atrophy		
		DOA plus	125250	Childhood or early adulthood	AD	Multiple deletions	Optic atrophy with deafness, ophthalmoplegia, myopathy, ataxia, and/or neuropathy		
POLG	DNA polymerase γ	Childhood myocerebrohepatopathy spectrum disorders		Infancy	AR	Depletion	Hypotonia, hepatopathy, developmental delay	*MIP1*	[22–39]
		MDDS 4A	203700	Early childhood	AR	Depletion	Alpers–Huttenlocher syndrome with encephalopathy, neuropathy, and hepatopathy		
		MDDS 4B	613662	Childhood to adulthood	AR	Depletion and multiple deletions	MNGIE with gastrointestinal dysmotility, myopathy, and neuropathy		
		Mitochondrial recessive ataxia syndrome	607459	Adolescence, early adulthood	AR	Multiple deletions	SANDO/SCAE, ANS, MEMSA with ataxia, neuropathy, encephalopathy, epilepsy and/or myopathy		

Table 1. *Cont.*

Human Gene	Protein Function	Disease	OMIM Number	Onset	Inheritance	mtDNA Alteration	Main Phenotype	Yeast Gene	Study in Yeast
		PEO, autosomal dominant 1	157640	Adulthood	AD	Multiple deletions	Ophthalmoplegia and myopathy		
		PEO, autosomal recessive	258450	Adolescence, adulthood	AR	Multiple deletions	Ophthalmoplegia		
POLG2	DNA polymerase γ accessory subunit	MDDS 16 (hepatic type)	618528	Infancy	AR	Depletion	Hepatophty	NP	/
		MDDS 16B	619425	Childood	AR	Depletion	Neuroophthalmic type		
		PEO, autosomal dominant 4	610131	Infancy to adulthood	AD	Multiple deletions	Myopathy and ophthalmoplegia		
RNASEH1	Ribonuclease H1	PEO, autosomal recessive 2	616479	Adulthood	AR	Multiple deletions	Ophthalmoplegia	*RNH1*	/
RRM2B	Ribonucleotide reductase, M2 B	MDDS 8A and 8B	612075	Infancy	AR	Depletion	Myopathy, encephalopathy and tubulopathy or MNGIE	*RNR2*	[40]
		PEO, autosomal recessive		Childhood	AR	Multiple deletions	Ophthalmoplegia and myopathy		
		PEO, autosomal dominant 5	613077	Adulthood	AD	Multiple deletions	Ophthalmoplegia and myopathy		
SLC25A21	Mitochondrial oxodicarboxylate carrier	MDDS 18	618811	Early childhood	AR	Depletion	Muscular atrophy and myopathy	*ODC1/ ODC2*	/
SLC25A4 (ANT1)	Mitochondrial ADP/ATP translocator	MDDS 12A (cardiomyopathic type)	617184	Neonatal	AD	Depletion	Myopathy and cardiomyopathy	*AAC2 (AAC1, AAC3)*	[41–49]
		MDDS 12B (cardiomyopathic type)	615418	Childhood	AR	Depletion and multiple deletions	Myopathy and cardiomyopathy		
		PEO, autosomal dominant 2	609283	Adulthood	AD	Multiple deletions	Ophthalmoplegia and myopathy		
SLC25A10 (DIC)	Mitochondrial dicarboxylate carrier	MDDS 19	618972	Infancy	AR	Depletion	Encephalopathy an hypotonia	*DIC1*	/

Table 1. *Cont.*

Human Gene	Protein Function	Disease	OMIM Number	Onset	Inheritance	mtDNA Alteration	Main Phenotype	Yeast Gene	Study in Yeast
SSBP1	Single-stranded DNA-binding protein 1	Optic atrophy 13	165510	Infancy to early adulthood	AD	Depletion	Optic atrophy	*RIM1*	/
SUCLA2	Succinyl-CoA ligase, β subunit	MDDS 5	612073	Infancy or early childhood	AR	Depletion	Encephalopathy and myopathy with or without methylmalonic aciduria	*LSC2*	/
SUCLG1	Succinyl-CoA ligase, α subunit	MDDS 9	245400	Neonatal period or infancy	AR	Depletion	Encephalopathy and myopathy with methylmalonic aciduria	*LSC1*	/
TFAM	Mitochondrial transcription factor 1	MDDS 15	617156	Neonatal	AR	Depletion	Hepatocerebral syndrome	*ABF2*	/
TOP3A	DNA topoisomerase III	PEO, autosomal recessive 5	618098	Adulthood	AR	Multiple deletions	Ophthalmoplegia and ataxia	*TOP3*	/
TK2	Mitochondrial thymidine kinase	PEO, autosomal recessive 3	617069	Mid-Adulthood	AR	Multiple deletions	Ophthalmoplegia and myopathy	NP	/
		MDDS 2	609560	Infancy or childhood	AR	Depletion	Myopathy,		
TWNK	Twinkle mtDNA helicase	MDDS 7 (hepatocerebral type), IOSCA	271245	Infancy	AR	Depletion	Ataxia, encephalopathy, and neuropathy	NP	/
		PEO, autosomal dominant 3	609286	Early adulthood	AD	Multiple deletions	Ophthalmoplegia and myopathy		
		Hepatocerebral MDMD		Neonatal or early infancy	AR	Depletion	Alpers-like with encephalopathy and hepatopathy		
TYMP	Thymidine phosphorylase	MDDS 1	603041	Adolescence to adulthood	AR	Depletion and multiple deletions	MNGIE with gastrointestinal dysmotility, myopathy, and neuropathy	NP	/

Genes are reported if mtDNA depletion and/or mtDNA deletions are found in most affected patients. Pathologies are reported only if mtDNA depletion and/or deletions occur. Yeast studies are reported only if yeast was used to confirm the pathogenicity of the mutations found in patients. Abbreviations: PEO: progressive external ophthalmoplegia; DOA: dominant optic atrophy; MDMD: mitochondrial DNA maintenance defects; MDDS: Mitochondrial DNA depletion syndrome; NP: not present.

Since the discovery that mitochondrial diseases associated with mtDNA depletion and/or multiple deletions can be caused by mutations in nuclear genes, the yeast *Saccharomyces cerevisiae* has been used to model these mutations with different aims. Recently, novel mutations in genes previously associated with mitochondrial diseases and mutations in novel genes have been found in patients through whole-exome sequencing (WES) or whole-genome sequencing (WGS). When mutations are identified in a patient through these analyses, often the familiar history is not known, therefore a model system can be useful to "validate" mutations, i.e., to confirm that the variant is the cause of the disorder and not a single nucleotide polymorphism (SNP). In addition, as detailed below, model systems can be useful for understanding the molecular mechanisms through which the mutations exert their pathological effects as well as to find genetic methods or drugs able to rescue the detrimental effects.

Yeast mtDNA is longer compared to its mammalian counterpart: depending on the strain, the length can be 68 Kbp (in short strains) to 86 Kbp (in long strains). The mitochondrial genome currently used as a reference is that sequenced by Foury and coauthors from strain FY1679, isogenic to the reference strain S288C [50]. The yeast mitochondrial genome contains seven genes encoding for subunits of the respiratory complexes (*COB* for complex III; *COX1*, *COX2*, *COX3* for complex IV; *ATP6*, *ATP8*, and *ATP9* for complex V), one gene, *VAR1*, encoding for a subunit of the mitochondrial ribosome, two genes for the 15S rRNA and for the 21S rRNA, 24 genes for tRNAs, and one gene, *RPM1*, encoding an RNA subunit of the RNase P, which is involved in the processing of the mitochondrial pre-tRNAs. In addition, several genes encoding for maturases, endonucleases, and a reverse transcriptase are present, and some of them are located inside the introns of *COX1*, *COB*, and 21S rRNA genes, however their role has not been fully characterized (www.yeastgenome.org, accessed on 19 November 2021).

Yeast mtDNA is rich in AT-segments, especially in the intergenic regions, with small regions rich in GC-segments, among which four copies of regions called *ori* initially associate with mtDNA replication [51,52]. Until now, several hypotheses have been proposed in order to explain the mechanisms of mtDNA replication: according to the most recent findings, *S. cerevisiae* mtDNA should be replicated through a rolling circle mechanism in which the leading strand is primed by recombinational structures, and which results in the coexistence of circular molecules and linear concatemers of DNA [53–56] (reviewed in [57,58]).

As for its human counterpart, yeast cells contain several copies of mtDNA molecules. Although the exact number depends on several conditions, such as the carbon source added to the growth medium, the growth temperature, and the haploid/diploid status of yeast, it has been estimated that each cell contains 10–50 copies per nuclear genome to 50–200 ones [59–61]. As in human mitochondria, yeast mtDNA is packaged into protein–DNA complexes, called nucleoids, which are anchored to the IMM [62–64]. Several proteins are present in the nucleoids, among which proteins involved in replication, transcription, repair and recombination, heat shock proteins, and some enzymes of the Krebs cycle [60,62,64–67].

One of the most important characteristics of *S. cerevisiae* as a model to study mtDNA instability is its *petite* positivity, i.e., yeast can survive without mtDNA or with long deletions in it. In this case, energy is produced through alcoholic fermentation, provided that a fermentable carbon source is added in the medium. Due to the limited quantity of ATP produced through alcoholic fermentation and the inability to utilize the ethanol released in the medium, the colonies deriving from cells harboring these mutations have a smaller size and are called "*petite*". The *petite* phenotype can be caused by mutations in nuclear genes involved in the maintenance of the mtDNA integrity and are inherited either in a Mendelian way (*pet* mutants) [68] or directly by mtDNA mutations (cytoplasmic *petite* mutants) [69]. Cytoplasmic *petite* mutants, called more simply "*petites*", arise spontaneously at high frequency (around 1–10%, depending on the strain and on the growth condition), and can be devoid of mtDNA (*rho*^{0} cells) or carry long and multiple deletions of mtDNA (*rho*$^-$ cells); in *rho*$^-$ cells, the mtDNA often contains several repeats of the same sequence. [70–72]. Cells containing the whole mtDNA are called *rho*$^+$. It has been postulated that *rho*$^-$ cells are generated mainly by homologous recombination, which is highly active in mitochondria, between imperfect repeats [73,74]. Besides, most *rho*$^-$ mtDNA genomes are unstable and can result in the loss of mtDNA, making the cell *rho*^{0}. Mutations in nuclear genes can affect the *rho* status of the cells and influence the frequency of *petite* mutants. Among these genes, there are those involved in the replication, recombination, and repair of the mtDNA, but also in the maintenance, in the integrity, and in the inheritance of the mitochondrial genome (reviewed in [75]).

Contrary to mammals, heteroplasmy is just a transient condition in yeast. If two types of DNA molecules are present in a cell, for example, a *rho*$^+$ molecule and a *rho*$^-$ molecule, after a few generations, yeast become homoplasmic, i.e., there are two populations of cells, each with only a single type of mtDNA genome [70,76,77]. Although the exact mechanisms leading to homoplasmy in few generations are not fully understood, it has been hypothesized that several pathways could be involved, such as the positioning of the nucleoids in a different part of the cells, the transport of mitochondria in the buds, the concatemeric formation of mtDNA, and the asymmetric inheritance of mtDNA [55,78–82].

In this review, we focus on the use of yeast for studying pathological mutations in nuclear genes associated with mtDNA instability, underlying the methodologies used to construct the models and to study the phenotypes associated with mtDNA instability and the main findings to which yeast has contributed [50–69,71,72,74–84].

2. Creation of the Model and Techniques Used to Measure Instability of mtDNA

2.1. Construction of the Model Systems

As reported in Table 1, several human genes associated with mtDNA depletion and/or deletion pathologies are present and conserved in yeast. To evaluate the consequence of a given mutation, it must be considered whether to use, as a wild-type reference, the *S. cerevisiae* gene (homologous complementation), the human cDNA (heterologous complementation), or a chimeric construct between the two genes (chimeric complementation) [85]. Each of these complementation approaches has its respective pros and cons which should be considered. The use of heterologous complementation allows for the studying of any mutation under analysis, since the human gene is directly introduced in yeast. In addition, if a detrimental effect is observed, the allele is very likely pathogenic; on the contrary, if no effect is observed, the allele is likely either not pathogenic or hypomorphic, although it cannot be excluded that, in humans, the gene has a second function which cannot be studied in yeast. Despite these advantages, heterologous complementation is only possible in the case that the human cDNA complements the deletion of its yeast counterpart. Moreover, the complementation degree should be sufficient to evaluate the possible defective phenotypes associated with the mutant alleles, and the gene expression should not be too high to hide the detrimental effect of mutant alleles. If complementation does not occur, chimeric complementation can be exploited. As has been reported, the mitochondrial targeting sequence (MTS) can be different between yeast and animals [86]; in some cases, it is sufficient to replace a human MTS with a yeast one. Both the MTS of the corresponding yeast gene and a generic MTS can be used. After the processing of the MTS, the protein present in the mitochondria is equal to the human one. In other cases, it is necessary to replace a longer region of the human protein with that of yeast to allow complementation, creating a true chimeric protein. If chimeric complementation does not work, homologous complementation must be used. The advantage of using the yeast gene is that it can be cloned under its natural promoter, avoiding effects due to non-physiological expression. However, only amino acids which are conserved or lie in a conserved stretch between the human protein and the yeast protein can be studied. In the former case, the gene can be mutagenized to introduce the mutation, resulting in the amino acid substitution. In the latter case, if the amino acid lies in a conserved region that aligns unambiguously with its human counterpart, the yeast gene must at first be humanized, i.e., the amino acid present in the human wild-type gene must be introduced and, once it has been established that the "humanization" has no or minimal effects on the phenotype, the alleged pathogenic amino acidic variant can be introduced. The use of the homologous complementation is based on the hypothesis that, if an amino acid is conserved during evolution, or lies in a conserved region, it should carry out the same function in all the orthologous proteins. For this reason, if the substitution in the yeast gene results in an affected phenotype, it is very likely that the human substitution is also pathological; on the contrary, if no detrimental phenotype is observed in yeast, it cannot be excluded that it is not pathological in humans.

Although, in the case of homologous complementation, it is possible to introduce the mutation on the yeast genomic locus through specific techniques such as *Delitto Perfetto* or the CRISPR/Cas9 (reviewed in [87]); in most cases, the mutant strain is obtained by one-step gene disruption of the gene under analysis [88] and the transformation with wild-type or mutant alleles introduced in a cloning centromeric plasmid under its natural promoter. On the contrary, in the case of heterologous complementation, the human cDNA must be cloned in an expression vector, which can be centromeric (single copy) or episomal (multicopy), under a specific yeast promoter, and is then introduced in a null yeast mutant. The promoter can be either that of the orthologous yeast gene or, more often, an ad hoc yeast promoter. In general, a constitutive promoter, such as *CYC1* [89], *PGK*1 [90], *ADH1* [91]; an inducible promoter, such as *CUP1* [92] or *GAL1* [93]; or a highly regulatable promoter, such as TETOff or TETOn [89,94], are used. Cloning under such promoters can require a consensus sequence that optimizes the start of translation (Kozak sequence), which for yeast is (A/T)A(A/C)A(A/C)A, inserted just upstream of the start codon [95].

If the gene under analysis is fundamental for the maintenance of the mtDNA, its deletion in a haploid strain leads to mtDNA loss, thus resulting in a *rho*^{0} strain, in which reintroduction of the wild-type gene does not recover the presence of mtDNA. To overcome this problem, several techniques can be used to maintain the mtDNA before the introduction of the mutant allele, as reviewed in [85,96]. The most used technique is the plasmid shuffling strategy, wherein one-step gene disruption is performed in an *ura3* strain previously transformed with the yeast wild-type gene cloned in a centromeric plasmid harboring *URA3* as a selection marker. After the disruption of the chromosomal gene, the strain is transformed with the mutant allele cloned in a plasmid harboring a different selectable marker. Finally, the double transformant strain is treated with 5-fluoroorotic acid (5-FOA), a drug that is toxic for *URA3* strains. After this treatment, only strains that have lost the *URA3* plasmid and that contain the plasmid harboring the mutant allele can grow.

2.2. Evaluation of mtDNA Instability

Thanks to its *petite* positivity, the effects of nuclear mutant alleles on mtDNA stability can be measured through the determination of the *petite* frequency, which is the ratio between the number of *petite* colonies and the number of total colonies deriving from a population of cells. The higher the detrimental effects of the mutation on the maintenance of the integrity and on the stability of the mtDNA are, the higher the frequency of *petites* is. The *petite* frequency of a strain depends on two factors: the strain background, including the mutant allele under investigation, which affects the onset of *petites* per generation, and an extrinsic factor, which depends on the growth conditions and can influence both the onset of *petite* cells and the growth rates of *rho*$^+$ vs. *petite* cells [75]. In order to minimize the effects that interfere with the onset of the *petites*, a comparison between strains harboring the wild-type and the mutant alleles must be performed in the same genetic background, and in the same growth conditions. Although three main methods can be used to measure *petite* frequency (described in [83]), all the methods are based on a pre-growth in a medium supplemented with a non-fermentable carbon source, such as ethanol or glycerol, in order to minimize the initial percentage of *petites*, which cannot grow in such conditions. After this counterselection step, the cells are grown in liquid or, through replica plating, in solid medium in the presence of a fermentable carbon source, such as glucose, in order to allow the onset and the growth of *petite* cells. The growth is performed for several generations (at least 10–15), in order to reach a *petite* frequency that is constant and mainly due to the mutation present in the strain. Both these "bulk" methods offer the advantage that the onset of *petites* occurs independently several times, resulting in a rather constant frequency. Alternatively, a third method is based on measuring the *petite* frequency of cells deriving from single colonies; since, in this case, the number of *petites* in each colony is strongly influenced by the time of the onset of the first *petite* cell, the *petite* frequency of each colony is highly variable and thus the results must be analyzed as in a fluctuation test based on the median [84].

In order to evaluate whether a mutant allele results primarily either in the loss of mtDNA or in the onset of deletions of mtDNA, it is possible to discriminate whether the *petites* are *rho*^{0} or *rho*$^-$ cells, which mimics a situation either of depletion or of heteroplasmic mtDNA deletions, respectively, in human cells. Three main methods allow for the distinguishing of *rho*$^-$ and *rho*^{0} colonies. The first is based on crossing *petite* colonies of the mutant strain with tester strains of opposite mating type, which are harbor *mit*$^-$ in a single point mutation in a mitochondrial gene encoding for a respiratory complex subunit and thus are respiratory deficient (RD). Since recombination in yeast mitochondria occurs at a high rate, after crossing a *petite* colony with a battery of tester *mit*$^-$ strains, which should include at least mutants in *COB, COX2* e *COX3*, if at least one of the diploid strains is respiratory proficient (RP), it means that the tested colony retained an mtDNA fragment encompassing the *mit*$^-$ mutation of the tester strain and is thus *rho*$^-$. This is due to the fact that most of the *rho*$^-$ mutants retain at least one gene among *COB*, *COX2*, and *COX3* [97,98]. The pro of this technique is that several colonies can be analyzed at the same time, making more powerful statistical tests possible; however, this technique can be used only for haploid strains, and some *rho*$^-$ colonies can be categorized erroneously as *rho*^{0} if they do not contain any of the abovementioned genes. The second method is based on the extraction of mtDNA from single *petite* colonies, digestion with a restriction endonuclease and Southern Blot using an *ori* fragment as a probe, since at least one *ori* sequence is maintained in most of the spontaneous *rho*$^-$ cells [70]. The pro of this technique is that almost all the *rho*$^-$ colonies are identified correctly; however, the technique is time-consuming and can be applied only in a limited number of clones. The third method is based on the staining of mtDNA with a specific probe, such as 4′,6-Diamidine-2′-phenylindole dihydrochloride (DAPI), which, in yeast, binds both nuclear DNA and mtDNA, and on the subsequent visualization of cells through a confocal or an epifluorescence microscope. The mtDNA appears as small spots in the periphery of the cell, which are not present in *rho*^{0} cells. The pros and cons of this method are similar to those of the previous method.

2.3. Evaluation of the mtDNA Levels

Independent from the increase in the *petite* frequency, the mutant alleles of genes involved in the maintenance of the mtDNA can result in a decrease in the mtDNA levels. In this case, the strain is RP, but the mtDNA copy number is decreased. The mtDNA levels can be measured through quantitative PCR (qPCR), amplifying a region of one of the mitochondrial protein genes as a target and a region of the nuclear DNA, such as *ACT1*, as a reference [99]. For example, mtDNA levels can be measured using oligos amplifying a region of *COX1*, such as qCOX1Fw (CTACAGATACAGCATTTCCAAGA) and qCOX1Rv (GTGCCTGAATAGATGATAATGGT) [100]. A comparison of the mtDNA/nuclear DNA ratio between strains transformed with the wild-type allele and with the mutant allele allows for the evaluating of whether the mutation is associated with a decrease in the mtDNA levels. It must be underlined that, unless the mutant is *petite*-negative, such as the *aac2Δ* strain, this analysis should not be performed in the presence of a fermentable carbon source, especially if the mutant allele is associated with an increase in the *petite* frequency. In such conditions, *rho*$^-$ cells can be present, invalidating the results. As a matter of fact, several *rho*$^-$ cells contain an mtDNA fragment repeated several times, and thus the mtDNA levels could be erroneously misestimated. The analysis should thus be performed by growing the strain in a nonfermentable carbon source, where only *rho*$^+$ strains with intact mtDNA molecules can grow.

3. Genes Studied in Yeast

3.1. *MPV17/SYM1*

An intriguing protein necessary for mtDNA maintenance is MPV17, whose function was puzzling and elusive for a long time and is still not yet completely understood. Originally, MPV17 was considered as a peroxisomal membrane protein [101,102], however, in 2006, it was clearly demonstrated that it is mitochondrially localized [16].

To date, it is known that the human *MPV17* gene encodes a small hydrophobic protein of 176 amino acids embedded in the IMM and characterized by four predicted hydrophobic transmembrane domains and short hydrophilic stretches in the intermembrane space (IMS) and matrix regions [16,103].

Mutations in *MPV17* were initially identified in three families with hepatocerebral mtDNA depletion syndrome [16] and in probands with Navajo neurohepatopathy, an autosomal recessive multisystem disorder [104]. So far, 41 pathogenic variants have been reported in the *MPV17* gene on The Human Gene Mutation Database (HGMD) (http://www.hgmd.cf.ac.uk/ac/index.php, accessed on 20 October 2021). Although the clinical presentations associated with *MPV17* mutations are highly variable, hepatopathy and neurologic abnormalities are the most recurrent clinical features [105]. Symptoms generally occur in the first months of life or in infancy, although cases of adult-onset neuropathy and leukoencephalopathy or progressive external ophthalmoplegia (PEO), characterized by multiple mitochondrial DNA deletions rather than depletion, have been reported [106–108].

Based on sequence homology (48% similarity and 32% identity), as well as the similar organization of transmembrane domains, the ortholog of *MPV17* in the yeast *S. cerevisiae* is *SYM1* (Stress-inducible Yeast Mpv17), a gene identified in 2004 which is induced by heat stress and necessary for growth on ethanol at 37 °C [109]. Furthermore, the expression of human *MPV17* in *sym1Δ* null mutant complemented the phenotype, demonstrating functional homology [16,109].

Given the functional conservation, yeast was first used to demonstrate a causative role between the alleged pathological mutation identified for the first time in *MPV17* and the disease [16]. The temperature-sensitive OXPHOS phenotype of the *sym1Δ* yeast strain was rescued by re-expressing the wild-type *SYM1* gene, whereas very limited correction was observed by expressing $sym1^{R51Q}$, a variant harboring the mutation equivalent to human R50Q, and no correction was obtained with $sym1^{R51W}$ and $sym1^{N172K}$, the equivalent to human R50W and N166K, thus validating the pathogenic significance of the human mutations [16,17]. In agreement with this observation, the human R50Q, the equivalent to the yeast R51Q mutation, is associated with a milder phenotype [16]. Four additional missense mutations (G24W, P104L, A168D, and S176F, the equivalent to human mutations G24W, P98L, A162D, and S170F, respectively), localized in different protein domains, were studied in yeast, demonstrating deleterious effects for all mutations regarding OXPHOS metabolism and mtDNA stability, measured as the frequency of *petite* colonies [17]. To elucidate the molecular consequences of the mutations, protein stability and localization were assessed. The results obtained showed that all the Sym1 mutant proteins correctly localize into the mitochondria, indicating that no mutation compromises the mitochondrial target sequence, which is still unknown, and suggest that protein instability is the main molecular mechanism underlying the pathology for mutation G24W (hG24W) [17]. Studies based on blue native gel electrophoresis have demonstrated that Sym1 takes part in vivo within a high molecular weight complex, whose composition is still unknown [110]. This offered another hint for investigation; in fact, it was suggested that, for most of the mutations studied in yeast, the pathogenicity is related to the inability to form a fully assembled functional complex [17]. Notably, in both cell cultures and mouse tissues, MPV17 is part of a high molecular weight complex of unknown composition, which is essential for mtDNA maintenance in the liver of an *MPV17* knockout (KO) mouse model [111].

Beside the validation and the comprehension of the molecular mechanisms underlying the disease, yeast was also used to attempt to elucidate Sym1 protein function. In addition to the role of oxidative growth and of mtDNA stability in stressing conditions, it was demonstrated that Sym1 is necessary for glycogen storage and mitochondrial morphology [110]. In fact, ultrastructural analyses have shown that the mitochondria of the null mutant are spherical with flattened or absent mitochondrial *cristae* and some mitochondria show electron-dense bodies [110]. Similar results were obtained on mitochondria from *Mpv17*$^{-/-}$ KO mouse [112] and zebrafish [113]. Furthermore, several observations point out a defective Krebs cycle in the *sym1Δ* strain, confirmed by a reduction in succinate dehydrogenase (SDH) activity. This finding is in agreement with the reduction in glycogen accumulation, a process dependent on gluconeogenesis and therefore regulated by the anaplerotic flow of the intermediates of the Krebs cycle from the mitochondria to the cytosol [110]. Interestingly, patients with *MPV17* mutations suffer from severe, sometimes fatal, hypoglycemic crises [16,114]. The data obtained in yeast suggest that these are due to a glycogen deficiency in the liver, providing a possible explanation of this clinical phenotype [110].

The reconstitution of purified Sym1 into lipid bilayers and electrophysiological measurements demonstrated that Sym1 forms a membrane pore in the IMM, whose diameter is large enough to enable the passage of metabolites, the nature of which is not yet known [115]. Similar results were obtained with recombinant MPV17, revealing that it forms a non-selective channel with a pore diameter of 1.8 nm and suggesting the role of MPV17 as a $\Delta\psi$m-modulating channel that contributes to mitochondrial homeostasis [116]. However, the physiological role of the channel and the nature of the cargo remain elusive.

Although biochemical functions of Sym1/MPV17 remain mainly unknown, it appears to be essential for mtDNA copy number maintenance, since the loss of the function of this protein causes mtDNA depletion in patients [16] and in *MPV17* KO mice [112], and mtDNA instability in *S. cerevisiae* [16,17,110]. However, the role of MPV17 in mtDNA maintenance is not yet completely understood and several hypotheses have been proposed. The enhanced reactive oxygen species (ROS) production observed in the glomeruli of *Mpv17* KO mice suggests an involvement of MPV17 in the regulation of ROS levels [117], even if it is not clear whether the ROS increase is a consequence of an impaired OXPHOS process resulting from the reduction in mitochondrial DNA content or the cause of the mtDNA damage [118]. A decrease in mitochondrial deoxynucleoside triphosphate (dNTP) pool, observed in the liver mitochondria of rat tissues and fibroblasts derived from patients with mutations in the *MPV17* gene, and the demonstration that supplementation of dNTPs prevents and rescues mtDNA depletion in patients' fibroblasts, strongly suggest that the insufficient availability of mitochondrial dNTPs is the principal cause of mtDNA depletion [119]. At a molecular level, it seems that MPV17 supports the mitochondrial purine salvage pathway, since a decreased expression of enzymes involved in this pathway was observed in the *Mpv17* KO mouse and in patient-derived fibroblasts [119]. The hypothesis that MPV17 may be involved in mitochondrial nucleotide metabolism is also supported by the observation that the deficiency of *MPV17* orthologous gene in zebrafish results in a strong reduction in pigment cell iridophores, mainly constituted by guanine [120]. It has been proposed that the lack of MPV17 leads to a reduction in the uptake of guanosine or its phosphate derivatives, resulting in mitochondrial dysfunction and in iridophore death. Moreover, iridophore and melanophore loss in zebrafish embryos can be caused by the chemical inhibition of pyrimidine de novo synthesis [121]. Interestingly, it has been demonstrated that supplementation with dNTPs and pyrimidine precursors as orotic acid leads to a significant increase in both iridophore number and mtDNA content in *mpv17*$^{-/-}$ zebrafish mutants, thus linking the loss of MPV17 to pyrimidine de novo synthesis [113]. Furthermore, MPV17 deficiency in HeLa cells has been shown to be associated with a reduction in folate levels and with an increase in the uracil level, a marker of impaired deoxythymidine monophosphate (dTMP) synthesis, without compromising either de novo or salvage pathway. This suggests that MPV17 can provide another dTMP source and prevents uracil misincorporation in mtDNA [122], which could lead to DNA strand breaks and genome instability [123]. On the other hand, in *S. cerevisiae*, Sym1 has been related to a homeostatic control of tricarboxylic acid cycle (TCA) intermediates, such as oxalacetate and α-ketoglutarate [110].

Very recently, it was shown that the yeast mitochondria of *sym1Δ* mutant displayed a significant decrease in the levels of all the dNTPs, suggesting that, as in the *Mpv17*$^{-/-}$ mouse and in *MPV17*-deficient human fibroblasts, the instability of mtDNA in *sym1Δ* yeast mitochondria is associated with a decrease in dNTPs levels, providing strong evidence that the cause of the mtDNA deletion/depletion in Sym1 deficient cells is a shortage of precursors for DNA synthesis [40]. Furthermore, it was demonstrated that the nucleotide reduction is not limited to the mitochondrial compartment but is instead extended to the entire cell compartment. The whole-cell dNTP pool decrease could be due to a reduced ATP production caused by the OXPHOS impairment; however, OXPHOS defect and mtDNA instability occur independently in *sym1Δ* yeast cells [110]. Another tempting explanation links this dNTP shortage to the Sym1 proposed function, as TCA cycle intermediates homeostatic control [40]. Indeed, it was previously reported that the OXPHOS defect of *sym1Δ* strain was rescued by the overexpression of mitochondrial transporters of TCA intermediates (*YMC1* and *ODC1*), or by the supplementation of different amino acids (glutamate, aspartate, glutamine, or asparagine) produced from TCA intermediates and all precursors of nucleotides synthesis [110]. Finally, as yeast lacks a deoxynucleotide salvage pathway, the mitochondrial dNTP pool is exclusively dependent on cytosolic dNTPs transport; as such, the reduction in the cytosolic dNTP pool could be reflected in a decrease in mitochondrial nucleotides.

After 15 years of intense studies, many steps forward have been made to understand the role of MPV17/Sym1 in the cell: it has been shown that it forms a non-selective channel, and it has been clearly demonstrated that it is involved in maintaining a correct level of dNTPs in mitochondria and, consequently, in the stability of mtDNA; however, the nature of the cargo and the exact function of MPV17/Sym1 remain to be clarified.

3.2. *MRM2/MRM2*

MRM2 (Mitochondrial rRNA Methyltransferase 2, also known as *RRMJ2* or *FTSJ2*) encodes for a mitochondrial 2′-O-ribose methyltransferase, which is required for a correct maturation of mitochondrial rRNA. In particular, MRM2 mediates the methylation of U(1369) located in the A-loop of the 16S rRNA, the core of the large mitochondrial ribosome subunit [124–126]. Recently, a structural approach provided essential insights into the last steps of the large mitoribosomal subunit biogenesis pathway [127]; however, the exact role of the RNA modifications is largely unknown.

U1369 is in the peptidyl transferase center and is implicated in the interaction of the ribosome with a tRNA in the aminoacyl(A)-site. 2′-O-ribose methylation of uridine 1369 was shown to be critical for proper mitochondrial translation and, consequently, for mitochondrial respiratory function [126]. In fact, siRNA knockdown cell lines showed a reduction in mitochondrial protein synthesis flanked by a reduction in the oxygen consumption rate (OCR) [126]. However, further studies are required to elucidate the exact functional role of the U1369 modification. In bacteria, 2′-O-ribose methylation of the A-loop uridine is not only required for proper ribosome biogenesis, but also plays a role in regulating translational accuracy [128]. Interestingly, MRM2 was found to be in the proximity of mtDNA nucleoids in mouse cell-cultures [124].

In 2017, a pathological mutation in *MRM2* was associated with childhood-onset of rapidly progressive encephalomyopathy and stroke-like episodes [18], underling the importance of rRNA methylation for mitoribosomal function. Multiple OXPHOS defects and a decreased mtDNA copy number were detected in muscle homogenate [18]. On the contrary, primary fibroblasts derived from the patient did not recapitulate the mitochondrial phenotypes, possibly due to tissue-specificity, pushing the authors to use the yeast *S. cerevisiae* to study the pathogenicity, thanks to the presence of the ortholog gene *MRM2*. It was previously demonstrated that not only is the human MRM2 highly similar to yeast Mrm2 on the sequence level, but that it also retained a functional homology during ribosome biogenesis [126]. The yeast Mrm2 protein is responsible for the 2′-O-ribose methylation of U2791 in 21S rRNA [129], which is equivalent to human U1369 of 16S [126]. Deletion of the *MRM2* gene led to a thermosensitive oxidative growth defect and rapid loss of mtDNA [129].

Notably, the yeast model expressing the G259R variant, the equivalent to human substitution G189R, showed a significant reduction in respiratory activity at the permissive temperature (28 °C) that was further exacerbated at the stressing temperature (37 °C), thus validating the pathological significance of this mutation [18]. Furthermore, through a radioactively labeled reverse transcriptase primer extension (RT-PEx) reaction analysis, it was shown that the substitution of the conserved residue results in a diminished Um2791 modification in yeast mitochondrial 21S [18], thus providing an insight into the molecular mechanism underlying the dysfunction.

3.3. *OPA1/MGM1*

OPA1 (Optic Atrophy 1) encodes for a mitochondrial dynamin-like GTPase mainly involved in mitochondrial fusion, *cristae* integrity, mtDNA stability, and copy number maintenance [130–133]. Beside these roles, OPA1 is also implicated in apoptosis regulation [134,135], in mitochondrial quality control [136,137], and in mitochondrial homeostasis regulation [19,138,139].

OPA1 is composed of an N-terminal MTS followed by a transmembrane domain (TM), a coiled-coil domain and three highly conserved dynamin constituents: a GTPase domain, a middle domain, and a coiled-coil GTPase effector domain (GED) [140]. Alternative splicing of exons 4, 4b, and 5b leads to eight OPA1 variants with a tissue-specific pattern of expression [141]. These variants are required to finely tune and to provide flexibility of mitochondrial dynamics under different cellular conditions [140]. The cleavage of the MTS produces long isoforms collectively called long-OPA1 (l-OPA1), which are anchored to the IMM and are essential for mitochondrial fusion and *cristae* organization [142,143]. Through a mechanism known as alternative topogenesis, about half of the long isoforms are then subjected to a proteolytic process in the rhomboid cleavage region (RCR) located before the GTPase domain to generate the short isoforms, called short-OPA1 (s-OPA1), which are devoid of the TM segment [144] and localized in the IMS. In addition, it is known that the short forms interact with some subunits of the mitochondrial contact site and *cristae* junction organization system (MICOS) to maintain the integrity of *cristae* junctions [140,145]. A specific ratio of l- and s-forms (2 long:2 short or 1 long:2 short) is necessary for an efficient mitochondrial fusion; in fact, an unbalancing toward the l-forms leads to an increase in mitochondrial network fragmentation [140].

OPA1 mutations are associated with dominant optic atrophy, one of the most common inherited optic neuropathies, which is characterized by the degeneration of the retinal ganglion cells (RGCs) and by an insidious onset of visual impairment in childhood, with moderate to severe loss of visual acuity [21,146–148]. To date, hundreds of pathological mutations have been identified (https://databases.lovd.nl/shared/variants/OPA1/unique, accessed on 20 October 2021) as the cause of Dominant Optic Atrophy (DOA). *OPA1* mutations associated with a DOA cluster mostly in the GTPase domain and are caused mainly by substitutions and by single deletions and insertions generating haploinsufficiency [149,150]. Some DOA patients that harbor missense *OPA1* mutations with a severe dominant-negative effect due to the interference of the mutant variant with the wild-type one display a syndromic form of DOA named DOA plus or DOA+. In these patients, which represent about 30% of those with DOA, optic atrophy in childhood is followed by chronic progressive external ophthalmoplegia (PEO), ataxia, sensorineural deafness, sensory-motor neuropathy, myopathy, and mtDNA multiple deletions in adult life [130,131].

The *OPA1* role in mtDNA maintenance has been clarified thanks to the functional homology with the orthologous gene *MGM1* (Mitochondrial Genome Maintenance) of the yeast *S. cerevisiae*. Mgm1 is a dynamin family member protein that was first discovered in *S. cerevisiae* in a genetic screening for nuclear genes required for the maintenance of mtDNA [151]. Subsequent studies indicated an additional role for Mgm1 in IMM fusion and in the formation and maintenance of *cristae* structure [152]. As for OPA1, Mgm1 exists in two forms: long (l-Mgm1) and short (s-Mgm1). The first is an integral IMM protein that spans the inner membrane, thanks to an N-terminal transmembrane segment (TD), while the second is a short soluble isoform present in the IMS [153]. A 1:1 ratio of l- to s-forms is fundamental for correct mitochondrial morphology and function; in fact, the overexpression of l-Mgm1, increasing this ratio, leads to mtDNA loss and mitochondrial network fragmentation [154]. The *mgm1Δ* mutant strain, which contains highly fragmented mitochondria because of the absence of the fusion machinery, is unable to segregate mtDNA and, after a few generations, loses mtDNA and becomes RD.

As for many genes causing mtDNA deletions, the use of patients' fibroblasts for studying mtDNA instability has some limitations, and model systems are needed [155]. Besides mammalian cell models and animal models, in recent years, yeast has been proven to be a useful model for evaluating in short times the pathogenicity and the dominance of novel mutations [155]. Although OPA1 and Mgm1 have a similar predicted structure, the sequence identity is limited to approximately 20%, and *OPA1* cannot complement the deletion of *MGM1*. To overcome this problem, which would prevent the study of most mutations, a chimeric complementation approach was used to study the effect of several pathogenic *OPA1* mutations in yeast [20,21]. Six different chimeras with different portions of Mgm1 and OPA1 regions were constructed, and complementation studies demonstrated that one of the chimeras, named *CHIM3*, was able to partially complement the *mgm1Δ* OXPHOS phenotypes. Furthermore, the aberrant mitochondrial morphology of *mgm1Δ*, due to a mitochondrial fusion deficit and not to the rho^0 condition, was partially rescued when *CHIM3* was expressed [20]. *CHIM3* encodes for the MPS, the TM, and the RCR of Mgm1 fused to the catalytic region of OPA1 (GTPase domain, middle domain, and GTPase effector domain), and was cloned under the TEToff promoter in a single copy vector [20]. A similar construct was used in [21].

To validate *CHIM3* as a suitable model with which to study *OPA1* pathological mutations, three well-known missense mutations were introduced in *CHIM3*: I382M associated to DOA but is also present in healthy subjects and was proposed to be a phenotypic modifier [156–158]; R445H associated to DOA plus [159]; and K468E associated to DOA [160]. In agreement with the severe pathological role of R445H and K468E substitutions, the corresponding yeast mutant strains showed a deficient respiratory phenotype and lacked tubular mitochondria. Moreover, the ratio between the s- and the l- forms is altered in both mutant strains, suggesting that the processing of l- and s-forms is impaired and that an alteration of the ratio between the l- and s-OPA1 could also be the cause of the disease. In contrast, the strain carrying $chim3^{I382M}$ mutant was able to complement the deletion of *MGM1*, showing only a 1.5-fold increase in the *petite* frequency, a slight decrease in oxidative growth, and a 20% reduction in respiratory rate as well as of some respiratory complex activities [20], supporting the hypothesis that I382M acts as phenotypic modifier, contributing to the worsening of the phenotype when in compound with another mutation [157].

Moreover, diploid heteroallelic models, harboring a wild-type allele and a mutant allele, also provide a useful approach for detecting the mechanism of dominance, i.e., loss-of-function (that, in humans, is associated with dominance by haploinsufficiency) or gain-of-function [19,20]. In this regard, heteroallelic strains expressing the I382M or K468E mutations and one copy of wild-type *CHIM3* ($CHIM3/chim3^{I382M}$ or $CHIM3/chim3^{K468E}$) displayed a normal oxidative phenotype as with the homoallelic strain (*CHIM3/CHIM3*), indicating a loss of function mutations, while the heteroallelic strain carrying the R445H mutations ($CHIM3/chim3^{R445H}$), associated to DOA plus, showed an oxidative growth defect indicative of a dominant negative mutation. Further studies, carried out using these yeast models, revealed that most of the missense mutations associated with DOA or DOA plus in the haploid strain cause the inability to grow on oxidative carbon sources due to the loss of mtDNA, whereas, in the diploid strain, they can cause a reduction in oxidative growth, suggesting that the mutant variants interfere with the activity of wild-type protein and are partially dominant-negative. It has been proposed that the severity of the phenotype as defined by pure DOA or DOA plus depends either on the degree of interference of the mutant protein on wild-type one or other factors such as the co-presence of modifying variants in other genes as well as environmental factors or age/gender of the patients [19]. Finally, the fact that the relative amount of l- and s-Mgm1 was altered in most mutant strains indicates that an increased ratio could be used as an additional indicator of the pathogenicity of the mutations.

3.4. *POLG/MIP1*

POLG encodes for the DNA polymerase γ, the catalytic subunit of the only mitochondrial replicase identified in animal mitochondria [161,162]. This enzyme was identified in 1970 as an RNA-dependent DNA polymerase in human Hela cells and represents only 1–5% of the total cellular DNA polymerase activity [163,164]. It is involved in replication, recombination, and the repair of mtDNA [165–169]. In mammals, the DNA polymerase γ acts as a complex containing three subunits: a catalytic subunit of 140 kDa encoded by *POLG* and two accessory subunits of 55 kDa encoded by *POLG2*, that enhance the processivity of the enzyme [170]. The catalytic subunit is composed of three domains: the N-terminal domain (residues 170-440), with a $3' \rightarrow 5'$ exonuclease activity involved in the proofreading activity; the C-terminal domain (residues 440–475 and 785–1239), characterized by polymerase activity responsible of the synthesis of mtDNA and a spacer region (residues 475–785) [169,171–174]. The spacer region is divided into two subdomains: the former is responsible for the intrinsic processivity whereas the latter, which interacts with the POLG2 subunit, is responsible for the enhancement of the processivity [175–177].

To date, more than 300 pathogenic mutations of *POLG* have been reported (http://tools.niehs.nih.gov/polg/, accessed on 20 October 2021), and pathologies caused by these mutations can be divided into two main groups: pathologies associated with mtDNA depletion and pathologies associated with multiple mtDNA deletions. The former leads to diseases with infancy to childhood-onset and is characterized by the involvement of many tissues and organs, whereas the latter leads to diseases with adolescence to adulthood-onset and is characterized by the involvement of a limited number of tissues (reviewed in [178]). Among the diseases characterized by mtDNA depletion, there is the lethal childhood myocerebrohepatopathy (MCHS), which affects infants and is characterized by development delay, myopathy, hepatic failure, pancreatitis, acidosis, and occurs generally in the first months of life [179]; the Alpers–Huttenlocher syndrome (AHS), characterized by childhood-onset, and is responsible for severe encephalopathy with liver failure and intractable epilepsy [180]; a MNGIE-like disease that has a variable onset and is characterized by gastrointestinal dysmotility, ptosis, myopathy, and sensory neuropathy [181]. Other syndromes associated with mtDNA deletions and characterized by epilepsy and ataxia include MERRF, MELAS, MEMSA, SCAE, and SANDO [182–185]. Adult-onset PEO is the most frequent mitochondrial pathology characterized by multiple mtDNA deletions associated to mutations in *POLG* [22,169]. Whereas all the previous pathologies are inherited through an autosomal recessive inheritance, PEO can be recessive (arPEO) and is characterized by the progressive weakness of the extraocular muscles which determine ptosis and ophthalmoparesis, or dominant (adPEO), and is characterized by axonal neuropathy, ataxia, depression, parkinsonism, and hypogonadism [186]. Most of the substitutions causing adPEO are in the polymerase domain, while recessive mutations are found throughout the whole protein. Altogether, mutations in *POLG* represent the main cause of mitochondrial diseases with Mendelian inheritance [178]. Due to the high frequency of mutations and SNPs in *POLG*, several patients harbor three or more mutations and SNPs, and, for this reason, it is often difficult to understand which mutation(s) are the cause of the pathology and whether the mutation(s) are dominant or recessive.

MIP1 (MItochondrial Polymerase 1) is the orthologous gene of *POLG* in the yeast *S. cerevisiae* and encodes for a protein of 140 kDa [187]. Mip1 shows a 45% similarity with its human counterpart. However, the similarity is not homogeneously distributed along the protein, but it is higher in the exonuclease and in the polymerase domain [35,187]. Mip1 is also characterized by a C-terminal extension (CTE), which is specific to fungal polymerase γ, and is variable among the species and reaches the maximum length in species of the *Saccharomyces* genus, where it plays a key role in balancing the exonuclease and the polymerase activities [188–190].

Yeast has proven to be an excellent genetic system to obtain information and validate in vivo the pathogenicity of *POLG* mutations (reviewed in [187]). Several substitutions found in patients affected by *POLG*-related diseases have been introduced in the corresponding position of Mip1 in a *mip1Δ* strain. Most of the mutations have been studied through homologous complementation and, since the deletion of the chromosomal *MIP1* gene leads to the loss of mtDNA, the introduction of *mip1* mutant alleles has been obtained principally by plasmid shuffling on 5-FOA. A few mutations have been studied through chimeric complementation by replacing *MIP1* with human *POLG* and *POLG2* containing the MTS of Mip1 [35]. Besides the determination of the *petite* frequency, discrimination of the nature of the *petites*, and measurement of the mtDNA levels, yeast has been used to evaluate whether substitutions are associated with an increase in mtDNA point mutations and to specific biochemical defects. Various information has been obtained by using yeast models (reviewed in [187]): (i) The higher the *petite* frequency associated with a *MIP1* mutations is, the higher the *rho*^{0} frequency among the *petite* colonies is, indicating that the most severe mutations result primarily in the loss of mtDNA; (ii) Some mutations increase the frequency of mtDNA point mutations, but this increase does not correlate with the increase in the *petite* frequency, indicating that the mechanisms are different. Thus, the increase in the point mutations does not seem to be involved in the development of the pathology. However, a decrease in the replication fidelity may influence the progression of the disease. Interestingly, several mutations in the exonuclease domain strongly increase the *petite* frequency, indicating that the exonuclease domain interacts with the polymerase domain and is fundamental for a proper replication process and to avoid deletions; (iii) Exposure to the exogenous base-alkylating agent methyl methanesulfonate increased the frequency of mtDNA point mutations when mutations in *MIP1* are present, suggesting that mutagenic molecules may negatively affect the progression of the pathology; (iv) Oxidized bases, which are produced in the mitochondria mainly by the presence of ROS, may play a role in the increase in mtDNA instability when some *mip1* mutant alleles are present. Indeed, in several *mip1* mutant strains, the *petite* frequency is decreased by the treatment of antioxidant molecules such as lipoic acid and MitoQ; (v) When more than two mutations were present in patients, yeast allowed to discriminate which mutations are the cause of the pathology; (vi) Yeast allowed to discriminate whether the heredity of a mutation is recessive or dominant; (vii) Moreover, thanks to the results obtained in yeast, two hypotheses can explain the dominance of some mutations in the polymerase domain. Several mutant variants have a similar DNA binding affinity but have no polymerase activity. The mutant variant binds the DNA with the same affinity, thus blocking the replication and preventing at the same time the binding of the wild-type enzyme. Alternatively, some mutant variants may directly introduce lesions to mtDNA, such as oxidized bases; (viii) It has been found that some SNPs are not neutral, but rather phenotypic modifiers, which can worsen the phenotype associated with a mutation; (ix) Although most of the SNPs are neutral and do not affect the *petite* frequency and the frequency of mtDNA point mutations, mutant variants harboring some of these SNPs are much more sensitive to the nucleoside analogs reverse transcriptase inhibitors (NRTIs) used to treat HIV infection, such as stavudine and zalcitabine, or valproic acid, used to treat epilepsy; (x) Almost all the pathological mutations are associated with an increase in the *petite* frequency. In addition, different mutations have a different severity, and the phenotypic defect on mtDNA stability induced by a mutation grossly parallelizes with the severity of the pathology, making yeast a suitable model for predicting the severity of a pathology; (xi) Mutations can affect the function of Mip1 in several ways: some mutations reduce the protein stability, especially at higher temperatures, other mutations reduce the catalytic activity, the processivity, the DNA binding affinity, the affinity for the incoming dNTPs, or the specific constant, whereas other mutations cause an increase in the ratio between the exonuclease activity and the polymerase activity. On the contrary, the exonuclease activity is only slightly or not at all reduced, even in the case of mutations in the exonuclease domain, indicating that the defects of mtDNA replication are not due to defects of the proofreading activity.

Moreover, it was observed that the overexpression of *RNR1* or the deletion of *SML1*, whose function is reported in Section 3.5, reduces the *petite* frequency in most mutant strains harboring pathological mutations by increasing the concentration of dNTPs [23,27,29,30,191]. The lower the mtDNA instability is, the greater the effect of the *RNR1* overexpression or the *SML1* deletion is, indicating that mutant variants that retain most of their catalytic activity benefit more by the increase in the concentration of the dNTPs. The rescuing activity exerted by the dNTPs observed initially in yeast explains the observation that supplementation to the myotubes of patients harboring mutations in *POLG* with specific concentrations of dNMPs, which are easily converted to dNTPs, lead to an almost complete normalization of the mtDNA levels [192].

The increase in the *petite* frequency and the point mtDNA mutability caused by Mip1 mutations are rescued also by the overexpression of DNA polymerase ζ, whose subunits are encoded by *REV3* and *REV7* genes, and by the deoxycytidyl transferase, encoded by the *REV1* gene. Both enzymes are involved in the error-prone translesion synthesis and localize also in mitochondria both in yeast and, though limited to Rev3, in humans [30,193,194]. It was shown that *MIP1* mutations rescued by DNA polymerase ζ overexpression are not recovered by the treatment with antioxidant molecules and vice versa, suggesting two different mechanisms of rescue. The decrease in *petite* frequency induced by overexpression of DNA polymerase ζ could be due to its capacity to partially replace Mip1 variants which mainly decrease the catalytic activity, whereas the decrease in the mtDNA instability induced by antioxidant drugs could be due to a decrease in the concentration of oxidized bases that can be incorporated by other Mip1 variants.

3.5. *RRM2B/RNR2*

RRM2B encodes for the Ribonucleotide Reductase Regulatory TP53 Inducible Subunit M2B, which is able to associate with the large subunit R1 to form an active ribonucleotide reductase complex (RNR), which catalyzes the synthesis of deoxyribonucleoside diphosphates from ribonucleotides diphosphates. Its expression is essential for DNA repair and mtDNA synthesis in postmitotic cells, since, in cycling cells, another small subunit called R2 interacts with R1 to form another type of RNR complex [195]. Due to the key role of the RNR complex in the biosynthesis of dNTPs, the limited availability of DNA building blocks for mtDNA synthesis results in mtDNA a maintenance defect when *RRM2B* is deficient. Since the first identification of an *RRM2B* mutation by Bourdon et al. in 2007, about 40 different mutations have been described until now [196]. The clinical manifestations associated with *RRM2B* mutations are very heterogeneous both in terms of severity of symptoms and the age of onset. Mutations in *RRM2B* can result in mtDNA depletion associated with a severe and childhood-onset multisystemic disease [130,197–201] or in the accumulation of multiple mtDNA deletions associated with a milder adult-onset PEO [202]. In the first case, pediatric patients manifest muscle hypotonia and weakness, often associated with severe respiratory distress; the disease progresses very quickly and causes death in a few months from the onset of symptoms [203]. In the second case, patients are clinically characterized by ptosis and weakness of the extraocular muscles [202].

The disease pathogenesis can be caused by mutations that affect amino acids involved in iron binding (crucial for the catalytic activity of the enzyme) amino acids that are essential for the conformation and stability of the active site, or amino acids that allow the interaction of R2 with an R1 subunit, thus interfering with RNR assembly [197,204,205].

In yeast, the large subunit R1 of RNR is encoded by *RNR1* and *RNR3*, while the small subunit R2 is encoded by *RNR2* and *RNR4*. As in other organisms, RNR is a heterotetramer containing two R1 subunits and two R2 subunits. The main isoform present in yeast is (Rnr1)2-Rnr2-Rnr4; whereas expression of *RNR2–4* is constant during the cell cycle, *RNR1* is upregulated before the S phase [206–211]. Rnr1 is thus the limiting factor for the assembly of the RNR complex. In addition, Rnr1 is inhibited by Sml1, a protein that binds Rnr1, blocking its assembly with the R2 subunits [212,213]. Concerning the R2 subunits, Rnr2 has a catalytic role, while Rnr4 folds correctly and stabilizes the radical-storing Rnr2 in an Rnr2-Rnr4 heterodimer which localizes in the nucleus during most of the cell cycle but undergoes cytoplasm redistribution during the S-phase in order to form the full RNR complex [214–217]. Whereas deletion of *RNR2* makes the strain inviable, the deletion of *RNR1* or *RNR4* makes the strain viable but devoid of mtDNA (www.yeastgenome.org).

Human RRM2B and its yeast orthologue Rnr2 share 55% identity and 78% similarity. In order to create a yeast model useful for the search of suppressors or drugs capable of rescuing the phenotype of mutations in *RNR2*, four mutations, equivalent to human substitutions R41P, I224R, M282I, and L317V were introduced in *RNR2*. All the mutant variants increase both the frequency of *petite* colonies and, among these, the frequency of *rho*^{0} colonies, at 37 °C, although at different extents. For all the mutant strains, the increase in the *petite* frequency was partially rescued by the overexpression of *RNR1* and, to a lesser extent, of *RNR4*. Combined with the fact that mutant strains are not fully devoid of mtDNA, and thus Rnr2 mutant variants retain part of their catalytic activity, this observation suggests that an increase in the RNR levels and/or its stabilization could improve the synthesis of the dNTPs, despite the detrimental substitution present in Rnr2 (Baruffini and Dallabona, personal observations).

The most detrimental mutation was L362V, equivalent to human L317V, which increased the *petite* frequency to ~60% [40]. The mutant strain harboring this mutation was used to evaluate the capability of some drugs to reduce the mtDNA instability, as reported in Section 4.

3.6. *SLC25A4 (ANT1)/AAC2*

ANT1 (Adenine Nucleotide Translocase), also known as *SLC25A4*, encodes for an ADP/ATP transporter, one of the most abundant proteins of IMM. Although the primary function of ANT1 is fully understood, the link with mtDNA instability is highly debated and far from being clarified.

In humans, there are four ANT isoforms with no functional difference but differential tissue expression [218,219]. *ANT1* is highly expressed in post-mitotic cells and encodes for the most abundant isoform in heart and muscle [220]; *ANT2* (*SLC25A5*) is expressed at a low level in differentiated tissues and is particularly abundant in proliferative, undifferentiated cells [221]; *ANT3* (*SLC25A6*) is ubiquitously expressed at variable levels depending on oxidative metabolism; *ANT4* is exclusively expressed in liver, testis, and brain [219].

ANT1 belongs to the large family of mitochondrial carriers [222,223], presenting a six alpha-helices transmembrane domain that forms a nucleotide translocation channel [224]. Although ANT1 has been thought to function as homodimers for decades [225–227], more recent observations have challenged this point of view and a monomeric functional unit was proposed [228–230]. Its primary function is to mediate the 1:1 exchange of ATP and ADP across the membrane, importing cytosolic ADP into the mitochondrial matrix to fuel the ATP production by ATP synthase (Complex V) and exporting to the cytosol the ATP produced by the OXPHOS process and is necessary to support all cellular activities. However, in vitro experiments show that this direction is achieved when cells possess a normal membrane potential that drives productive transport (ATPout, ADPin). If the membrane potential is altered, ANT can mediate non-productive ADP/ADP or ATP/ATP exchange or counterproductive (ATPin, ADPout) transport [231].

However, the role of ANT is not limited to ADP/ATP transport; in fact, it has a role in the regulation of the mitochondrial permeability transition pore [232,233] and mitochondria-mediated apoptosis [234], in mitophagy [235], and in proton loss across the IMM, thus mildly uncoupling the membrane and avoiding the hyperpolarization and overproduction of ROS [236].

The first disease-linked substitution in the *ANT1* gene (A114P) was identified more than two decades ago [47], associated with adPEO. Since then, a total of ten missense mutations have been identified in *ANT1*: five (A90D, L98P, D104G, A114P, and V289M) are dominant and were found in patients affected by adPEO, clinically characterized by ptosis and impairment of eye movements [47,237–240]; two (A123D and R236P) are recessive loss-of-function mutations and have been found in subjects affected by mitochondrial myopathy and cardiomyopathy [41,241]; three are de novo dominant mutations associated with severe non-adPEO disease (R80H and R235G) [49] or with a mild myopathy (K33Q) (King et al., 2018). Despite the wide spectrum of clinical presentations, all patients share a molecular feature, i.e., mtDNA defects in affected tissues. Patients affected by adPEO and patients carrying homozygous recessive mutations present an accumulation of multiple mtDNA deletions whereas patients carrying the de novo mutations show mtDNA depletion.

The assessment of the impact of *ANT1* dominant mutations in mammals is hindered by the absence of suitable models. In fact, the most used human cell lines, fibroblasts, express the *ANT1* gene at a very low level, and its exogenous expression induces apoptotic cell death [47,242]. Contrariwise, mouse myotubes express naturally high levels of *ANT1*, therefore it was possible to exogenously express mutant alleles. This model was used to assess the effect of some dominant mutations on ANT1 transport activity, demonstrating a decreased ADP/ATP exchange function and abnormal translocator reversal potential; however, no mtDNA deletions or depletion were observed [243].

The *ANT1* gene is highly conserved in all eukaryotic organisms, including the yeast *S. cerevisiae*. In yeast three genes coding for the mitochondrial ATP/ADP carrier (*AAC1*, *AAC2*, *AAC3*) have been identified [244–247], and, among these, *AAC2*, encoding the major isoform of the translocator and the only one required for respiratory growth [245], is considered the yeast ortholog of human *ANT1*. Worth considering is that the *aac2*Δ null mutant is *petite*-negative [248], i.e., it is not viable in the absence of mtDNA. Contrariwise to *rho*$^+$ respiring cells, in which mitochondrial membrane potential is generated by respiratory complexes proton pumping, in cells lacking mtDNA (*rho*^{0} cells) in which respiration is impaired, a minimal membrane potential is maintained by the electrogenic exchange of cytosolic ATP (containing four negative charges) for mitochondrial ADP (containing three negative charges) by means of Aac2 reversing the transport direction [249,250]. Even if respiration and mitochondrial DNA are dispensable in *S. cerevisiae*, the mitochondrial membrane potential is essential for viability and therefore the absence of both Aac2 function and mtDNA (and thus proton pumping activity) leads to lethality.

The presence of the *ANT1* ortholog gene in yeast gave the possibility to exploit yeast as a model system; in fact, studies on the pathogenic mechanism of *ANT1* mutations were mostly carried out in this organism.

Homologous complementation studies in which the mutant allele was introduced into the null mutant allowed to analyze the effects of each pathogenic mutation on mitochondrial functions, for example through oxidative growth analysis, oxygen consumption measurements, respiratory cytochromes content quantification, ADP/ATP transport activity analyses [41,43,47,49]. In addition, for a non-conserved mutation, a y*AAC2*/h*ANT1* chimeric construct was used [44]. As the haploid yeast with nonfunctional Aac2 is *petite*-negative, a direct measurement of mtDNA instability is not possible. In an attempt to demonstrate a link between the pathological mutation A123D and mtDNA instability, Palmieri and colleagues proposed a viability test as an indirect measure, under the assumption that a lethal phenotype could result from deletion(s) or loss of mtDNA. Indeed, yeast expressing the equivalent *aac2*^{A137D} mutant allele showed a very severe viability loss [41]. Interestingly, treatment with two well-known antioxidant drugs, dihydrolipoic acid and

N-acetyl cysteine, partially rescued the viability decrease in the mutant strain, suggesting that oxidative stress could have a role in the pathogenesis of mtDNA damage [41].

However, since most mutations are dominant, the best model for studying these mutations must be considered as a strain carrying both a wild-type and mutant copy of the gene. Such a model was created by introducing mutations into a wild-type strain, thus mimicking human heterozygous condition, giving rise to heteroallelic haploid strains. The evaluation of these strains allowed clarifying some aspects, including the effect on mtDNA stability. For example, the phenotypic characterization of heteroallelic strains carrying $aac2^{R96H}$ or $aac2^{R252G}$ alleles not only served to demonstrate that the mutations are dominant, but also suggested that the dominance of R80H and R235G in *ANT1* is likely due to gain-of-function [48]. Notably, all the heteroallelic strains so far studied recapitulate the mtDNA instability found in patients, as demonstrated by an increase in the frequency of *petite* colonies [43,48].

Extensive studies were also carried out in yeast in an attempt to clarify the link between ANT1/Aac2 dysfunction and mtDNA instability, leading to different hypotheses. The ANT1 mutant variant may have a reduced activity in ATP/ADP translocation across the IMM that subsequently causes an imbalance in the mitochondrial deoxynucleotide pool. Consequently, it would affect the accuracy of mtDNA replication, thereby leading to the accumulation of mutant mtDNA [243]. Altered intramitochondrial ATP levels in the matrix due to reverse ADP/ATP exchange or a defect in nucleotide transport could cause electron transport stalling, increased ROS production, and consequent oxidative mtDNA damage [41]. Formation of an unregulated channel on the IMM, induced by a mutated form of Ant1, rather than a defect in ATP/ADP translocation, could be the primary pathogenic factor in human adPEO. Accumulation of mtDNA mutations may be a consequence of the loss of mtDNA replication precursors following membrane permeabilization [42]. More recently, it was proposed that misfolding and propensity to form large aggregates by Aac2 mutant could generate stress on the membrane, affecting mitochondrial biogenesis, particularly causing severe damage to the electron transport chain assembly and mtDNA integrity [45,251,252]. Nevertheless, the experiments performed so far are not yet conclusive, and none of these hypotheses seem to be totally clarifyied for now. It is not to be excluded that different mechanisms could coexist, giving a contribution to the pathophysiological mechanism.

4. Use of the Yeast Models for the Identification of Drugs by Means of a Drug Repurposing Approach

Currently, no approved treatments for mitochondrial diseases associated with mtDNA depletion and multiple deletions are available [253]. In recent years, yeast has been used as a model organism for the development of phenotypic screenings to identify molecules capable of suppressing mitochondrial disease-related phenotypes, including mtDNA instability. In 2011, an assay using the yeast *S. cerevisiae* to rapidly identify potential drugs that could be used to treat mitochondrial diseases was developed. This assay, called drug drop test, allows a high throughput analysis of chemical libraries in a reasonably short time. This approach exploits the respiratory-deficient phenotype of specific strains harboring mutations that affect the OXPHOS activity. The mutant strain is plated onto a medium containing a non-fermentable carbon source on which paper disks are placed; a different molecule is then deposited on each disk, using a disk for the negative control (the solvent in which the molecules are dissolved). The plate thus prepared is incubated at the temperature of interest and growth is monitored for several days. Since each drug diffuses from the disk forming a gradient, the great advantage of this technique is that it allows evaluating the effect of the tested molecules at different and continuous concentrations, avoiding the need to determine the effective and non-toxic dilutions. In particular, a molecule gives rise to a halo of growth around the filter if it is able to rescue the phenotype caused by the mutation harbored by the strain [254].

This approach can be used either to test chemical libraries containing novel drugs (natural and/or synthetic) or drugs previously approved by specific agencies such as the Food and Drug Administration (FDA) and the European Medicines Agency (EMA) for the treatment of other diseases. In the latter case, a drug repurposing approach, that is the investigation of existing drugs for new diseases, is used. This approach allows for a considerable reduction in times and costs, since the safety and pharmacological parameters have been already assessed [255]. It is based on the fact that several drugs have secondary targets or, alternatively, that the primary target inhibited by the drug participates in a pathway involved in the onset of the disease. Due to these characteristics, drug repurposing is an optimal approach, especially in the field of rare diseases, including mitochondrial ones [256].

Concerning mitochondrial diseases associated with mtDNA instability, the drug repurposing approach has been used to find drugs able to rescue the oxidative growth defects of *MIP1*, *MGM1*, *AAC2*, *SYM1*, and *RNR2* mutant strains.

The $mip1^{G651S}$ mutant, carrying the mutation equivalent to *POLG* G848S, shows a defective oxidative growth compared to the wild-type *MIP1* at a non-permissive temperature (37 °C) due to the high instability of the mtDNA (*petite* frequency > 99%). Through the screening of ~1500 drugs, clofilium tosylate, an anti-arrhythmic drug, was found as a positive hit [257] able to decrease the *petite* frequency of a plethora of *mip1* mutant strains, to increase the protein levels of wild-type and mutant variants, and to partially restore the respiratory activity of the mutant strains. Further analyses showed that clofilium tosylate is able to rescue the phenotypes caused by mutations/deletions in *POLG* in the worm *Caenorhabditis elegans*, in zebrafish, and in patients' fibroblasts, making it suitable as a potential treatment for *POLG*-associated pathologies [257,258].

The $mgm1^{I322M}$ mutant, harboring the hypomorphic mutation equivalent to I382M mutation in *OPA1*, shows a limited oxidative growth associated with both a strong defect in the maintenance of mtDNA (*petite* frequency > 95%) and a low respiratory activity (~5% compared to wild-type strain) at a non-permissive temperature (37° C). Through the screening of ~2600 drugs of two different chemical libraries, among which a library containing 1018 FDA-approved drugs and a library containing 1596 molecules with a high degree of chemical diversity, 42 positive hits were identified. Among these, 26 molecules were also able to partially recover the thermosensitive growth defect of a chimeric mutant strain harboring the *OPA1* S616L pathological mutation. Further analyses showed that six of them, able to decrease the mtDNA mutability in yeast, were able to improve the pathological phenotype of $opa1^{-/-}$ mouse embryo fibroblasts expressing the human *OPA1* isoform carrying two mutations, the R445H or D603H, associated with DOA plus and DOA, respectively. Interestingly, different drugs rescue different defects induced by mutations in *OPA1*, such as mitochondrial morphology, cell viability, or energetics, suggesting thatthe rescuing mechanisms of each drug are different. Analysis performed on DOA patients' fibroblasts allowed to select the most promising molecule, tolfenamic acid, to be translated in a clinical trial for DOA or other neurodegeneration linked with *OPA1* mutations [259].

The haploid $aac2^{M114P}$ mutant, which carries the mutation equivalent to *ANT1* L98P found in adPEO patients, exhibits a defective respiratory growth, making it possible to search for molecules that restore this defect. The screening of 1018 FDA-approved compounds led to the identification of five positive hits, able to bring the level of oxygen consumption rate of the $aac2^{M114P}$ mutant strain to the wild-type level. Furthermore, these molecules were able to restore the respiratory activity and to reduce the mtDNA instability in the heteroallelic $AAC2/aac2^{M114P}$ strain, which mimics the human heterozygous condition of adPEO patients, and in the heteroallelic strain carrying the R96H mutation equivalent to the de novo dominant missense mutation R80H, associated with a more severe disease, thus expanding the possible applications for the treatment. Positive results on two drugs, albeit preliminary, were also obtained in *C. elegans*, indicating that these drugs identified in yeast are also beneficial in a multi-organ animal model and can be potentially applied to humans [260].

The thermosensitive yeast mutant $sym1^{R51W}$, harboring the mutation equivalent to *MPV17* R50W, was exploited to identify potential therapeutic molecules for MDDS caused by mutations in *MPV17*. Through screening on a library containing 1018 FDA-approved drugs and on other six molecules previously identified as positive on another yeast model of MDDS, 10 drugs were found as positive hits, being able to rescue the OXPHOS growth defect and to reduce the mtDNA instability of the *sym1*^{R51W} mutant strain. Rescue of the growth defect and a decrease in the *petite* frequency were also obtained in the absence of Sym1 protein, thus suggesting that the observed beneficial effect was due to a bypass mechanism. Further analysis showed that the decrease in mtDNA instability obtained with all ten drugs was associated with an increase in mitochondrial dNTP pools, especially of dTTP, providing evidence that the reduced availability of DNA synthesis precursors is the cause of the mtDNA maintenance defect in Sym1 deficiency. Some of these molecules were also able to reduce mtDNA instability on another MDDS yeast model, characterized by mutations in *MIP1* or *RNR2*, likely increasing the levels of dNTPs [40]. Although it is necessary to test these molecules on mammalian cells and model organisms, the identification of molecules capable of reducing the mtDNA instability in different MDDS yeast models makes them a starting point for developing drugs for the treatment of diseases caused by mutations in different genes but all resulting in mtDNA synthesis defects.

5. Conclusions

In the last ten years, most of the novel pathological mutations were found through whole-exome sequencing or through whole-genome sequencing. Sometimes, only the exome or the genome of the proband are sequenced, and, besides the pathological mutations, other mutations are often identified. It is thus fundamental to evaluate whether the putative pathological mutation is the cause of the disease or not, and the use of specific models in which just a single mutation is introduced, such as yeast, can be helpful. Because of the conservation of genes and pathways during evolution, the study of human genetic diseases associated with mtDNA depletion and multiple deletions has also been directly addressed in the model organism *Saccharomyces cerevisiae*. Thanks to the use of yeast as a model system, the causal relationship between pathologies associated with mtDNA instability and novel nuclear mutations has been established or confirmed for tens of mutations. Besides validation, yeast has been also used to determine the pathogenic mechanisms behind these mutations. In this regard, it must be underlined that yeast also has some limitations, among which a different size of the mtDNA, partial differences in the replication process, and the fact that, if two mtDNA molecules are present, the patients are mainly heteroplasmic, whereas yeast is homoplasmic. Despite these differences, pathological mechanisms identified in yeast have been observed in mutant cells derived from patients or in biochemical assays (reviewed in [261]).

To our knowledge, mutations in approximately 25 nuclear genes are associated with diseases characterized by depletion and/or multiple deletions. More than 15 genes are conserved in yeast, although, for some of them, such as *TFAM/ABF2* or *SSBP1/RIM1*, the protein similarity is low. Mutations in six genes have been validated in yeast, mainly through homologous complementation (*ANT1/AAC2, MRM2/MRM2, MPV71/SYM1, POLG/MIP1, RRM2B/RNR2*) or, to a lesser extent, through chimeric complementation (*ANT1/AAC2, OPA1/MGM1, POLG/MIP1*).

A major challenge regarding mitochondrial diseases associated with mtDNA depletion and multiple deletions is the availability of pharmacological treatments. For some of these diseases, preclinical studies have been performed or are ongoing [262–267]. Recently, an open-label clinical study showed that the administration of deoxynucleoside monophosphates and deoxynucleosides to 16 children affected by *TK2*-related disease improved, in most cases, the health condition [268]. However, the identification of therapeutic molecules effective on a broad spectrum of mitochondrial pathologies would be critical. From this point of view, yeast disease models have proven to be useful, thanks to the existence of numerous disease models and to the technique reported above, which allows for the analysis of many molecules in short time.

Author Contributions: Conceptualization, P.G., E.B. and C.D.; writing, A.I.G., C.C.B., M.M., G.d.P., P.G., E.B. and C.D.; table preparation, A.I.G. and C.C.B.; supervision, E.B. and C.D.; funding acquisition, E.B. All authors have read and agreed to the published version of the manuscript.

Funding: This research was funded by "Fondazione Telethon", grant number GGP19287A (E.B.). A.I.G. is supported by a fellowship partially funded by "Fondazione Telethon", grant number GGP19287A. C.C.B. and G.d.P. are supported by a fellowship of the Italian Ministry of Health, grant number GR-2016-02361449 (E.B).

Institutional Review Board Statement: Not applicable.

Informed Consent Statement: Not applicable.

Conflicts of Interest: The authors declare no conflict of interest.

References

1. Nass, M.M.; Nass, S. Intramitochondrial fibers with DNA characteristics. I. Fixation and electron staining reactions. *J. Cell Biol.* **1963**, *19*, 593–611. [CrossRef]
2. Viscomi, C.; Zeviani, M. MtDNA-Maintenance Defects: Syndromes and Genes. *J. Inherit. Metab. Dis.* **2017**, *40*, 587–599. [CrossRef]
3. Rusecka, J.; Kaliszewska, M.; Bartnik, E.; Tońska, K. Nuclear Genes Involved in Mitochondrial Diseases Caused by Instability of Mitochondrial DNA. *J. Appl. Genet.* **2018**, *59*, 43–57. [CrossRef]
4. Chen, H.; Vermulst, M.; Wang, Y.E.; Chomyn, A.; Prolla, T.A.; McCaffery, J.M.; Chan, D.C. Mitochondrial Fusion Is Required for MtDNA Stability in Skeletal Muscle and Tolerance of MtDNA Mutations. *Cell* **2010**, *141*, 280–289. [CrossRef]
5. El-Hattab, A.W.; Craigen, W.J.; Scaglia, F. Mitochondrial DNA Maintenance Defects. *Biochim. Biophys. Acta Mol. Basis Dis.* **2017**, *1863*, 1539–1555. [CrossRef]
6. Koopman, W.J.H.; Distelmaier, F.; Smeitink, J.A.M.; Willems, P.H.G.M. OXPHOS Mutations and Neurodegeneration. *EMBO J.* **2013**, *32*, 9–29. [CrossRef]
7. Filograna, R.; Mennuni, M.; Alsina, D.; Larsson, N.-G. Mitochondrial DNA Copy Number in Human Disease: The More the Better? *FEBS Lett.* **2021**, *595*, 976–1002. [CrossRef]
8. Holt, I.J.; He, J.; Mao, C.-C.; Boyd-Kirkup, J.D.; Martinsson, P.; Sembongi, H.; Reyes, A.; Spelbrink, J.N. Mammalian Mitochondrial Nucleoids: Organizing an Independently Minded Genome. *Mitochondrion* **2007**, *7*, 311–321. [CrossRef]
9. Kukat, C.; Wurm, C.A.; Spåhr, H.; Falkenberg, M.; Larsson, N.-G.; Jakobs, S. Super-Resolution Microscopy Reveals That Mammalian Mitochondrial Nucleoids Have a Uniform Size and Frequently Contain a Single Copy of MtDNA. *Proc. Natl. Acad. Sci. USA* **2011**, *108*, 13534–13539. [CrossRef]
10. Bonekamp, N.A.; Larsson, N.-G. SnapShot: Mitochondrial Nucleoid. *Cell* **2018**, *172*, 388-388.e1. [CrossRef]
11. Robberson, D.L.; Kasamatsu, H.; Vinograd, J. Replication of Mitochondrial DNA. Circular Replicative Intermediates in Mouse L Cells. *Proc. Natl. Acad. Sci. USA* **1972**, *69*, 737–741. [CrossRef]
12. Falkenberg, M. Mitochondrial DNA Replication in Mammalian Cells: Overview of the Pathway. *Essays Biochem.* **2018**, *62*, 287–296. [CrossRef]
13. Taylor, R.W.; Turnbull, D.M. Mitochondrial DNA Mutations in Human Disease. *Nat. Rev. Genet.* **2005**, *6*, 389–402. [CrossRef]
14. Sciacco, M.; Bonilla, E.; Schon, E.A.; DiMauro, S.; Moraes, C.T. Distribution of Wild-Type and Common Deletion Forms of MtDNA in Normal and Respiration-Deficient Muscle Fibers from Patients with Mitochondrial Myopathy. *Hum. Mol. Genet.* **1994**, *3*, 13–19. [CrossRef]
15. Basel, D. Mitochondrial DNA Depletion Syndromes. *Clin. Perinatol.* **2020**, *47*, 123–141. [CrossRef]
16. Spinazzola, A.; Viscomi, C.; Fernandez-Vizarra, E.; Carrara, F.; D'Adamo, P.; Calvo, S.; Marsano, R.M.; Donnini, C.; Weiher, H.; Strisciuglio, P.; et al. MPV17 Encodes an Inner Mitochondrial Membrane Protein and Is Mutated in Infantile Hepatic Mitochondrial DNA Depletion. *Nat. Genet.* **2006**, *38*, 570–575. [CrossRef]
17. Gilberti, M.; Baruffini, E.; Donnini, C.; Dallabona, C. Pathological Alleles of MPV17 Modeled in the Yeast *Saccharomyces cerevisiae* Orthologous Gene SYM1 Reveal Their Inability to Take Part in a High Molecular Weight Complex. *PLoS ONE* **2018**, *13*, e0205014. [CrossRef]
18. Garone, C.; D'Souza, A.R.; Dallabona, C.; Lodi, T.; Rebelo-Guiomar, P.; Rorbach, J.; Donati, M.A.; Procopio, E.; Montomoli, M.; Guerrini, R.; et al. Defective Mitochondrial RRNA Methyltransferase MRM2 Causes MELAS-like Clinical Syndrome. *Hum. Mol. Genet.* **2017**, *26*, 4257–4266. [CrossRef]
19. Del Dotto, V.; Fogazza, M.; Musiani, F.; Maresca, A.; Aleo, S.J.; Caporali, L.; La Morgia, C.; Nolli, C.; Lodi, T.; Goffrini, P.; et al. Deciphering OPA1 Mutations Pathogenicity by Combined Analysis of Human, Mouse and Yeast Cell Models. *Biochim. Biophys. Acta Mol. Basis Dis.* **2018**, *1864*, 3496–3514. [CrossRef]
20. Nolli, C.; Goffrini, P.; Lazzaretti, M.; Zanna, C.; Vitale, R.; Lodi, T.; Baruffini, E. Validation of a MGM1/OPA1 Chimeric Gene for Functional Analysis in Yeast of Mutations Associated with Dominant Optic Atrophy. *Mitochondrion* **2015**, *25*, 38–48. [CrossRef]
21. Nasca, A.; Rizza, T.; Doimo, M.; Legati, A.; Ciolfi, A.; Diodato, D.; Calderan, C.; Carrara, G.; Lamantea, E.; Aiello, C.; et al. Not Only Dominant, Not Only Optic Atrophy: Expanding the Clinical Spectrum Associated with OPA1 Mutations. *Orphanet J. Rare Dis.* **2017**, *12*, 89. [CrossRef]

22. Stuart, G.R.; Santos, J.H.; Strand, M.K.; Van Houten, B.; Copeland, W.C. Mitochondrial and Nuclear DNA Defects in *Saccharomyces cerevisiae* with Mutations in DNA Polymerase Gamma Associated with Progressive External Ophthalmoplegia. *Hum. Mol. Genet.* **2006**, *15*, 363–374. [CrossRef]
23. Baruffini, E.; Lodi, T.; Dallabona, C.; Puglisi, A.; Zeviani, M.; Ferrero, I. Genetic and Chemical Rescue of the *Saccharomyces cerevisiae* Phenotype Induced by Mitochondrial DNA Polymerase Mutations Associated with Progressive External Ophthalmoplegia in Humans. *Hum. Mol. Genet.* **2006**, *15*, 2846–2855. [CrossRef]
24. Baruffini, E.; Ferrero, I.; Foury, F. Mitochondrial DNA Defects in *Saccharomyces cerevisiae* Caused by Functional Interactions between DNA Polymerase Gamma Mutations Associated with Disease in Human. *Biochim. Biophys. Acta* **2007**, *1772*, 1225–1235. [CrossRef]
25. Szczepanowska, K.; Foury, F. A Cluster of Pathogenic Mutations in the 3′-5′ Exonuclease Domain of DNA Polymerase Gamma Defines a Novel Module Coupling DNA Synthesis and Degradation. *Hum. Mol. Genet.* **2010**, *19*, 3516–3529. [CrossRef]
26. Stricker, S.; Prüss, H.; Horvath, R.; Baruffini, E.; Lodi, T.; Siebert, E.; Endres, M.; Zschenderlein, R.; Meisel, A. A Variable Neurodegenerative Phenotype with Polymerase Gamma Mutation. *J. Neurol. Neurosurg. Psychiatry* **2009**, *80*, 1181–1182. [CrossRef]
27. Stumpf, J.D.; Bailey, C.M.; Spell, D.; Stillwagon, M.; Anderson, K.S.; Copeland, W.C. Mip1 Containing Mutations Associated with Mitochondrial Disease Causes Mutagenesis and Depletion of MtDNA in *Saccharomyces cerevisiae*. *Hum. Mol. Genet.* **2010**, *19*, 2123–2133. [CrossRef]
28. Stewart, J.D.; Horvath, R.; Baruffini, E.; Ferrero, I.; Bulst, S.; Watkins, P.B.; Fontana, R.J.; Day, C.P.; Chinnery, P.F. Polymerase γ Gene POLG Determines the Risk of Sodium Valproate-Induced Liver Toxicity. *Hepatology* **2010**, *52*, 1791–1796. [CrossRef]
29. Baruffini, E.; Horvath, R.; Dallabona, C.; Czermin, B.; Lamantea, E.; Bindoff, L.; Invernizzi, F.; Ferrero, I.; Zeviani, M.; Lodi, T. Predicting the Contribution of Novel POLG Mutations to Human Disease through Analysis in Yeast Model. *Mitochondrion* **2011**, *11*, 182–190. [CrossRef]
30. Baruffini, E.; Serafini, F.; Ferrero, I.; Lodi, T. Overexpression of DNA Polymerase Zeta Reduces the Mitochondrial Mutability Caused by Pathological Mutations in DNA Polymerase Gamma in Yeast. *PLoS ONE* **2012**, *7*, e34322. [CrossRef]
31. Stumpf, J.D.; Copeland, W.C. The Exonuclease Activity of the Yeast Mitochondrial DNA Polymerase γ Suppresses Mitochondrial DNA Deletions between Short Direct Repeats in *Saccharomyces cerevisiae*. *Genetics* **2013**, *194*, 519–522. [CrossRef]
32. Stumpf, J.D.; Copeland, W.C. MMS Exposure Promotes Increased MtDNA Mutagenesis in the Presence of Replication-Defective Disease-Associated DNA Polymerase γ Variants. *PLoS Genet.* **2014**, *10*, e1004748. [CrossRef]
33. Kaliszewska, M.; Kruszewski, J.; Kierdaszuk, B.; Kostera-Pruszczyk, A.; Nojszewska, M.; Łusakowska, A.; Vizueta, J.; Sabat, D.; Lutyk, D.; Lower, M.; et al. Yeast Model Analysis of Novel Polymerase Gamma Variants Found in Patients with Autosomal Recessive Mitochondrial Disease. *Hum. Genet.* **2015**, *134*, 951–966. [CrossRef]
34. Hoyos-Gonzalez, N.; Trasviña-Arenas, C.H.; Degiorgi, A.; Castro-Lara, A.Y.; Peralta-Castro, A.; Jimenez-Sandoval, P.; Diaz-Quezada, C.; Lodi, T.; Baruffini, E.; Brieba, L.G. Modeling of Pathogenic Variants of Mitochondrial DNA Polymerase: Insight into the Replication Defects and Implication for Human Disease. *Biochim. Biophys. Acta Gen. Subj.* **2020**, *1864*, 129608. [CrossRef]
35. Qian, Y.; Kachroo, A.H.; Yellman, C.M.; Marcotte, E.M.; Johnson, K.A. Yeast Cells Expressing the Human Mitochondrial DNA Polymerase Reveal Correlations between Polymerase Fidelity and Human Disease Progression. *J. Biol. Chem.* **2014**, *289*, 5970–5985. [CrossRef]
36. Qian, Y.; Ziehr, J.L.; Johnson, K.A. Alpers Disease Mutations in Human DNA Polymerase Gamma Cause Catalytic Defects in Mitochondrial DNA Replication by Distinct Mechanisms. *Front. Genet.* **2015**, *6*, 135. [CrossRef]
37. Baruffini, E.; Ferrari, J.; Dallabona, C.; Donnini, C.; Lodi, T. Polymorphisms in DNA Polymerase γ Affect the MtDNA Stability and the NRTI-Induced Mitochondrial Toxicity in *Saccharomyces cerevisiae*. *Mitochondrion* **2015**, *20*, 52–63. [CrossRef]
38. Baruffini, E.; Lodi, T. Construction and Validation of a Yeast Model System for Studying in Vivo the Susceptibility to Nucleoside Analogues of DNA Polymerase Gamma Allelic Variants. *Mitochondrion* **2010**, *10*, 183–187. [CrossRef]
39. Spinazzola, A.; Invernizzi, F.; Carrara, F.; Lamantea, E.; Donati, A.; Dirocco, M.; Giordano, I.; Meznaric-Petrusa, M.; Baruffini, E.; Ferrero, I.; et al. Clinical and Molecular Features of Mitochondrial DNA Depletion Syndromes. *J. Inherit. Metab. Dis.* **2009**, *32*, 143–158. [CrossRef]
40. Di Punzio, G.; Gilberti, M.; Baruffini, E.; Lodi, T.; Donnini, C.; Dallabona, C. A Yeast-Based Repurposing Approach for the Treatment of Mitochondrial DNA Depletion Syndromes Led to the Identification of Molecules Able to Modulate the dNTP Pool. *Int. J. Mol. Sci.* **2021**, *22*, 12223. [CrossRef]
41. Palmieri, L.; Alberio, S.; Pisano, I.; Lodi, T.; Meznaric-Petrusa, M.; Zidar, J.; Santoro, A.; Scarcia, P.; Fontanesi, F.; Lamantea, E.; et al. Complete Loss-of-Function of the Heart/Muscle-Specific Adenine Nucleotide Translocator Is Associated with Mitochondrial Myopathy and Cardiomyopathy. *Hum. Mol. Genet.* **2005**, *14*, 3079–3088. [CrossRef]
42. Chen, X.J. Induction of an Unregulated Channel by Mutations in Adenine Nucleotide Translocase Suggests an Explanation for Human Ophthalmoplegia. *Hum. Mol. Genet.* **2002**, *11*, 1835–1843. [CrossRef]
43. Fontanesi, F.; Palmieri, L.; Scarcia, P.; Lodi, T.; Donnini, C.; Limongelli, A.; Tiranti, V.; Zeviani, M.; Ferrero, I.; Viola, A.M. Mutations in AAC2, Equivalent to Human AdPEO-Associated ANT1 Mutations, Lead to Defective Oxidative Phosphorylation in *Saccharomyces cerevisiae* and Affect Mitochondrial DNA Stability. *Hum. Mol. Genet.* **2004**, *13*, 923–934. [CrossRef]

44. Lodi, T.; Bove, C.; Fontanesi, F.; Viola, A.M.; Ferrero, I. Mutation D104G in ANT1 Gene: Complementation Study in *Saccharomyces cerevisiae* as a Model System. *Biochem. Biophys. Res. Commun.* **2006**, *341*, 810–815. [CrossRef]
45. Liu, Y.; Wang, X.; Chen, X.J. Misfolding of Mutant Adenine Nucleotide Translocase in Yeast Supports a Novel Mechanism of Ant1-Induced Muscle Diseases. *Mol. Biol. Cell* **2015**, *26*, 1985–1994. [CrossRef]
46. Wang, X.; Salinas, K.; Zuo, X.; Kucejova, B.; Chen, X.J. Dominant Membrane Uncoupling by Mutant Adenine Nucleotide Translocase in Mitochondrial Diseases. *Hum. Mol. Genet.* **2008**, *17*, 4036–4044. [CrossRef]
47. Kaukonen, J.; Juselius, J.K.; Tiranti, V.; Kyttälä, A.; Zeviani, M.; Comi, G.P.; Keränen, S.; Peltonen, L.; Suomalainen, A. Role of Adenine Nucleotide Translocator 1 in MtDNA Maintenance. *Science* **2000**, *289*, 782–785. [CrossRef]
48. Dallabona, C.; Baruffini, E.; Goffrini, P.; Lodi, T. Dominance of Yeast Aac2R96H and Aac2R252G Mutations, Equivalent to Pathological Mutations in Ant1, Is Due to Gain of Function. *Biochem. Biophys. Res. Commun.* **2017**, *493*, 909–913. [CrossRef]
49. Thompson, K.; Majd, H.; Dallabona, C.; Reinson, K.; King, M.S.; Alston, C.L.; He, L.; Lodi, T.; Jones, S.A.; Fattal-Valevski, A.; et al. Recurrent De Novo Dominant Mutations in SLC25A4 Cause Severe Early-Onset Mitochondrial Disease and Loss of Mitochondrial DNA Copy Number. *Am. J. Hum. Genet.* **2016**, *99*, 1405. [CrossRef]
50. Foury, F.; Roganti, T.; Lecrenier, N.; Purnelle, B. The Complete Sequence of the Mitochondrial Genome of *Saccharomyces cerevisiae*. *FEBS Lett.* **1998**, *440*, 325–331. [CrossRef]
51. Blanc, H.; Dujon, B. Replicator Regions of the Yeast Mitochondrial DNA Responsible for Suppressiveness. *Proc. Natl. Acad. Sci. USA* **1980**, *77*, 3942–3946. [CrossRef] [PubMed]
52. De Zamaroczy, M.; Marotta, R.; Faugeron-Fonty, G.; Goursot, R.; Mangin, M.; Baldacci, G.; Bernardi, G. The Origins of Replication of the Yeast Mitochondrial Genome and the Phenomenon of Suppressivity. *Nature* **1981**, *292*, 75–78. [CrossRef] [PubMed]
53. Maleszka, R.; Skelly, P.J.; Clark-Walker, G.D. Rolling Circle Replication of DNA in Yeast Mitochondria. *EMBO J.* **1991**, *10*, 3923–3929. [CrossRef]
54. Ling, F.; Shibata, T. Recombination-Dependent MtDNA Partitioning: In Vivo Role of Mhr1p to Promote Pairing of Homologous DNA. *EMBO J.* **2002**, *21*, 4730–4740. [CrossRef]
55. Ling, F.; Shibata, T. Mhr1p-Dependent Concatemeric Mitochondrial DNA Formation for Generating Yeast Mitochondrial Homoplasmic Cells. *Mol. Biol. Cell* **2004**, *15*, 310–322. [CrossRef] [PubMed]
56. Prasai, K.; Robinson, L.C.; Scott, R.S.; Tatchell, K.; Harrison, L. Evidence for Double-Strand Break Mediated Mitochondrial DNA Replication in *Saccharomyces cerevisiae*. *Nucleic Acids Res.* **2017**, *45*, 7760–7773. [CrossRef]
57. Chen, X.J.; Clark-Walker, G.D. Unveiling the Mystery of Mitochondrial DNA Replication in Yeasts. *Mitochondrion* **2018**, *38*, 17–22. [CrossRef]
58. Ling, F.; Yoshida, M. Rolling-Circle Replication in Mitochondrial DNA Inheritance: Scientific Evidence and Significance from Yeast to Human Cells. *Genes* **2020**, *11*, 514. [CrossRef]
59. Dujon, B.; Slonimski, P.P.; Weill, L. Mitochondrial Genetics IX: A Model for Recombination and Segregation of Mitochondrial Genomes in *Saccharomyces cerevisiae*. *Genetics* **1974**, *78*, 415–437. [CrossRef]
60. Solieri, L. Mitochondrial Inheritance in Budding Yeasts: Towards an Integrated Understanding. *Trends Microbiol.* **2010**, *18*, 521–530. [CrossRef]
61. Hori, A.; Yoshida, M.; Shibata, T.; Ling, F. Reactive Oxygen Species Regulate DNA Copy Number in Isolated Yeast Mitochondria by Triggering Recombination-Mediated Replication. *Nucleic Acids Res.* **2009**, *37*, 749–761. [CrossRef] [PubMed]
62. Chen, X.J.; Butow, R.A. The Organization and Inheritance of the Mitochondrial Genome. *Nat. Rev. Genet.* **2005**, *6*, 815–825. [CrossRef] [PubMed]
63. Lipinski, K.A.; Kaniak-Golik, A.; Golik, P. Maintenance and Expression of the S. Cerevisiae Mitochondrial Genome—From Genetics to Evolution and Systems Biology. *Biochim. Biophys. Acta* **2010**, *1797*, 1086–1098. [CrossRef]
64. Kucej, M.; Butow, R.A. Evolutionary Tinkering with Mitochondrial Nucleoids. *Trends Cell. Biol.* **2007**, *17*, 586–592. [CrossRef] [PubMed]
65. Diffley, J.F.; Stillman, B. A Close Relative of the Nuclear, Chromosomal High-Mobility Group Protein HMG1 in Yeast Mitochondria. *Proc. Natl. Acad. Sci. USA* **1991**, *88*, 7864–7868. [CrossRef] [PubMed]
66. Westermann, B. Mitochondrial Inheritance in Yeast. *Biochim. Biophys. Acta* **2014**, *1837*, 1039–1046. [CrossRef]
67. Miyakawa, I. Organization and Dynamics of Yeast Mitochondrial Nucleoids. *Proc. Jpn. Acad. Ser. B Phys. Biol. Sci.* **2017**, *93*, 339–359. [CrossRef]
68. Sherman, F. Respiration-Deficient Mutants of Yeast. I. Genetics. *Genetics* **1963**, *48*, 375–385. [CrossRef]
69. Ephrussi, B.; Slonimski, P.P. Subcellular Units Involved in the Synthesis of Respiratory Enzymes in Yeast. *Nature* **1955**, *176*, 1207–1208. [CrossRef]
70. Dujon, B. Mitochondrial genetics and functions. In *The Molecular Biology of the Yeast Saccharomyces. Life Cycle and Inheritance*; Cold Spring Harbor Laboratory Press: New York, NY, USA, 1981; pp. 505–635.
71. Ling, F.; Hori, A.; Shibata, T. DNA Recombination-Initiation Plays a Role in the Extremely Biased Inheritance of Yeast [Rho-] Mitochondrial DNA That Contains the Replication Origin Ori5. *Mol. Cell. Biol.* **2007**, *27*, 1133–1145. [CrossRef]
72. Dujon, B. Mitochondrial Genetics Revisited. *Yeast* **2020**, *37*, 191–205. [CrossRef] [PubMed]
73. Lazowska, J.; Slonimski, P.P. Site-Specific Recombination in "Petite Colony" Mutants of *Saccharomyces cerevisiae*. I. Electron Microscopic Analysis of the Organization of Recombinant DNA Resulting from End to End Joining of Two Mitochondrial Segments. *Mol. Gen. Genet.* **1977**, *156*, 163–175. [CrossRef]

74. Gaillard, C.; Strauss, F.; Bernardi, G. Excision Sequences in the Mitochondrial Genome of Yeast. *Nature* **1980**, *283*, 218–220. [CrossRef]
75. Contamine, V.; Picard, M. Maintenance and Integrity of the Mitochondrial Genome: A Plethora of Nuclear Genes in the Budding Yeast. *Microbiol. Mol. Biol. Rev.* **2000**, *64*, 281–315. [CrossRef] [PubMed]
76. Birky, C.W. The Inheritance of Genes in Mitochondria and Chloroplasts: Laws, Mechanisms, and Models. *Annu. Rev. Genet.* **2001**, *35*, 125–148. [CrossRef] [PubMed]
77. Shibata, T.; Ling, F. DNA Recombination Protein-Dependent Mechanism of Homoplasmy and Its Proposed Functions. *Mitochondrion* **2007**, *7*, 17–23. [CrossRef]
78. Berger, K.H.; Yaffe, M.P. Mitochondrial DNA Inheritance in *Saccharomyces cerevisiae*. *Trends Microbiol.* **2000**, *8*, 508–513. [CrossRef]
79. Strausberg, R.L.; Perlman, P.S. The Effect of Zygotic Bud Position on the Transmission of Mitochondrial Genes in *Saccharomyces cerevisiae*. *Mol. Gen. Genet.* **1978**, *163*, 131–144. [CrossRef]
80. Nunnari, J.; Marshall, W.F.; Straight, A.; Murray, A.; Sedat, J.W.; Walter, P. Mitochondrial Transmission during Mating in *Saccharomyces cerevisiae* Is Determined by Mitochondrial Fusion and Fission and the Intramitochondrial Segregation of Mitochondrial DNA. *Mol. Biol. Cell* **1997**, *8*, 1233–1242. [CrossRef]
81. Okamoto, K.; Perlman, P.S.; Butow, R.A. The Sorting of Mitochondrial DNA and Mitochondrial Proteins in Zygotes: Preferential Transmission of Mitochondrial DNA to the Medial Bud. *J. Cell Biol.* **1998**, *142*, 613–623. [CrossRef]
82. Azpiroz, R.; Butow, R.A. Patterns of Mitochondrial Sorting in Yeast Zygotes. *Mol. Biol. Cell* **1993**, *4*, 21–36. [CrossRef]
83. Baruffini, E.; Ferrero, I.; Foury, F. In Vivo Analysis of MtDNA Replication Defects in Yeast. *Methods* **2010**, *51*, 426–436. [CrossRef]
84. Lea, D.E.; Coulson, C.A. The Distribution of the Numbers of Mutants in Bacterial Populations. *J. Genet.* **1949**, *49*, 264–285. [CrossRef]
85. Ceccatelli Berti, C.; di Punzio, G.; Dallabona, C.; Baruffini, E.; Goffrini, P.; Lodi, T.; Donnini, C. The Power of Yeast in Modelling Human Nuclear Mutations Associated with Mitochondrial Diseases. *Genes* **2021**, *12*, 300. [CrossRef] [PubMed]
86. Fukasawa, Y.; Tsuji, J.; Fu, S.-C.; Tomii, K.; Horton, P.; Imai, K. MitoFates: Improved Prediction of Mitochondrial Targeting Sequences and Their Cleavage Sites. *Mol. Cell Proteom.* **2015**, *14*, 1113–1126. [CrossRef] [PubMed]
87. Fraczek, M.G.; Naseeb, S.; Delneri, D. History of Genome Editing in Yeast. *Yeast* **2018**, *35*, 361–368. [CrossRef] [PubMed]
88. Rothstein, R.J. One-Step Gene Disruption in Yeast. *Methods Enzymol.* **1983**, *101*, 202–211. [CrossRef]
89. Garí, E.; Piedrafita, L.; Aldea, M.; Herrero, E. A Set of Vectors with a Tetracycline-Regulatable Promoter System for Modulated Gene Expression in *Saccharomyces cerevisiae*. *Yeast* **1997**, *13*, 837–848. [CrossRef]
90. Graham, I.R.; Chambers, A. Constitutive Expression Vectors: PGK. *Methods Mol. Biol.* **1997**, *62*, 159–169. [CrossRef]
91. Palmer, E.A.; Kruse, K.B.; McCracken, A.A. A Yeast Expression Vector and Leucine Selection in *Escherichia coli* to Aid in the Identification of Novel Genes. *Plasmid* **2001**, *46*, 57–59. [CrossRef]
92. Mascorro-Gallardo, J.O.; Covarrubias, A.A.; Gaxiola, R. Construction of a CUP1 Promoter-Based Vector to Modulate Gene Expression in *Saccharomyces cerevisiae*. *Gene* **1996**, *172*, 169–170. [CrossRef]
93. Gueldener, U.; Heinisch, J.; Koehler, G.J.; Voss, D.; Hegemann, J.H. A Second Set of LoxP Marker Cassettes for Cre-Mediated Multiple Gene Knockouts in Budding Yeast. *Nucleic Acids Res.* **2002**, *30*, e23. [CrossRef]
94. Bellí, G.; Garí, E.; Piedrafita, L.; Aldea, M.; Herrero, E. An Activator/Repressor Dual System Allows Tight Tetracycline-Regulated Gene Expression in Budding Yeast. *Nucleic Acids Res.* **1998**, *26*, 942–947. [CrossRef]
95. Li, J.; Liang, Q.; Song, W.; Marchisio, M.A. Nucleotides Upstream of the Kozak Sequence Strongly Influence Gene Expression in the Yeast *S. cerevisiae*. *J. Biol. Eng.* **2017**, *11*, 25. [CrossRef] [PubMed]
96. Figuccia, S.; Degiorgi, A.; Ceccatelli Berti, C.; Baruffini, E.; Dallabona, C.; Goffrini, P. Mitochondrial Aminoacyl-TRNA Synthetase and Disease: The Yeast Contribution for Functional Analysis of Novel Variants. *Int. J. Mol. Sci.* **2021**, *22*, 4524. [CrossRef] [PubMed]
97. Fukuhara, H.; Wesolowski, M. Genetics and Biogenesis of Mitochondria. In *Mitochondria*; De Gruyter: Berlin, Germany, 1977; pp. 123–131.
98. Mathews, S.; Schweyen, R.J.; Kaudewitz, F. Genetics and Biogenesis of Mitochondria. In *Mitochondria*; De Gruyter: Berlin, Germany, 1977; pp. 133–139.
99. Gonzalez-Hunt, C.P.; Rooney, J.P.; Ryde, I.T.; Anbalagan, C.; Joglekar, R.; Meyer, J.N. PCR-Based Analysis of Mitochondrial DNA Copy Number, Mitochondrial DNA Damage, and Nuclear DNA Damage. *Curr. Protoc. Toxicol.* **2016**, *67*, 20.11. [CrossRef]
100. Taylor, S.D.; Zhang, H.; Eaton, J.S.; Rodeheffer, M.S.; Lebedeva, M.A.; O'rourke, T.W.; Siede, W.; Shadel, G.S. The Conserved Mec1/Rad53 Nuclear Checkpoint Pathway Regulates Mitochondrial DNA Copy Number in *Saccharomyces cerevisiae*. *Mol. Biol. Cell* **2005**, *16*, 3010–3018. [CrossRef]
101. Weiher, H.; Noda, T.; Gray, D.A.; Sharpe, A.H.; Jaenisch, R. Transgenic Mouse Model of Kidney Disease: Insertional Inactivation of Ubiquitously Expressed Gene Leads to Nephrotic Syndrome. *Cell* **1990**, *62*, 425–434. [CrossRef]
102. Zwacka, R.M.; Reuter, A.; Pfaff, E.; Moll, J.; Gorgas, K.; Karasawa, M.; Weiher, H. The Glomerulosclerosis Gene Mpv17 Encodes a Peroxisomal Protein Producing Reactive Oxygen Species. *EMBO J.* **1994**, *13*, 5129–5134. [CrossRef]
103. Wong, L.-J.C.; Brunetti-Pierri, N.; Zhang, Q.; Yazigi, N.; Bove, K.E.; Dahms, B.B.; Puchowicz, M.A.; Gonzalez-Gomez, I.; Schmitt, E.S.; Truong, C.K.; et al. Mutations in the MPV17 Gene Are Responsible for Rapidly Progressive Liver Failure in Infancy. *Hepatology* **2007**, *46*, 1218–1227. [CrossRef]

104. Karadimas, C.L.; Vu, T.H.; Holve, S.A.; Chronopoulou, P.; Quinzii, C.; Johnsen, S.D.; Kurth, J.; Eggers, E.; Palenzuela, L.; Tanji, K.; et al. Navajo Neurohepatopathy Is Caused by a Mutation in the MPV17 Gene. *Am. J. Hum. Genet.* **2006**, *79*, 544–548. [CrossRef]
105. AlSaman, A.; Tomoum, H.; Invernizzi, F.; Zeviani, M. Hepatocerebral Form of Mitochondrial DNA Depletion Syndrome Due to Mutation in MPV17 Gene. *Saudi J. Gastroenterol.* **2012**, *18*, 285–289. [CrossRef] [PubMed]
106. Blakely, E.L.; Butterworth, A.; Hadden, R.D.M.; Bodi, I.; He, L.; McFarland, R.; Taylor, R.W. MPV17 Mutation Causes Neuropathy and Leukoencephalopathy with Multiple MtDNA Deletions in Muscle. *Neuromuscul. Disord.* **2012**, *22*, 587–591. [CrossRef] [PubMed]
107. Garone, C.; Rubio, J.C.; Calvo, S.E.; Naini, A.; Tanji, K.; Dimauro, S.; Mootha, V.K.; Hirano, M. MPV17 Mutations Causing Adult-Onset Multisystemic Disorder with Multiple Mitochondrial DNA Deletions. *Arch. Neurol* **2012**, *69*, 1648–1651. [CrossRef]
108. Sommerville, E.W.; Chinnery, P.F.; Gorman, G.S.; Taylor, R.W. Adult-Onset Mendelian PEO Associated with Mitochondrial Disease. *J. Neuromuscul. Dis.* **2014**, *1*, 119–133. [CrossRef] [PubMed]
109. Trott, A.; Morano, K.A. SYM1 Is the Stress-Induced *Saccharomyces cerevisiae* Ortholog of the Mammalian Kidney Disease Gene Mpv17 and Is Required for Ethanol Metabolism and Tolerance during Heat Shock. *Eukaryot. Cell* **2004**, *3*, 620–631. [CrossRef]
110. Dallabona, C.; Marsano, R.M.; Arzuffi, P.; Ghezzi, D.; Mancini, P.; Zeviani, M.; Ferrero, I.; Donnini, C. Sym1, the Yeast Ortholog of the MPV17 Human Disease Protein, Is a Stress-Induced Bioenergetic and Morphogenetic Mitochondrial Modulator. *Hum. Mol. Genet.* **2010**, *19*, 1098–1107. [CrossRef]
111. Bottani, E.; Giordano, C.; Civiletto, G.; Di Meo, I.; Auricchio, A.; Ciusani, E.; Marchet, S.; Lamperti, C.; d'Amati, G.; Viscomi, C.; et al. AAV-Mediated Liver-Specific MPV17 Expression Restores MtDNA Levels and Prevents Diet-Induced Liver Failure. *Mol. Ther.* **2014**, *22*, 10–17. [CrossRef]
112. Viscomi, C.; Spinazzola, A.; Maggioni, M.; Fernandez-Vizarra, E.; Massa, V.; Pagano, C.; Vettor, R.; Mora, M.; Zeviani, M. Early-Onset Liver MtDNA Depletion and Late-Onset Proteinuric Nephropathy in Mpv17 Knockout Mice. *Hum. Mol. Genet.* **2009**, *18*, 12–26. [CrossRef]
113. Martorano, L.; Peron, M.; Laquatra, C.; Lidron, E.; Facchinello, N.; Meneghetti, G.; Tiso, N.; Rasola, A.; Ghezzi, D.; Argenton, F. The Zebrafish Orthologue of the Human Hepatocerebral Disease Gene MPV17 Plays Pleiotropic Roles in Mitochondria. *Dis. Model. Mech.* **2019**, *12*, dmm037226. [CrossRef]
114. Parini, R.; Furlan, F.; Notarangelo, L.; Spinazzola, A.; Uziel, G.; Strisciuglio, P.; Concolino, D.; Corbetta, C.; Nebbia, G.; Menni, F.; et al. Glucose Metabolism and Diet-Based Prevention of Liver Dysfunction in MPV17 Mutant Patients. *J. Hepatol.* **2009**, *50*, 215–221. [CrossRef] [PubMed]
115. Reinhold, R.; Krüger, V.; Meinecke, M.; Schulz, C.; Schmidt, B.; Grunau, S.D.; Guiard, B.; Wiedemann, N.; van der Laan, M.; Wagner, R.; et al. The Channel-Forming Sym1 Protein Is Transported by the TIM23 Complex in a Presequence-Independent Manner. *Mol. Cell. Biol.* **2012**, *32*, 5009–5021. [CrossRef] [PubMed]
116. Antonenkov, V.D.; Isomursu, A.; Mennerich, D.; Vapola, M.H.; Weiher, H.; Kietzmann, T.; Hiltunen, J.K. The Human Mitochondrial DNA Depletion Syndrome Gene MPV17 Encodes a Non-Selective Channel That Modulates Membrane Potential. *J. Biol. Chem.* **2015**, *290*, 13840–13861. [CrossRef] [PubMed]
117. Binder, C.J.; Weiher, H.; Exner, M.; Kerjaschki, D. Glomerular Overproduction of Oxygen Radicals in Mpv17 Gene-Inactivated Mice Causes Podocyte Foot Process Flattening and Proteinuria: A Model of Steroid-Resistant Nephrosis Sensitive to Radical Scavenger Therapy. *Am. J. Pathol.* **1999**, *154*, 1067–1075. [CrossRef]
118. Löllgen, S.; Weiher, H. The Role of the Mpv17 Protein Mutations of Which Cause Mitochondrial DNA Depletion Syndrome (MDDS): Lessons from Homologs in Different Species. *Biol. Chem.* **2015**, *396*, 13–25. [CrossRef]
119. Dalla Rosa, I.; Cámara, Y.; Durigon, R.; Moss, C.F.; Vidoni, S.; Akman, G.; Hunt, L.; Johnson, M.A.; Grocott, S.; Wang, L.; et al. MPV17 Loss Causes Deoxynucleotide Insufficiency and Slow DNA Replication in Mitochondria. *PLoS Genet.* **2016**, *12*, e1005779. [CrossRef]
120. Krauss, J.; Astrinidis, P.; Frohnhöfer, H.G.; Walderich, B.; Nüsslein-Volhard, C. Erratum: Transparent, a Gene Affecting Stripe Formation in Zebrafish, Encodes the Mitochondrial Protein Mpv17 That Is Required for Iridophore Survival. *Biol. Open* **2013**, *2*, 979. [CrossRef]
121. White, Y.A.R.; Woods, D.C.; Wood, A.W. A Transgenic Zebrafish Model of Targeted Oocyte Ablation and de Novo Oogenesis. *Dev. Dyn.* **2011**, *240*, 1929–1937. [CrossRef]
122. Alonzo, J.R.; Venkataraman, C.; Field, M.S.; Stover, P.J. The Mitochondrial Inner Membrane Protein MPV17 Prevents Uracil Accumulation in Mitochondrial DNA. *J. Biol. Chem.* **2018**, *293*, 20285–20294. [CrossRef]
123. Blount, B.C.; Mack, M.M.; Wehr, C.M.; MacGregor, J.T.; Hiatt, R.A.; Wang, G.; Wickramasinghe, S.N.; Everson, R.B.; Ames, B.N. Folate Deficiency Causes Uracil Misincorporation into Human DNA and Chromosome Breakage: Implications for Cancer and Neuronal Damage. *Proc. Natl. Acad. Sci. USA* **1997**, *94*, 3290–3295. [CrossRef]
124. Lee, K.-W.; Okot-Kotber, C.; LaComb, J.F.; Bogenhagen, D.F. Mitochondrial Ribosomal RNA (RRNA) Methyltransferase Family Members Are Positioned to Modify Nascent RRNA in Foci near the Mitochondrial DNA Nucleoid. *J. Biol. Chem.* **2013**, *288*, 31386–31399. [CrossRef] [PubMed]
125. Lee, K.-W.; Bogenhagen, D.F. Assignment of 2′-O-Methyltransferases to Modification Sites on the Mammalian Mitochondrial Large Subunit 16 S Ribosomal RNA (RRNA). *J. Biol. Chem.* **2014**, *289*, 24936–24942. [CrossRef]

126. Rorbach, J.; Boesch, P.; Gammage, P.A.; Nicholls, T.J.J.; Pearce, S.F.; Patel, D.; Hauser, A.; Perocchi, F.; Minczuk, M. MRM2 and MRM3 Are Involved in Biogenesis of the Large Subunit of the Mitochondrial Ribosome. *Mol. Biol. Cell* **2014**, *25*, 2542–2555. [CrossRef] [PubMed]
127. Cipullo, M.; Gesé, G.V.; Khawaja, A.; Hällberg, B.M.; Rorbach, J. Structural Basis for Late Maturation Steps of the Human Mitoribosomal Large Subunit. *Nat. Commun.* **2021**, *12*, 3673. [CrossRef] [PubMed]
128. Widerak, M.; Kern, R.; Malki, A.; Richarme, G. U2552 Methylation at the Ribosomal A-Site Is a Negative Modulator of Translational Accuracy. *Gene* **2005**, *347*, 109–114. [CrossRef] [PubMed]
129. Pintard, L.; Bujnicki, J.M.; Lapeyre, B.; Bonnerot, C. MRM2 Encodes a Novel Yeast Mitochondrial 21S RRNA Methyltransferase. *EMBO J.* **2002**, *21*, 1139–1147. [CrossRef] [PubMed]
130. Amati-Bonneau, P.; Valentino, M.L.; Reynier, P.; Gallardo, M.E.; Bornstein, B.; Boissière, A.; Campos, Y.; Rivera, H.; de la Aleja, J.G.; Carroccia, R.; et al. OPA1 Mutations Induce Mitochondrial DNA Instability and Optic Atrophy "plus" Phenotypes. *Brain* **2008**, *131*, 338–351. [CrossRef]
131. Hudson, G.; Amati-Bonneau, P.; Blakely, E.L.; Stewart, J.D.; He, L.; Schaefer, A.M.; Griffiths, P.G.; Ahlqvist, K.; Suomalainen, A.; Reynier, P.; et al. Mutation of OPA1 Causes Dominant Optic Atrophy with External Ophthalmoplegia, Ataxia, Deafness and Multiple Mitochondrial DNA Deletions: A Novel Disorder of MtDNA Maintenance. *Brain* **2008**, *131*, 329–337. [CrossRef]
132. Elachouri, G.; Vidoni, S.; Zanna, C.; Pattyn, A.; Boukhaddaoui, H.; Gaget, K.; Yu-Wai-Man, P.; Gasparre, G.; Sarzi, E.; Delettre, C.; et al. OPA1 Links Human Mitochondrial Genome Maintenance to MtDNA Replication and Distribution. *Genome Res.* **2011**, *21*, 12–20. [CrossRef]
133. Spiegel, R.; Saada, A.; Flannery, P.J.; Burté, F.; Soiferman, D.; Khayat, M.; Eisner, V.; Vladovski, E.; Taylor, R.W.; Bindoff, L.A.; et al. Fatal Infantile Mitochondrial Encephalomyopathy, Hypertrophic Cardiomyopathy and Optic Atrophy Associated with a Homozygous OPA1 Mutation. *J. Med. Genet.* **2016**, *53*, 127–131. [CrossRef]
134. Olichon, A.; Baricault, L.; Gas, N.; Guillou, E.; Valette, A.; Belenguer, P.; Lenaers, G. Loss of OPA1 Perturbates the Mitochondrial Inner Membrane Structure and Integrity, Leading to Cytochrome c Release and Apoptosis. *J. Biol. Chem.* **2003**, *278*, 7743–7746. [CrossRef]
135. Frezza, C.; Cipolat, S.; Martins de Brito, O.; Micaroni, M.; Beznoussenko, G.V.; Rudka, T.; Bartoli, D.; Polishuck, R.S.; Danial, N.N.; De Strooper, B.; et al. OPA1 Controls Apoptotic Cristae Remodeling Independently from Mitochondrial Fusion. *Cell* **2006**, *126*, 177–189. [CrossRef]
136. Carelli, V.; Musumeci, O.; Caporali, L.; Zanna, C.; La Morgia, C.; Del Dotto, V.; Porcelli, A.M.; Rugolo, M.; Valentino, M.L.; Iommarini, L.; et al. Syndromic Parkinsonism and Dementia Associated with OPA1 Missense Mutations. *Ann. Neurol.* **2015**, *78*, 21–38. [CrossRef] [PubMed]
137. Liao, C.; Ashley, N.; Diot, A.; Morten, K.; Phadwal, K.; Williams, A.; Fearnley, I.; Rosser, L.; Lowndes, J.; Fratter, C.; et al. Dysregulated Mitophagy and Mitochondrial Organization in Optic Atrophy Due to OPA1 Mutations. *Neurology* **2017**, *88*, 131–142. [CrossRef] [PubMed]
138. Chan, D.C. Mitochondrial Dynamics and Its Involvement in Disease. *Annu. Rev. Pathol.* **2020**, *15*, 235–259. [CrossRef] [PubMed]
139. Giacomello, M.; Pyakurel, A.; Glytsou, C.; Scorrano, L. The Cell Biology of Mitochondrial Membrane Dynamics. *Nat. Rev. Mol. Cell. Biol.* **2020**, *21*, 204–224. [CrossRef] [PubMed]
140. Del Dotto, V.; Mishra, P.; Vidoni, S.; Fogazza, M.; Maresca, A.; Caporali, L.; McCaffery, J.M.; Cappelletti, M.; Baruffini, E.; Lenaers, G.; et al. OPA1 Isoforms in the Hierarchical Organization of Mitochondrial Functions. *Cell Rep.* **2017**, *19*, 2557–2571. [CrossRef]
141. Olichon, A.; Elachouri, G.; Baricault, L.; Delettre, C.; Belenguer, P.; Lenaers, G. OPA1 Alternate Splicing Uncouples an Evolutionary Conserved Function in Mitochondrial Fusion from a Vertebrate Restricted Function in Apoptosis. *Cell Death Differ.* **2007**, *14*, 682–692. [CrossRef]
142. Ishihara, N.; Fujita, Y.; Oka, T.; Mihara, K. Regulation of Mitochondrial Morphology through Proteolytic Cleavage of OPA1. *EMBO J.* **2006**, *25*, 2966–2977. [CrossRef] [PubMed]
143. Anand, R.; Wai, T.; Baker, M.J.; Kladt, N.; Schauss, A.C.; Rugarli, E.; Langer, T. The I-AAA Protease YME1L and OMA1 Cleave OPA1 to Balance Mitochondrial Fusion and Fission. *J. Cell Biol.* **2014**, *204*, 919–929. [CrossRef]
144. MacVicar, T.; Langer, T. OPA1 Processing in Cell Death and Disease—The Long and Short of It. *J. Cell Sci.* **2016**, *129*, 2297–2306. [CrossRef]
145. Ding, C.; Wu, Z.; Huang, L.; Wang, Y.; Xue, J.; Chen, S.; Deng, Z.; Wang, L.; Song, Z.; Chen, S. Mitofilin and CHCHD6 Physically Interact with Sam50 to Sustain Cristae Structure. *Sci. Rep.* **2015**, *5*, 16064. [CrossRef] [PubMed]
146. Alexander, C.; Votruba, M.; Pesch, U.E.; Thiselton, D.L.; Mayer, S.; Moore, A.; Rodriguez, M.; Kellner, U.; Leo-Kottler, B.; Auburger, G.; et al. OPA1, Encoding a Dynamin-Related GTPase, Is Mutated in Autosomal Dominant Optic Atrophy Linked to Chromosome 3q28. *Nat. Genet.* **2000**, *26*, 211–215. [CrossRef]
147. Delettre, C.; Lenaers, G.; Griffoin, J.M.; Gigarel, N.; Lorenzo, C.; Belenguer, P.; Pelloquin, L.; Grosgeorge, J.; Turc-Carel, C.; Perret, E.; et al. Nuclear Gene OPA1, Encoding a Mitochondrial Dynamin-Related Protein, Is Mutated in Dominant Optic Atrophy. *Nat. Genet.* **2000**, *26*, 207–210. [CrossRef] [PubMed]
148. Lenaers, G.; Hamel, C.; Delettre, C.; Amati-Bonneau, P.; Procaccio, V.; Bonneau, D.; Reynier, P.; Milea, D. Dominant Optic Atrophy. *Orphanet J. Rare Dis.* **2012**, *7*, 46. [CrossRef]

149. Olichon, A.; Guillou, E.; Delettre, C.; Landes, T.; Arnauné-Pelloquin, L.; Emorine, L.J.; Mils, V.; Daloyau, M.; Hamel, C.; Amati-Bonneau, P.; et al. Mitochondrial Dynamics and Disease, OPA1. *Biochim. Biophys. Acta* **2006**, *1763*, 500–509. [CrossRef]
150. Ferré, M.; Bonneau, D.; Milea, D.; Chevrollier, A.; Verny, C.; Dollfus, H.; Ayuso, C.; Defoort, S.; Vignal, C.; Zanlonghi, X.; et al. Molecular Screening of 980 Cases of Suspected Hereditary Optic Neuropathy with a Report on 77 Novel OPA1 Mutations. *Hum. Mutat.* **2009**, *30*, E692–E705. [CrossRef]
151. Jones, B.A.; Fangman, W.L. Mitochondrial DNA Maintenance in Yeast Requires a Protein Containing a Region Related to the GTP-Binding Domain of Dynamin. *Genes Dev.* **1992**, *6*, 380–389. [CrossRef] [PubMed]
152. Meeusen, S.; DeVay, R.; Block, J.; Cassidy-Stone, A.; Wayson, S.; McCaffery, J.M.; Nunnari, J. Mitochondrial Inner-Membrane Fusion and Crista Maintenance Requires the Dynamin-Related GTPase Mgm1. *Cell* **2006**, *127*, 383–395. [CrossRef]
153. Herlan, M.; Vogel, F.; Bornhovd, C.; Neupert, W.; Reichert, A.S. Processing of Mgm1 by the Rhomboid-Type Protease Pcp1 Is Required for Maintenance of Mitochondrial Morphology and of Mitochondrial DNA. *J. Biol. Chem.* **2003**, *278*, 27781–27788. [CrossRef]
154. Zick, M.; Duvezin-Caubet, S.; Schäfer, A.; Vogel, F.; Neupert, W.; Reichert, A.S. Distinct Roles of the Two Isoforms of the Dynamin-like GTPase Mgm1 in Mitochondrial Fusion. *FEBS Lett.* **2009**, *583*, 2237–2243. [CrossRef]
155. Del Dotto, V.; Carelli, V. Dominant Optic Atrophy (DOA): Modeling the Kaleidoscopic Roles of OPA1 in Mitochondrial Homeostasis. *Front. Neurol.* **2021**, *12*, 681326. [CrossRef]
156. Schaaf, C.P.; Blazo, M.; Lewis, R.A.; Tonini, R.E.; Takei, H.; Wang, J.; Wong, L.-J.; Scaglia, F. Early-Onset Severe Neuromuscular Phenotype Associated with Compound Heterozygosity for OPA1 Mutations. *Mol. Genet. Metab.* **2011**, *103*, 383–387. [CrossRef]
157. Bonifert, T.; Karle, K.N.; Tonagel, F.; Batra, M.; Wilhelm, C.; Theurer, Y.; Schoenfeld, C.; Kluba, T.; Kamenisch, Y.; Carelli, V.; et al. Pure and Syndromic Optic Atrophy Explained by Deep Intronic OPA1 Mutations and an Intralocus Modifier. *Brain* **2014**, *137*, 2164–2177. [CrossRef]
158. Carelli, V.; Sabatelli, M.; Carrozzo, R.; Rizza, T.; Schimpf, S.; Wissinger, B.; Zanna, C.; Rugolo, M.; La Morgia, C.; Caporali, L.; et al. "Behr Syndrome" with OPA1 Compound Heterozygote Mutations. *Brain* **2015**, *138*, e321. [CrossRef] [PubMed]
159. Amati-Bonneau, P.; Odent, S.; Derrien, C.; Pasquier, L.; Malthiéry, Y.; Reynier, P.; Bonneau, D. The Association of Autosomal Dominant Optic Atrophy and Moderate Deafness May Be Due to the R445H Mutation in the OPA1 Gene. *Am. J. Ophthalmol.* **2003**, *136*, 1170–1171. [CrossRef]
160. Pesch, U.E.; Leo-Kottler, B.; Mayer, S.; Jurklies, B.; Kellner, U.; Apfelstedt-Sylla, E.; Zrenner, E.; Alexander, C.; Wissinger, B. OPA1 Mutations in Patients with Autosomal Dominant Optic Atrophy and Evidence for Semi-Dominant Inheritance. *Hum. Mol. Genet.* **2001**, *10*, 1359–1368. [CrossRef] [PubMed]
161. Bolden, A.; Noy, G.P.; Weissbach, A. DNA Polymerase of Mitochondria Is a Gamma-Polymerase. *J. Biol. Chem.* **1977**, *252*, 3351–3356. [CrossRef]
162. Kaguni, L.S. DNA Polymerase Gamma, the Mitochondrial Replicase. *Annu. Rev. Biochem.* **2004**, *73*, 293–320. [CrossRef]
163. Loeb, L.A.; Liu, P.K.; Fry, M. DNA Polymerase-Alpha: Enzymology, Function, Fidelity, and Mutagenesis. *Prog. Nucleic Acid Res. Mol. Biol.* **1986**, *33*, 57–110. [CrossRef] [PubMed]
164. Kornberg, A.; Baker, T.A. *DNA Replication*; University Science Books: Melville, NY, USA, 2005; ISBN 978-1-891389-44-3.
165. Ropp, P.A.; Copeland, W.C. Cloning and Characterization of the Human Mitochondrial DNA Polymerase, DNA Polymerase Gamma. *Genomics* **1996**, *36*, 449–458. [CrossRef] [PubMed]
166. Pinz, K.G.; Bogenhagen, D.F. Efficient Repair of Abasic Sites in DNA by Mitochondrial Enzymes. *Mol. Cell. Biol.* **1998**, *18*, 1257–1265. [CrossRef] [PubMed]
167. Pinz, K.G.; Bogenhagen, D.F. Characterization of a Catalytically Slow AP Lyase Activity in DNA Polymerase Gamma and Other Family A DNA Polymerases. *J. Biol. Chem.* **2000**, *275*, 12509–12514. [CrossRef]
168. Pinz, K.G.; Bogenhagen, D.F. The Influence of the DNA Polymerase Gamma Accessory Subunit on Base Excision Repair by the Catalytic Subunit. *DNA Rep.* **2006**, *5*, 121–128. [CrossRef] [PubMed]
169. Longley, M.J.; Prasad, R.; Srivastava, D.K.; Wilson, S.H.; Copeland, W.C. Identification of 5′-Deoxyribose Phosphate Lyase Activity in Human DNA Polymerase Gamma and Its Role in Mitochondrial Base Excision Repair in Vitro. *Proc. Natl. Acad. Sci. USA* **1998**, *95*, 12244–12248. [CrossRef]
170. Yakubovskaya, E.; Chen, Z.; Carrodeguas, J.A.; Kisker, C.; Bogenhagen, D.F. Functional Human Mitochondrial DNA Polymerase Gamma Forms a Heterotrimer. *J. Biol. Chem.* **2006**, *281*, 374–382. [CrossRef]
171. Kunkel, T.A.; Soni, A. Exonucleolytic Proofreading Enhances the Fidelity of DNA Synthesis by Chick Embryo DNA Polymerase-Gamma. *J. Biol. Chem.* **1988**, *263*, 4450–4459. [CrossRef]
172. Kunkel, T.A.; Mosbaugh, D.W. Exonucleolytic Proofreading by a Mammalian DNA Polymerase. *Biochemistry* **1989**, *28*, 988–995. [CrossRef]
173. Ito, J.; Braithwaite, D.K. Yeast Mitochondrial DNA Polymerase Is Related to the Family A DNA Polymerases. *Nucleic Acids Res.* **1990**, *18*, 6716. [CrossRef]
174. Graziewicz, M.A.; Longley, M.J.; Copeland, W.C. DNA Polymerase Gamma in Mitochondrial DNA Replication and Repair. *Chem. Rev.* **2006**, *106*, 383–405. [CrossRef]
175. Fan, L.; Sanschagrin, P.C.; Kaguni, L.S.; Kuhn, L.A. The Accessory Subunit of MtDNA Polymerase Shares Structural Homology with Aminoacyl-TRNA Synthetases: Implications for a Dual Role as a Primer Recognition Factor and Processivity Clamp. *Proc. Natl. Acad. Sci. USA* **1999**, *96*, 9527–9532. [CrossRef] [PubMed]

176. Fan, L.; Kaguni, L.S. Multiple Regions of Subunit Interaction in Drosophila Mitochondrial DNA Polymerase: Three Functional Domains in the Accessory Subunit. *Biochemistry* **2001**, *40*, 4780–4791. [CrossRef] [PubMed]
177. Lee, Y.-S.; Kennedy, W.D.; Yin, Y.W. Structural Insight into Processive Human Mitochondrial DNA Synthesis and Disease-Related Polymerase Mutations. *Cell* **2009**, *139*, 312–324. [CrossRef] [PubMed]
178. Rahman, S.; Copeland, W.C. POLG-Related Disorders and Their Neurological Manifestations. *Nat. Rev. Neurol.* **2019**, *15*, 40–52. [CrossRef]
179. Hikmat, O.; Tzoulis, C.; Chong, W.K.; Chentouf, L.; Klingenberg, C.; Fratter, C.; Carr, L.J.; Prabhakar, P.; Kumaraguru, N.; Gissen, P.; et al. The Clinical Spectrum and Natural History of Early-Onset Diseases Due to DNA Polymerase Gamma Mutations. *Genet. Med.* **2017**, *19*, 1217–1225. [CrossRef]
180. Harding, B.N. Progressive Neuronal Degeneration of Childhood with Liver Disease (Alpers-Huttenlocher Syndrome): A Personal Review. *J. Child. Neurol.* **1990**, *5*, 273–287. [CrossRef]
181. Tang, S.; Dimberg, E.L.; Milone, M.; Wong, L.-J.C. Mitochondrial Neurogastrointestinal Encephalomyopathy (MNGIE)-like Phenotype: An Expanded Clinical Spectrum of POLG1 Mutations. *J. Neurol.* **2012**, *259*, 862–868. [CrossRef]
182. Van Goethem, G.; Mercelis, R.; Löfgren, A.; Seneca, S.; Ceuterick, C.; Martin, J.J.; Van Broeckhoven, C. Patient Homozygous for a Recessive POLG Mutation Presents with Features of MERRF. *Neurology* **2003**, *61*, 1811–1813. [CrossRef]
183. Deschauer, M.; Tennant, S.; Rokicka, A.; He, L.; Kraya, T.; Turnbull, D.M.; Zierz, S.; Taylor, R.W. MELAS Associated with Mutations in the POLG1 Gene. *Neurology* **2007**, *68*, 1741–1742. [CrossRef]
184. Anagnostou, M.-E.; Ng, Y.S.; Taylor, R.W.; McFarland, R. Epilepsy Due to Mutations in the Mitochondrial Polymerase Gamma (POLG) Gene: A Clinical and Molecular Genetic Review. *Epilepsia* **2016**, *57*, 1531–1545. [CrossRef]
185. Hanisch, F.; Kornhuber, M.; Alston, C.L.; Taylor, R.W.; Deschauer, M.; Zierz, S. SANDO Syndrome in a Cohort of 107 Patients with CPEO and Mitochondrial DNA Deletions. *J. Neurol. Neurosurg. Psychiatry* **2015**, *86*, 630–634. [CrossRef] [PubMed]
186. Cohen, B.H.; Chinnery, P.F.; Copeland, W.C. POLG-Related Disorders. In *GeneReviews®*; Adam, M.P., Ardinger, H.H., Pagon, R.A., Wallace, S.E., Bean, L.J., Mirzaa, G., Amemiya, A., Eds.; University of Washington Seattle: Seattle, WA, USA, 1993.
187. Lodi, T.; Dallabona, C.; Nolli, C.; Goffrini, P.; Donnini, C.; Baruffini, E. DNA Polymerase γ and Disease: What We Have Learned from Yeast. *Front. Genet.* **2015**, *6*, 106. [CrossRef]
188. Young, M.J.; Theriault, S.S.; Li, M.; Court, D.A. The Carboxyl-Terminal Extension on Fungal Mitochondrial DNA Polymerases: Identification of a Critical Region of the Enzyme from *Saccharomyces cerevisiae*. *Yeast* **2006**, *23*, 101–116. [CrossRef] [PubMed]
189. Viikov, K.; Jasnovidova, O.; Tamm, T.; Sedman, J. C-Terminal Extension of the Yeast Mitochondrial DNA Polymerase Determines the Balance between Synthesis and Degradation. *PLoS ONE* **2012**, *7*, e33482. [CrossRef]
190. Trasviña-Arenas, C.H.; Hoyos-Gonzalez, N.; Castro-Lara, A.Y.; Rodriguez-Hernandez, A.; Sanchez-Sandoval, M.E.; Jimenez-Sandoval, P.; Ayala-García, V.M.; Díaz-Quezada, C.; Lodi, T.; Baruffini, E.; et al. Amino and Carboxy-Terminal Extensions of Yeast Mitochondrial DNA Polymerase Assemble Both the Polymerization and Exonuclease Active Sites. *Mitochondrion* **2019**, *49*, 166–177. [CrossRef]
191. Lecrenier, N.; Foury, F. Overexpression of the RNR1 Gene Rescues *Saccharomyces cerevisiae* Mutants in the Mitochondrial DNA Polymerase-Encoding MIP1 Gene. *Mol. Gen. Genet.* **1995**, *249*, 1–7. [CrossRef]
192. Bulst, S.; Holinski-Feder, E.; Payne, B.; Abicht, A.; Krause, S.; Lochmüller, H.; Chinnery, P.F.; Walter, M.C.; Horvath, R. In Vitro Supplementation with Deoxynucleoside Monophosphates Rescues Mitochondrial DNA Depletion. *Mol. Genet. Metab.* **2012**, *107*, 95–103. [CrossRef] [PubMed]
193. Zhang, H.; Chatterjee, A.; Singh, K.K. *Saccharomyces cerevisiae* Polymerase Zeta Functions in Mitochondria. *Genetics* **2006**, *172*, 2683–2688. [CrossRef]
194. Singh, B.; Li, X.; Owens, K.M.; Vanniarajan, A.; Liang, P.; Singh, K.K. Human REV3 DNA Polymerase Zeta Localizes to Mitochondria and Protects the Mitochondrial Genome. *PLoS ONE* **2015**, *10*, e0140409. [CrossRef]
195. Pontarin, G.; Ferraro, P.; Bee, L.; Reichard, P.; Bianchi, V. Mammalian Ribonucleotide Reductase Subunit P53R2 Is Required for Mitochondrial DNA Replication and DNA Repair in Quiescent Cells. *Proc. Natl. Acad. Sci. USA* **2012**, *109*, 13302–13307. [CrossRef]
196. Finsterer, J.; Zarrouk-Mahjoub, S. Phenotypic and Genotypic Heterogeneity of RRM2B Variants. *Neuropediatrics* **2018**, *49*, 231–237. [CrossRef]
197. Bornstein, B.; Area, E.; Flanigan, K.M.; Ganesh, J.; Jayakar, P.; Swoboda, K.J.; Coku, J.; Naini, A.; Shanske, S.; Tanji, K.; et al. Mitochondrial DNA Depletion Syndrome Due to Mutations in the RRM2B Gene. *Neuromuscul. Disord.* **2008**, *18*, 453–459. [CrossRef]
198. Kollberg, G.; Darin, N.; Benan, K.; Moslemi, A.-R.; Lindal, S.; Tulinius, M.; Oldfors, A.; Holme, E. A Novel Homozygous RRM2B Missense Mutation in Association with Severe MtDNA Depletion. *Neuromuscul. Disord.* **2009**, *19*, 147–150. [CrossRef] [PubMed]
199. Acham-Roschitz, B.; Plecko, B.; Lindbichler, F.; Bittner, R.; Mache, C.J.; Sperl, W.; Mayr, J.A. A Novel Mutation of the RRM2B Gene in an Infant with Early Fatal Encephalomyopathy, Central Hypomyelination, and Tubulopathy. *Mol. Genet. Metab.* **2009**, *98*, 300–304. [CrossRef] [PubMed]
200. Stojanovic, V.; Mayr, J.A.; Sperl, W.; Barišić, N.; Doronjski, A.; Milak, G. Infantile Peripheral Neuropathy, Deafness, and Proximal Tubulopathy Associated with a Novel Mutation of the RRM2B Gene: Case Study. *Croat. Med. J.* **2013**, *54*, 579–584. [CrossRef] [PubMed]

201. Kropach, N.; Shkalim-Zemer, V.; Orenstein, N.; Scheuerman, O.; Straussberg, R. Novel RRM2B Mutation and Severe Mitochondrial DNA Depletion: Report of 2 Cases and Review of the Literature. *Neuropediatrics* **2017**, *48*, 456–462. [CrossRef]
202. Pitceathly, R.D.S.; Smith, C.; Fratter, C.; Alston, C.L.; He, L.; Craig, K.; Blakely, E.L.; Evans, J.C.; Taylor, J.; Shabbir, Z.; et al. Adults with RRM2B-Related Mitochondrial Disease Have Distinct Clinical and Molecular Characteristics. *Brain* **2012**, *135*, 3392–3403. [CrossRef]
203. Lim, A.Z.; McFarland, R.; Taylor, R.W.; Gorman, G.S. RRM2B Mitochondrial DNA Maintenance Defects. In *GeneReviews®*; Adam, M.P., Ardinger, H.H., Pagon, R.A., Wallace, S.E., Bean, L.J., Mirzaa, G., Amemiya, A., Eds.; University of Washington Seattle: Seattle, WA, USA, 1993.
204. Bourdon, A.; Minai, L.; Serre, V.; Jais, J.-P.; Sarzi, E.; Aubert, S.; Chrétien, D.; de Lonlay, P.; Paquis-Flucklinger, V.; Arakawa, H.; et al. Mutation of RRM2B, Encoding P53-Controlled Ribonucleotide Reductase (P53R2), Causes Severe Mitochondrial DNA Depletion. *Nat. Genet.* **2007**, *39*, 776–780. [CrossRef]
205. Pitceathly, R.D.S.; Fassone, E.; Taanman, J.-W.; Sadowski, M.; Fratter, C.; Mudanohwo, E.E.; Woodward, C.E.; Sweeney, M.G.; Holton, J.L.; Hanna, M.G.; et al. Kearns-Sayre Syndrome Caused by Defective R1/P53R2 Assembly. *J. Med. Genet.* **2011**, *48*, 610–617. [CrossRef]
206. Elledge, S.J.; Davis, R.W. Identification of the DNA Damage-Responsive Element of RNR2 and Evidence That Four Distinct Cellular Factors Bind It. *Mol. Cell. Biol.* **1989**, *9*, 5373–5386. [CrossRef]
207. Elledge, S.J.; Davis, R.W. DNA Damage Induction of Ribonucleotide Reductase. *Mol. Cell. Biol.* **1989**, *9*, 4932–4940. [CrossRef] [PubMed]
208. Elledge, S.J.; Davis, R.W. Two Genes Differentially Regulated in the Cell Cycle and by DNA-Damaging Agents Encode Alternative Regulatory Subunits of Ribonucleotide Reductase. *Genes Dev.* **1990**, *4*, 740–751. [CrossRef] [PubMed]
209. Huang, M.; Zhou, Z.; Elledge, S.J. The DNA Replication and Damage Checkpoint Pathways Induce Transcription by Inhibition of the Crt1 Repressor. *Cell* **1998**, *94*, 595–605. [CrossRef]
210. Wang, P.J.; Chabes, A.; Casagrande, R.; Tian, X.C.; Thelander, L.; Huffaker, T.C. Rnr4p, a Novel Ribonucleotide Reductase Small-Subunit Protein. *Mol. Cell. Biol.* **1997**, *17*, 6114–6121. [CrossRef]
211. Sanvisens, N.; de Llanos, R.; Puig, S. Function and Regulation of Yeast Ribonucleotide Reductase: Cell Cycle, Genotoxic Stress, and Iron Bioavailability. *Biomed. J.* **2013**, *36*, 51–58. [CrossRef] [PubMed]
212. Zhao, X.; Muller, E.G.; Rothstein, R. A Suppressor of Two Essential Checkpoint Genes Identifies a Novel Protein That Negatively Affects DNTP Pools. *Mol. Cell* **1998**, *2*, 329–340. [CrossRef]
213. Chabes, A.; Domkin, V.; Thelander, L. Yeast Sml1, a Protein Inhibitor of Ribonucleotide Reductase. *J. Biol. Chem.* **1999**, *274*, 36679–36683. [CrossRef]
214. Chabes, A.; Domkin, V.; Larsson, G.; Liu, A.; Graslund, A.; Wijmenga, S.; Thelander, L. Yeast Ribonucleotide Reductase Has a Heterodimeric Iron-Radical-Containing Subunit. *Proc. Natl. Acad. Sci. USA* **2000**, *97*, 2474–2479. [CrossRef]
215. Sommerhalter, M.; Voegtli, W.C.; Perlstein, D.L.; Ge, J.; Stubbe, J.; Rosenzweig, A.C. Structures of the Yeast Ribonucleotide Reductase Rnr2 and Rnr4 Homodimers. *Biochemistry* **2004**, *43*, 7736–7742. [CrossRef]
216. Yao, R.; Zhang, Z.; An, X.; Bucci, B.; Perlstein, D.L.; Stubbe, J.; Huang, M. Subcellular Localization of Yeast Ribonucleotide Reductase Regulated by the DNA Replication and Damage Checkpoint Pathways. *Proc. Natl. Acad. Sci. USA* **2003**, *100*, 6628–6633. [CrossRef]
217. An, X.; Zhang, Z.; Yang, K.; Huang, M. Cotransport of the Heterodimeric Small Subunit of the *Saccharomyces cerevisiae* Ribonucleotide Reductase between the Nucleus and the Cytoplasm. *Genetics* **2006**, *173*, 63–73. [CrossRef] [PubMed]
218. Doerner, A.; Pauschinger, M.; Badorff, A.; Noutsias, M.; Giessen, S.; Schulze, K.; Bilger, J.; Rauch, U.; Schultheiss, H.P. Tissue-Specific Transcription Pattern of the Adenine Nucleotide Translocase Isoforms in Humans. *FEBS Lett.* **1997**, *414*, 258–262. [CrossRef] [PubMed]
219. Dolce, V.; Scarcia, P.; Iacopetta, D.; Palmieri, F. A Fourth ADP/ATP Carrier Isoform in Man: Identification, Bacterial Expression, Functional Characterization and Tissue Distribution. *FEBS Lett.* **2005**, *579*, 633–637. [CrossRef]
220. Stepien, G.; Torroni, A.; Chung, A.B.; Hodge, J.A.; Wallace, D.C. Differential Expression of Adenine Nucleotide Translocator Isoforms in Mammalian Tissues and during Muscle Cell Differentiation. *J. Biol. Chem.* **1992**, *267*, 14592–14597. [CrossRef]
221. Chevrollier, A.; Loiseau, D.; Reynier, P.; Stepien, G. Adenine Nucleotide Translocase 2 Is a Key Mitochondrial Protein in Cancer Metabolism. *Biochim. Biophys. Acta* **2011**, *1807*, 562–567. [CrossRef] [PubMed]
222. Palmieri, F. The Mitochondrial Transporter Family (SLC25): Physiological and Pathological Implications. *Pflugers Arch.* **2004**, *447*, 689–709. [CrossRef] [PubMed]
223. Palmieri, F. Mitochondrial Transporters of the SLC25 Family and Associated Diseases: A Review. *J. Inherit. Metab. Dis.* **2014**, *37*, 565–575. [CrossRef]
224. Pebay-Peyroula, E.; Dahout-Gonzalez, C.; Kahn, R.; Trézéguet, V.; Lauquin, G.J.-M.; Brandolin, G. Structure of Mitochondrial ADP/ATP Carrier in Complex with Carboxyatractyloside. *Nature* **2003**, *426*, 39–44. [CrossRef]
225. Riccio, P.; Aquila, H.; Klingenberg, M. Purification of the Carboxy-Atractylate Binding Protein from Mitochondria. *FEBS Lett.* **1975**, *56*, 133–138. [CrossRef]
226. Hackenberg, H.; Klingenberg, M. Molecular Weight and Hydrodynamic Parameters of the Adenosine 5′-Diphosphate–Adenosine 5′-Triphosphate Carrier in Triton X-100. *Biochemistry* **1980**, *19*, 548–555. [CrossRef]

227. Block, M.R.; Zaccaï, G.; Lauquin, G.J.; Vignais, P.V. Small Angle Neutron Scattering of the Mitochondrial ADP/ATP Carrier Protein in Detergent. *Biochem. Biophys. Res. Commun.* **1982**, *109*, 471–477. [CrossRef]
228. Bamber, L.; Harding, M.; Butler, P.J.G.; Kunji, E.R.S. Yeast Mitochondrial ADP/ATP Carriers Are Monomeric in Detergents. *Proc. Natl. Acad. Sci. USA* **2006**, *103*, 16224–16229. [CrossRef]
229. Bamber, L.; Harding, M.; Monné, M.; Slotboom, D.-J.; Kunji, E.R.S. The Yeast Mitochondrial ADP/ATP Carrier Functions as a Monomer in Mitochondrial Membranes. *Proc. Natl. Acad. Sci. USA* **2007**, *104*, 10830–10834. [CrossRef]
230. Bamber, L.; Slotboom, D.-J.; Kunji, E.R.S. Yeast Mitochondrial ADP/ATP Carriers Are Monomeric in Detergents as Demonstrated by Differential Affinity Purification. *J. Mol. Biol.* **2007**, *371*, 388–395. [CrossRef] [PubMed]
231. Krämer, R.; Klingenberg, M. Enhancement of Reconstituted ADP, ATP Exchange Activity by Phosphatidylethanolamine and by Anionic Phospholipids. *FEBS Lett.* **1980**, *119*, 257–260. [CrossRef]
232. Zoratti, M.; Szabò, I. The Mitochondrial Permeability Transition. *Biochim. Biophys. Acta* **1995**, *1241*, 139–176. [CrossRef]
233. Kokoszka, J.E.; Waymire, K.G.; Levy, S.E.; Sligh, J.E.; Cai, J.; Jones, D.P.; MacGregor, G.R.; Wallace, D.C. The ADP/ATP Translocator Is Not Essential for the Mitochondrial Permeability Transition Pore. *Nature* **2004**, *427*, 461–465. [CrossRef] [PubMed]
234. Marzo, I.; Brenner, C.; Kroemer, G. The Central Role of the Mitochondrial Megachannel in Apoptosis: Evidence Obtained with Intact Cells, Isolated Mitochondria, and Purified Protein Complexes. *Biomed. Pharmacother.* **1998**, *52*, 248–251. [CrossRef]
235. Hoshino, A.; Wang, W.-J.; Wada, S.; McDermott-Roe, C.; Evans, C.S.; Gosis, B.; Morley, M.P.; Rathi, K.S.; Li, J.; Li, K.; et al. The ADP/ATP Translocase Drives Mitophagy Independent of Nucleotide Exchange. *Nature* **2019**, *575*, 375–379. [CrossRef]
236. Brand, M.D.; Esteves, T.C. Physiological Functions of the Mitochondrial Uncoupling Proteins UCP2 and UCP3. *Cell Metab.* **2005**, *2*, 85–93. [CrossRef]
237. Napoli, L.; Bordoni, A.; Zeviani, M.; Hadjigeorgiou, G.M.; Sciacco, M.; Tiranti, V.; Terentiou, A.; Moggio, M.; Papadimitriou, A.; Scarlato, G.; et al. A Novel Missense Adenine Nucleotide Translocator-1 Gene Mutation in a Greek AdPEO Family. *Neurology* **2001**, *57*, 2295–2298. [CrossRef]
238. Komaki, H.; Goto, Y. ANT1, twinkle, POLG mutation. *Nihon Rinsho* **2002**, *60*, 353–356.
239. Siciliano, G.; Tessa, A.; Petrini, S.; Mancuso, M.; Bruno, C.; Grieco, G.S.; Malandrini, A.; DeFlorio, L.; Martini, B.; Federico, A.; et al. Autosomal Dominant External Ophthalmoplegia and Bipolar Affective Disorder Associated with a Mutation in the ANT1 Gene. *Neuromuscul. Disord.* **2003**, *13*, 162–165. [CrossRef]
240. Deschauer, M.; Hudson, G.; Müller, T.; Taylor, R.W.; Chinnery, P.F.; Zierz, S. A Novel ANT1 Gene Mutation with Probable Germline Mosaicism in Autosomal Dominant Progressive External Ophthalmoplegia. *Neuromuscul. Disord.* **2005**, *15*, 311–315. [CrossRef] [PubMed]
241. Körver-Keularts, I.M.L.W.; de Visser, M.; Bakker, H.D.; Wanders, R.J.A.; Vansenne, F.; Scholte, H.R.; Dorland, L.; Nicolaes, G.A.F.; Spaapen, L.M.J.; Smeets, H.J.M.; et al. Two Novel Mutations in the *SLC25A4* Gene in a Patient with Mitochondrial Myopathy. In *JIMD Reports*; Springer: Berlin/Heidelberg, Germany, 2015; Volume 22, pp. 39–45. [CrossRef]
242. Bauer, M.K.; Schubert, A.; Rocks, O.; Grimm, S. Adenine Nucleotide Translocase-1, a Component of the Permeability Transition Pore, Can Dominantly Induce Apoptosis. *J. Cell Biol.* **1999**, *147*, 1493–1502. [CrossRef] [PubMed]
243. Kawamata, H.; Tiranti, V.; Magrané, J.; Chinopoulos, C.; Manfredi, G. AdPEO Mutations in ANT1 Impair ADP-ATP Translocation in Muscle Mitochondria. *Hum. Mol. Genet.* **2011**, *20*, 2964–2974. [CrossRef] [PubMed]
244. Adrian, G.S.; McCammon, M.T.; Montgomery, D.L.; Douglas, M.G. Sequences Required for Delivery and Localization of the ADP/ATP Translocator to the Mitochondrial Inner Membrane. *Mol. Cell. Biol.* **1986**, *6*, 626–634. [CrossRef]
245. Lawson, J.E.; Douglas, M.G. Separate Genes Encode Functionally Equivalent ADP/ATP Carrier Proteins in *Saccharomyces cerevisiae*. Isolation and Analysis of AAC2. *J. Biol. Chem.* **1988**, *263*, 14812–14818. [CrossRef]
246. Kolarov, J.; Kolarova, N.; Nelson, N. A Third ADP/ATP Translocator Gene in Yeast. *J. Biol. Chem.* **1990**, *265*, 12711–12716. [CrossRef]
247. Drgon, T.; Sabová, L.; Gavurniková, G.; Kolarov, J. Yeast ADP/ATP Carrier (AAC) Proteins Exhibit Similar Enzymatic Properties but Their Deletion Produces Different Phenotypes. *FEBS Lett.* **1992**, *304*, 277–280. [CrossRef]
248. Kovácová, V.; Irmlerová, J.; Kovác, L. Oxidative Phosphorylatiion in Yeast. IV. Combination of a Nuclear Mutation Affecting Oxidative Phosphorylation with Cytoplasmic Mutation to Respiratory Deficiency. *Biochim. Biophys. Acta* **1968**, *162*, 157–163. [CrossRef]
249. Appleby, R.D.; Porteous, W.K.; Hughes, G.; James, A.M.; Shannon, D.; Wei, Y.H.; Murphy, M.P. Quantitation and Origin of the Mitochondrial Membrane Potential in Human Cells Lacking Mitochondrial DNA. *Eur. J. Biochem.* **1999**, *262*, 108–116. [CrossRef]
250. Dupont, C.H.; Mazat, J.P.; Guerin, B. The Role of Adenine Nucleotide Translocation in the Energization of the Inner Membrane of Mitochondria Isolated from Rho + and Rho Degree Strains of *Saccharomyces cerevisiae*. *Biochem. Biophys. Res. Commun.* **1985**, *132*, 1116–1123. [CrossRef]
251. Liu, Y.; Chen, X.J. Adenine Nucleotide Translocase, Mitochondrial Stress, and Degenerative Cell Death. *Oxid. Med. Cell. Longev.* **2013**, *2013*, 146860. [CrossRef] [PubMed]
252. Coyne, L.P.; Chen, X.J. Consequences of Inner Mitochondrial Membrane Protein Misfolding. *Mitochondrion* **2019**, *49*, 46–55. [CrossRef] [PubMed]
253. Viscomi, C.; Zeviani, M. Strategies for Fighting Mitochondrial Diseases. *J. Intern. Med.* **2020**, *287*, 665–684. [CrossRef]

254. Couplan, E.; Aiyar, R.S.; Kucharczyk, R.; Kabala, A.; Ezkurdia, N.; Gagneur, J.; St Onge, R.P.; Salin, B.; Soubigou, F.; Le Cann, M.; et al. A Yeast-Based Assay Identifies Drugs Active against Human Mitochondrial Disorders. *Proc. Natl. Acad. Sci. USA* **2011**, *108*, 11989–11994. [CrossRef]
255. Talevi, A.; Bellera, C.L. Challenges and Opportunities with Drug Repurposing: Finding Strategies to Find Alternative Uses of Therapeutics. *Expert Opin. Drug Discov.* **2020**, *15*, 397–401. [CrossRef]
256. Pushpakom, S.; Iorio, F.; Eyers, P.A.; Escott, K.J.; Hopper, S.; Wells, A.; Doig, A.; Guilliams, T.; Latimer, J.; McNamee, C.; et al. Drug Repurposing: Progress, Challenges and Recommendations. *Nat. Rev. Drug Discov.* **2019**, *18*, 41–58. [CrossRef]
257. Pitayu, L.; Baruffini, E.; Rodier, C.; Rötig, A.; Lodi, T.; Delahodde, A. Combined Use of *Saccharomyces cerevisiae*, Caenorhabditis Elegans and Patient Fibroblasts Leads to the Identification of Clofilium Tosylate as a Potential Therapeutic Chemical against POLG-Related Diseases. *Hum. Mol. Genet.* **2016**, *25*, 715–727. [CrossRef]
258. Facchinello, N.; Laquatra, C.; Locatello, L.; Beffagna, G.; Brañas Casas, R.; Fornetto, C.; Dinarello, A.; Martorano, L.; Vettori, A.; Risato, G.; et al. Efficient Clofilium Tosylate-Mediated Rescue of POLG-Related Disease Phenotypes in Zebrafish. *Cell Death Dis.* **2021**, *12*, 100. [CrossRef]
259. Aleo, S.J.; Del Dotto, V.; Fogazza, M.; Maresca, A.; Lodi, T.; Goffrini, P.; Ghelli, A.; Rugolo, M.; Carelli, V.; Baruffini, E.; et al. Drug Repositioning as a Therapeutic Strategy for Neurodegenerations Associated with OPA1 Mutations. *Hum. Mol. Genet.* **2021**, *29*, 3631–3645. [CrossRef] [PubMed]
260. Di Punzio, G.; Di Noia, M.A.; Delahodde, A.; Sellem, C.; Donnini, C.; Palmieri, L.; Lodi, T.; Dallabona, C. A Yeast-Based Screening Unravels Potential Therapeutic Molecules for Mitochondrial Diseases Associated with Dominant ANT1 Mutations. *Int. J. Mol. Sci.* **2021**, *22*, 4461. [CrossRef] [PubMed]
261. Baile, M.G.; Claypool, S.M. The Power of Yeast to Model Diseases of the Powerhouse of the Cell. *Front. Biosci.* **2013**, *18*, 241–278. [CrossRef]
262. Taanman, J.-W.; Muddle, J.R.; Muntau, A.C. Mitochondrial DNA Depletion Can Be Prevented by DGMP and DAMP Supplementation in a Resting Culture of Deoxyguanosine Kinase-Deficient Fibroblasts. *Hum. Mol. Genet.* **2003**, *12*, 1839–1845. [CrossRef]
263. Rampazzo, C.; Miazzi, C.; Franzolin, E.; Pontarin, G.; Ferraro, P.; Frangini, M.; Reichard, P.; Bianchi, V. Regulation by Degradation, a Cellular Defense against Deoxyribonucleotide Pool Imbalances. *Mutat. Res.* **2010**, *703*, 2–10. [CrossRef]
264. Garone, C.; Garcia-Diaz, B.; Emmanuele, V.; Lopez, L.C.; Tadesse, S.; Akman, H.O.; Tanji, K.; Quinzii, C.M.; Hirano, M. Deoxypyrimidine Monophosphate Bypass Therapy for Thymidine Kinase 2 Deficiency. *EMBO Mol. Med.* **2014**, *6*, 1016–1027. [CrossRef]
265. Lopez-Gomez, C.; Levy, R.J.; Sanchez-Quintero, M.J.; Juanola-Falgarona, M.; Barca, E.; Garcia-Diaz, B.; Tadesse, S.; Garone, C.; Hirano, M. Deoxycytidine and Deoxythymidine Treatment for Thymidine Kinase 2 Deficiency. *Ann. Neurol.* **2017**, *81*, 641–652. [CrossRef]
266. Bulst, S.; Abicht, A.; Holinski-Feder, E.; Müller-Ziermann, S.; Koehler, U.; Thirion, C.; Walter, M.C.; Stewart, J.D.; Chinnery, P.F.; Lochmüller, H.; et al. In Vitro Supplementation with DAMP/DGMP Leads to Partial Restoration of MtDNA Levels in Mitochondrial Depletion Syndromes. *Hum. Mol. Genet.* **2009**, *18*, 1590–1599. [CrossRef]
267. Munro, B.; Horvath, R.; Müller, J.S. Nucleoside Supplementation Modulates Mitochondrial DNA Copy Number in the Dguok-/- Zebrafish. *Hum. Mol. Genet.* **2019**, *28*, 796–803. [CrossRef]
268. Domínguez-González, C.; Madruga-Garrido, M.; Mavillard, F.; Garone, C.; Aguirre-Rodríguez, F.J.; Donati, M.A.; Kleinsteuber, K.; Martí, I.; Martín-Hernández, E.; Morealejo-Aycinena, J.P.; et al. Deoxynucleoside Therapy for Thymidine Kinase 2-Deficient Myopathy. *Ann. Neurol.* **2019**, *86*, 293–303. [CrossRef] [PubMed]

MDPI
St. Alban-Anlage 66
4052 Basel
Switzerland
www.mdpi.com

Genes Editorial Office
E-mail: genes@mdpi.com
www.mdpi.com/journal/genes

www.ingramcontent.com/pod-product-compliance
Lightning Source LLC
Chambersburg PA
CBHW040742250926
51850CB00007BB/1379
* 9 7 8 3 0 3 6 5 9 8 0 3 1 *